Dana's Minerals
and How to Study Them

Dana's New Mineralogy: The System of Mineralogy of James Dwight Dana and Edward Salisbury Dana, Eighth Edition, 1997 (0-471-19310-0)
Entirely rewritten and greatly enlarged by
Richard V. Gaines
H. Catherine W. Skinner
Eugene E. Foord
Brian Mason
Abraham Rosensweig
With additional description by
Vandall T. King
And with illustrations by
Eric Dowty
Manual of Mineralogy (After James D. Dana), 21st Edition, 1993 (0-471-57452-X)
Cornelis Klein and Cornelius S. Hurlbut, Jr.
Manual of Mineralogy (After James D. Dana): Exercises in Crystallography, Mineralogy, and Hand-Specimen Petrology, 21st Edition, Revised, 1993 (0-471-00042-6)
Cornelis Klein
Mineralogy Tutorials: Interactive Instruction on CD-ROM, 1995 (0-471-10996-7)
Cornelis Klein and The S. M. Stoller Corporation

Dana's Minerals and How to Study Them

(After Edward Salisbury Dana)

Fourth Edition

Cornelius S. Hurlbut, Jr.
W. Edwin Sharp

John Wiley & Sons, Inc.

New York / Chichester / Weinheim / Brisbane / Singapore / Toronto

A word of CAUTION is necessary: Many of the tests in this book involve intense heat, caustic acids, noxious and poisonous fumes, and even poisonous minerals. These procedures are, of course, not to be undertaken by children, teenagers, or adults without proper supervision and understanding of precautions and the dangers involved.

This text is printed on acid-free paper.

Copyright © 1998 by John Wiley & Sons, Inc.

All rights reserved. Published simultaneously in Canada.

Reproduction or translation of any part of this work beyond that permitted by Section 107 or 108 of the 1976 United States Copyright Act without the permission of the copyright owner is unlawful. Requests for permission or further information should be addressed to the Permissions Department, John Wiley & Sons, Inc., 605 Third Avenue, New York, NY 10158-0012.

This publication is designed to provide accurate and authoritative information in regard to the subject matter covered. It is sold with the understanding that the publisher is not engaged in rendering legal, accounting, or other professional services. If legal advice or other expert assistance is required, the services of a competent professional person should be sought.

Library of Congress Cataloging-in-Publication Data
Hurlbut, Cornelius Searle, 1906–
    Dana's minerals and how to study them : (after Edward Salisbury Dana). — 4th ed. / Cornelius S. Hurlbut, Jr., W. Edwin Sharp.
        p.   cm.
    Rev. ed. of: Minerals and how to study them / Edward Salisbury Dana. 3rd ed. 1949.
    Includes index.
    ISBN 0-471-15677-9 (paper : alk. paper)
    1. Mineralogy. 2. Crystallography. I. Sharp, W. Edwin. II. Dana, Edward Salisbury, 1849–1935. Minerals and how to study them. III. Title.
QE365.H845   1997
549'.1—dc21   97-21330

Printed in the United States of America

10 9 8 7 6 5 4 3 2 1

Contents

Preface

The first edition of this book was written by Edward Salisbury Dana over a century ago. It was the first of its kind to be written for the amateur mineralogist and the mineral collector and as a textbook for the beginning student, whether in high school or in a course in adult education. In it, facts about minerals were presented in an informative and interesting manner. In the present edition an effort has been made to maintain that approach to mineralogy for a similiar audience. The presentation of material is from the general to the specific. That is, the first chapters consider the crystallographic, chemical, and physical properties of minerals and methods used in their determination. In Chapter 7 these properties are discussed as they relate to individual minerals.

In the 50 years since the publication of the last edition of this book, great strides have been made in several aspects of mineralogy: notably, crystallography, microscopy, crystal chemistry, and mineral chemistry. These advances have been made largely by use of techniques involving elaborate and sophisticated instruments, equipment that is very expensive and beyond the reach of the average person. These techniques are the basis of all current advanced mineralogy textbooks. However, this is done at the expense of the old, time-honored methods of mineral study and testing. We feel strongly that for the beginning student there is a place for the old, simple, easy-to-make tests. While informing the student of the existence of the new techniques and what can be accomplished with them, we have retained, in somewhat abbreviated form, much of the old, particularly the "blowpipe tests." These can be carried out in the laboratory or in the field with a minimum of equipment and with materials that are inexpensive and easily purchased. With most blowpipe tests, one determines the presence or absence of a given element. They thus have the advantage over many other tests of keeping the most important property of a mineral, its chemical composition, constantly before the student.

In Chapter 3 the general approach to crystallography remains the same but there is considerable updating of the nomenclature to conform to the changes that have taken place in the past 50 years. In the preceding edition, a few pages were devoted to crystal growing in the laboratory. Because such crystals are not minerals and because the discussion contributed little to an understanding of the manner in which minerals grow in nature, these pages have been deleted.

In Chapter 7 the properties, occurrence, and uses of about 150 minerals are described. Of the more than 3000 known minerals, these are the ones we consider to be most abundant and most important. A few mineral descriptions not in the preceding edition have been added, but several have been removed, keeping the total number about the same.

A new chapter has been added on mineral genesis. Also added is an appendix listing minerals in order of increasing specific gravity, which should be helpful in mineral determination.

We are indebted to Dr. Carl Francis, curator of the Harvard University Mineralogical Museum, for making minerals available for photographing. Most of the mineral photographs in this book are of Harvard specimens.

Cornelius S. Hurlbut, Jr.
W. Edwin Sharp

Cambridge, Massachusetts
Columbia, South Carolina

1

Minerals and Mineralogy

Minerals are the materials of which the earth is built. Although we live and move upon the earth, we know little about it by direct observation. It is possible, however, to measure its size and shape, to determine its density as a whole, and by seismic study to make assumptions as to its interior. But of the materials of which it is made we know little beyond those that form the surface upon which we walk. The miner digs down a little distance, the oil-well driller reaches down still deeper, and we can examine the materials that their work brings up. Perhaps we can go down with the miner and see the rocks in place, but the deepest mines are only a little over 3 kilometers in depth. Although this seems very deep to someone let down a shaft in a cage, it is only a little way compared to the 6500-kilometer distance to the center of the earth.

Fortunately, our knowledge is greatly increased by the study of rocks in a variety of places over the earth's surface. In the process of building mountains the earth has brought to the surface rocks that once were 50 or 60 kilometers below the surface, and certain kinds of volcanoes sometimes carry up fragments of rocks and minerals such as diamonds from depths as great as 200 kilometers. Careful study of these rocks gives us a better idea of the kinds of matter that occur at depth and the conditions under which they exist. Yet astronomers have calculated the density of the earth as a whole and found it to be nearly twice as great as that of the rocks at the surface. This suggests that the deep interior of the earth is made of very dense minerals, probably including metallic iron.

The mineralogist is limited to the study of the earth's crust that can be reach with a hammer. A mineral collection cannot extend much beyond this unless one includes, and indeed one does, those rare visitors from outer space—called *meteorites*—which occasionally fall to the earth. What do we learn from a study of this hard, rocky material that we find at the earth's surface? We find, in the first place, that it consists of different kinds of rocks, some of which have familiar names, such as granite, basalt, marble, sandstone, and slate. On closer examination we find that each rock consists of different substances, each having certain properties by which it can be recognized. It is to these individual substances that the name *mineral* is given.

Although difficult, it is possible to separate the minerals that make up hard rocks, but the sand of the seashore can easily be separated into various types of grains. All the grains of one type are chemically alike, with characteristic properties such as color, luster, and density which enable us to distinguish different types with comparative ease. Most of the grains are clear and glassy, hard enough to scratch glass: These are *quartz.* There are also black grains which, because they are heavy (high density), have been sorted out by wave action into little rifts on the white sand. Some of these jump to a magnet and for this reason are called *magnetite.* There are other black grains which the magnet does not attract; these are probably *ilmenite.* Some red, glassy grains of *garnet* may be present. There may be still others, depending on where the sand came from and what kind of rock has been ground by nature's mill and sorted by water to make the sand.

If a piece of granite is examined closely, it is possible to distinguish several kinds of minerals. There are hard glassy grains with irregular surfaces, which, like the greater part of the sand grains, are *quartz.* There are white, yellow, or pale flesh-red grains, also hard, but not as hard as quartz, which reflect the light from one or two smooth surfaces; these are *feldspar.* Then there is *mica,* which is either white and silvery or shiny black (in some granites both types are found) and which with the touch of a knife blade separates into very thin scales. In addition, there may be a few coal-black *tourmaline* grains, some bright red *garnets,* and smaller amounts of other minerals. If a cavity or open space can be found in the granite, it is often possible to find in it these same minerals, but in larger and more regular shapes called *crystals.*

If we examine a rock such as basalt from a lava flow that is fine-grained and compact, it will appear quite uniform to the unaided eye. But the skillful mineralogist can show us that it is, like the granite, a mineral aggregate but composed of tiny grains. This is done by making a thin section (a slice thin enough to be transparent) in which, when examined under a polarizing microscope, one can recognize a variety of minerals. In cavities in fine-grained rocks, well-formed minerals are frequently found that were deposited after the rocks had formed and differ from the minerals making up the massive rock.

In places we find rocks which examination shows are made entirely of grains of but one mineral, as, for example, the white marble of Vermont, in which the mineral is *calcite.* Another rock, quartzite, is composed of only *quartz.* Because they are made up of a single mineral, marble and quartzite are called *monomineralic* rocks.

The various substances that make up the rocky materials of the earth's crust and into which we can separate most rocks are called *minerals.* Each is a chemical element or compound with properties of hardness, density, luster, and color by which it may be distinguished. Moreover, if it is well crystallized, it has a definite shape that is also characteristic. Because to it belong all the properties that distinguish it from other minerals, it is called a *mineral species.*

In the study of minerals the student must learn the properties that distinguish one mineral from another, how these properties are determined, how minerals are classified, how they occur in nature, and something of their uses. The knowledge that mineralogists have accumulated through long years of study in both field and laboratory has been formulated in systematic form and makes up the science of *mineralogy*.

Since mineralogy includes not only the description of minerals but also the way in which they occur, the minerals with which they are associated and ultimately, their origin, it bears a definite relation to the broader science of *geology*. Geology is concerned with the history of the earth and all the changes through which it has gone as read in the record of the rocks. It is essential, therefore, that the geologist know something about mineralogy; and at the same time it is desirable for the mineralogist to know something about geology.

There is also a profitable exchange of knowledge between the chemist and mineralogist. With the exception of the gases of the atmosphere, minerals are the source of all the chemical elements as well as a wide variety of chemical compounds, and the industrial chemist and engineer learn from the mineralogist in what minerals, at what localities, and in what amounts they are to be found. Thus the mineralogist plays an important role in locating the minerals needed for many industrial materials and technological products.

A chemical classification of minerals is considered by most mineralogists today to be the best. It is from the chemist that the mineralogist learns not only how to classify minerals in chemical groups but also how to make chemical analyses to determine one of the fundamental properties of minerals—the chemical composition.

It is not to be inferred that before one undertakes a study of minerals, the student must be a geologist or a chemist, although for advanced work in mineralogy, a familiarity with these related sciences is essential. For the beginner, the reverse may be true and it will be discovered after a short while that one is learning some geology and chemistry as a by-product of the study of minerals.

Earlier we mentioned some of the properties of minerals in an effort to build a picture of a mineral in general terms, but before proceeding we should formulate a specific definition. This was done for us in 1995 by the Commission on New Minerals and Mineral Names of the International Mineralogical Association. The succinct definition of the commission and one with which most mineralogists concur is: *A mineral is an element or compound that is normally crystalline and which has been formed as the result of geological processes.* This definition includes with only minor exceptions all the substances described later in this book. Let us consider the implications of the definition.

For a mineral to be a *chemical element or compound* means that its composition can be expressed by a chemical formula. Some minerals have a definite composition, for example, *quartz:* silicon dioxide (SiO_2) and *galena:* lead sulfide (PbS). Other minerals have a range of compositions in which atoms of one element substitute for atoms of another in the crystal structure. For example, in the plagioclase feldspars there is a range in composition that extends from the sodium member, albite ($NaAlSi_3O_8$), to the calcium member, anorthite ($CaAl_2Si_2O_8$). Atoms of calcium can substitute for atoms of sodium in "all proportions" between pure albite and pure anorthite.

The restriction that minerals be crystalline excludes liquids and noncrystalline solids even though the solids are like minerals in chemistry and occurrence. These mineral-like substances, called mineraloids, are of two kinds. The *amorphous* mineraloids, such as opal, are solids that have never been crystalline but may be accepted as minerals if they can be proved to be true chemical compounds. The other type, called *metamict,* are substances that were crystalline when formed but whose crystallinity has been destroyed by radiation. Zircon is the only common mineral that has a metamict phase.

Formed as a result of geologic processes means that a mineral must be of natural occurrence; human beings cannot have taken part in its formation. Many compounds have been synthesized with properties identical to those of natural crystalline substances but cannot be considered minerals. Those that have been manufactured may be very difficult to distinguish from the natural. This is particularly true of the gem minerals, for their synthetic counterparts rival or even excel the natural substances in beauty and perfection of development. Diamond, ruby, emerald, sapphire, and many other gems have been synthesized and, strictly speaking, must be called synthetic diamond, synthetic ruby, and so on.

There are chemical compounds that have been produced entirely by biological processes without a geological component. For this reason, even if they closely resemble natural inorganic materials, the mineralogist usually excludes them from mineral collections. These substances include urinary calculi, the shell of the oyster and the pearl itself, the lime of the bones of animals, and the opal-like form of silica secreted by the growth of plants, such as the tabasheer found in the joints of bamboo. However, once accumulations of such biologically generated inorganic compounds have been compacted or recrystallized into a rock such as limestone, they have the necessary geological component and can be treated as minerals, as in any other rock.

However, the term *mineral* is not used in this restricted sense by everyone. Economists speaking of the *mineral resources* or *mineral wealth* of a country refer not only to the minerals of the mineralogist but also to coal and petroleum, which are of organic origin. When chemists or physicists need to describe a synthetic compound analogous to a given mineral, or biologists need to describe the nature of bone or shells, they often use mineral names.

2

Preliminary Hints on How to Study Minerals

We have seen that a mineral has certain properties of form, hardness, density, luster, and color that, collectively, characterize it and frequently enable us to separate it from other minerals. These are known as its *physical properties.* The most important property of a mineral, however, is its chemical composition, and directly dependent on it are its *chemical properties.* These two sets of properties are described in some detail in subsequent chapters; but first it is necessary to describe briefly how to study minerals if we wish to learn as much as possible about each species with the least effort.

The student must first use the eyes and other unaided senses in studying minerals; in other words, gain all the information possible about minerals by looking at and handling them. If this is done wisely, one will be surprised to find how keen the senses become and how much can be told merely by inspecting specimens. However, with experience, students find that this method can be carried only to a certain point and must recognize the importance of confirming first conclusions by more positive tests. The appearance of specimens of even the common species may, if depended upon alone, lead one considerably astray. The old saying that "all that glitters is not gold," and the names *fool's gold* and *false galena,* express the fact that the senses unassisted may readily be deceived.

The trained eye of the mineralogist will show at a glance whether a mineral has the regular geometrical shape of a crystal. It will also show whether a mineral has the natural, easy, smooth fracture of many crystalline substances, called *cleavage,* or only a rough irregular fracture. The eye will also perceive the color and note whether the mineral is transparent or opaque, as well as the peculiarities of luster that may be characteristic of a given mineral.

Touch will indicate whether the "feel" is greasy, as is true of talc and a few other very soft minerals. On "hefting" a mineral, one may recognize at once that it is heavy or light compared to familiar substances of the same appearance. The common minerals quartz, feldspar, and calcite have nearly the same density, and one can easily become so accustomed to them that a piece of gypsum seems light and a piece of barite seems heavy. Similarly, a piece of aluminum or, more particularly, magnesium seems very light because it is compared instinctively with apparently similar but much denser metals which we are more accustomed to handle.

Taste is a distinguishing characteristic in some minerals; *odor* is occasionally a useful property, such as the clayey odor given off by some minerals when they are breathed upon. It does, however, require some study and experience before the senses are so alert that all the properties to be noted are perceived at once and evaluated correctly. Everyone should strive toward this end, for one of the great benefits to be derived from a study of mineralogy is that it cultivates and stimulates the powers of observation.

When the senses have gleaned all that is possible, simple tests to aid them should be used. Touching the smooth surface of a mineral with the point of a knife serves to show whether it is relatively soft or hard. The color of the powder obtained by rubbing a mineral across a plate of unglazed porcelain, or scratching it with a knife, is called the *streak;* if the streak is quite different from the color of the surface, as it is in some minerals, it constitutes a very important property. Careful inspection of the mineral can suggest the presence of cleavage by reflection of light from their smooth surfaces; closer examination with a magnifying glass may reveal whether the cleavage is blocky, takes the form of splinters, or occurs as sheets.

If inspection of the specimen or the field tests mentioned fail to identify a mineral, some simple laboratory tests must be made. These include the density or specific gravity, use of a blowpipe, and a number of simple chemical tests to show the presence of certain elements. After these are made there are still others, which include the refined methods of the trained mineralogist with a precision goniometer for measuring crystal angles, a polarizing microscope for the study of optical properties, x-ray diffraction analysis for study of the internal structure, and accurate quantitative chemical analysis by use of x-ray fluoresence or with the electron microprobe. With these methods and others that are constantly being developed, most of nature's secrets may be learned and the properties of each mineral may be thoroughly studied.

SUGGESTIONS ON MAKING A MINERAL COLLECTION

A very important part of the study of minerals is the student's own collection, for all who wish to learn mineralogy must have a collection of their own to examine and experiment upon. It is very desirable that each school or college have a larger collection for reference and study, but this does not take the place of the individual collection, which will be studied, arranged, labeled, and handled over and over again until every specimen is perfectly familiar.

Furthermore, if possible, the student should obtain specimens by collecting them, and thus see the minerals in their natural surroundings. Even if one lives in a region that does not seem at first to afford many specimens, it is possible to find something that is worth keeping until better ones become available. Occasionally, one will have the opportunity to make trips to some of the noted localities, where there is a great variety of minerals. There is nothing more delightfully instructive and refreshing than to spend a day in the open air, with a good hammer in hand, a bag for the specimens, and plenty of soft paper to wrap them in.

The hammer should be of hard steel that will not chip on the edges; it may weigh a half-kilo to a kilo, the face should be square or slightly oblong, and the edges should be sharp, with the peen in the form of a point or a wedge, as shown in Fig. 2-1. A cold chisel or two, for working into cracks and crevices, will often be found

Fig. 2-1. Hammers and cold chisel. (Estwing; Rockford, Illinois.)

useful. Also, a small light hammer with a sharp edge for trimming specimens may prove valuable, for a blow from a heavy hammer will often shatter a specimen. Fragments from hammering on the rocks can yield sharp chips that can fly into the eyes or cut exposed skin; thus it is advisable to use safety goggles and gloves to protect the hands. Similarly, pants should be worn and hard-toed shoes, as rolling rocks can scrape exposed legs or crush toes.

It is better as a rule not to break crystals out of rock. A detached crystal of garnet is interesting when quite perfect, but in general the crystal is most interesting and instructive when in its own surroundings. The seller of minerals soon discovers this, and it is unfortunately not an uncommon trick for the local dealer to mount a loose crystal in a mass of rock in which it never belonged, thus to increase the value of the specimen and deceive the unwary purchaser.

If the specimen collected is nonfragile, it can simply be placed in a small plastic bag (about 17 by 15 centimeters) or wrapped in paper. To wrap the specimen, a sheet of paper is taken and a single fold made near the bottom so as to cover the specimen. Then folds are made over the specimen, first from one side and then the other. Finally, the specimen is rolled into the remaining sheet. It is critical that each specimen be labeled with the date and the location, as both are soon forgotten if more than one collecting site is visited that day or if placed aside even for a day or so. The label should consist of a stiff piece of paper such as that used for file cards and the information written using a soft pencil, permanent ink, or with a permanent marking pen. If the specimen is fragile, it should first be wrapped in soft paper, followed by wrapping in regular paper or placement in a small plastic bag. If especially fragile, the specimen must not be placed in the collecting bag with other specimens, where it might be crushed on the trip home.

The student is advised not to spend a great deal of money on buying specimens, particularly at any one time, for today many mineral localities are accessible by automobile, and it is possible to see minerals in place and to collect fine specimens of some minerals. Although many such sites are on public lands, there are many places where it is important to obtain permission from the landowner. By joining a local mineral club or society, one can learn not only about difficult-to-locate sites but can find out which are open and which require permission. In many cases local clubs have a working relationship with landowners and can obtain permission to visit sites not otherwise possible. Particularly popular localities may be open on certain days of the week for a small fee, and the owner in many cases will actively assist the novice.

However, a little money is by no means thrown away if judiciously expended from time to time, for it will serve to buy a few small characteristic specimens of the common species and pure fragments for simple tests. Fine specimens, especially of the rarer species, are now very expensive. Fortunately, sufficiently good specimens of the minerals that are important for the student to know well may be obtained for very little money.

It is better to collect, as far as possible, small specimens rather than large, such as will fit in a little paper tray 5 centimeters square, or 5 by 7 centimeters, or at most 7 by 7 centimeters. These trays are inexpensive and are very useful for the arrangement and preservation of a collection. If specimens are placed loose in a drawer, opening the drawer a few times will throw them into confusion and separate them from their labels; and sooner or later they will be badly injured. A depth of 1 centimeter is sufficient for the tray, but the drawers, if possible, should not be less than 6 or 8 centimeters deep. Alternatively, a layer of cotton batting may be used to line the bottom of the drawer; this will usually prevent the specimens from rolling. All the specimens in a collection should be carefully labeled, particularly as regards *locality.*

Even if considerable care is exercised, specimens may become separated from their labels. To prevent permanent separation, it is well to mark a number directly on the specimen as soon as it is acquired. The number with the locality and other pertinent data should be entered in a catalog. A good way to ensure permanence of the number is to paint on the mineral, in an inconspicuous place, a small area of white enamel and, when this is dry, to write a unique number on it in India ink.

Fortunately, most minerals the student is likely to begin collecting are quite stable and can be displayed for years while retaining the freshness they had when collected. However, dust can collect on them and mask their original beauty. Although in many cases this can be restored simply by blowing off the dust or washing with soap and water, it is actually better to place fine specimens in a cabinet with glass panels so that dust cannot accumulate on the surfaces. Even when specimens are stored in drawers, a cloth cover will help to maintain fresh surfaces. Special problems are associated with certain hydrous minerals, such as sylvite; some zeolites, such as lau-montite; and some borates, as they may gain or lose water. The bright, fresh surfaces of native silver and copper as well as those of most sulfides will tarnish in time. Particularly troubling is marcasite, many fine specimens of which crumble completely over time. For the beginner, the simplest solution to this difficulty is to store such specimens in sealed square glass canning jars. In time the student will learn of other effective methods for *cleaning and preserving minerals.*

Large, well-crystallized mineral specimens are usually not available to the average collector because of their scarcity, and such specimens of many minerals can be seen only in museums. There is a rule, however, to which there are few exceptions, that the smaller the crystal, the better it is formed and the more faces it possesses. Because of this fact, some collectors search for small, well-crystallized specimens that can be seen well only through a microscope. These specimens require very little space and are perfect for some city dwellers. They are mounted individually on pedestals, which in turn are cemented to the bottom of small boxes or trays. Figure 2-2 shows a *micromount* in position to be observed with a microscope, and Figure

Fig. 2-2. *Binocular microscope and micromount.*

Fig. 2-3. *Magnified micromount of rutile.*

2-3 is the same micromount magnified 20 times. When viewed in this manner, these tiny specimens rival or even surpass in beauty and perfection of development larger specimens of the same minerals.

Once the student has gained an overview of the wide variety of minerals to be collected, he or she may wish to specialize in some particular way. As illustrated above, one way is to specialize in collecting crystals as micromounts, for collecting large crystals is very difficult and expensive. Some might choose to collect minerals that fluoresce, others to collect crystals of a mineral such as quartz or calcite that has a large variety of forms, or to collect a particular mineral from as many different locations as possible, or to collect as many minerals as possible from a single locality.

3

Crystals and Crystallography

If you were to show a tray of common minerals to a seasoned mineralogist, most of the minerals could be named merely by looking at them. This can be done because the mineralogist has learned to recognize various properties that characterize each mineral. The color, luster, and shape of the crystals, all of which can be seen at a glance, may be sufficient to identify a specific mineral. Color, luster, and crystal form are but three of what are known as *physical properties.* Because of their characteristic shapes, crystal forms, when present, are probably the most informative of the several physical properties in mineral identification. The present chapter is concerned with crystals and their forms, habits, and aggregates.

STATES OF MATTER

Whether we explore the remote reaches of the universe with a powerful telescope or probe deep into the world of the very small with a powerful microscope, everything we see is made up of atoms of a relatively small number of elements. Atoms, atoms as ions, or groups of atoms called molecules exist in varying degrees of order known as *states of matter.* In the air we breath, in the helium that fills a balloon, the atoms and molecules are in a state of complete disorder and in constant motion, colliding with each other and with the walls of the container. This complete lack of order is called the *gaseous state.*

When gases are cooled down to a certain temperature, different for each gas, a small electrical attraction between the atoms or molecules causes them to approach each other as closely as possible, but they lack internal arrangement. They are still free to rotate or change position with each other, filling and taking the shape of their container. This is called the *liquid state.*

If heat energy is removed from a liquid, a temperature is reached, different for each liquid, called the *freezing point.* At this point a dramatic change takes place. The mobile liquid suddenly changes to a rigid solid and the liquid is said to have passed into the *solid state.* Under the proper conditions the solid may form as a crystal with smooth shining faces. The regular building of atoms takes place not only when a solid forms from a liquid, but a solid may form directly from a gas without going into the liquid state. If air is cooled sufficiently, the contained water vapor may form snowflakes, true crystalline solids. Figure 3-1 shows some of the many forms of snow crystals.

Fig. 3-1. Snow crystals.

The solid state is the crystalline state, quite different from gaseous and liquid states. It differs in that every atom or other building particle becomes locked into position by strong binding forces acting between them. Each atom has a definite environment; that is, it is the same for every other identical particle. They thus form a crystal lattice of orderly rows and each particle has definite neighbors at definite distances from it. There is a regular arrangement of atoms in three dimensions; that is, the neighbors of a given particle are not only beside it but above and below it as well. It is with such solids that the mineralogist is concerned, for minerals are true solids with a crystalline structure and will form crystals of regular geometric form under favorable conditions. A crystal is, therefore, the outward expression of the orderly internal arrangement of atoms.

This orderly arrangement is shown in the cubes of fluorite (Fig. 3-2), the six-sided prisms of quartz (Fig. 3-3), the 12-sided, 24-sided, or even more complex forms of

Fig. 3-2. *Fluorite, group of cubic crystals.*   **Fig. 3-3.** *Quartz crystal.*

garnet. Although these minerals may show other crystal faces, it is their "habit" to crystallize in the forms indicated. Thus *crystal habit* means the characteristic form assumed by a mineral, including the general shape and irregularities of growth. Even when a mineral does not show a regular external form, it may have an easy smooth fracture called *cleavage,* which indicates the orderly internal structure of a crystalline substance.

Efforts have long been made to grow crystals in the laboratory, and this has now become an important industrial art. Prior to the twentieth century, crystals of various salts were grown from aqueous solutions. Some instructions in this art are given in the third edition of this book. The modern art of crystal growth and *manufactured gems* began with the invention by Verneuil in 1902 of the flame fusion growth of ruby. These had the orderly structure of a crystalline material but lacked external crystal faces. During the following decades, a great effort was expended by many workers in crystal growing which proved to be highly successful. These include the growth of emerald from molten salts (1934), the hydrothermal growth in steel pressure vessels of quartz for oscillator plates (1950), and finally, the synthesis of diamond (1955) in large hydraulic rams that can reach pressures of 60,000 atmospheres and temperatures of 1200°C. Since then, diamond has been made by explosive shock and by vapor deposition. Today, most important gem minerals and many common minerals have been synthesized, many bounded by crystal faces.

Manufactured crystals cannot compete with the many beautiful and complex crystals of minerals with brilliant faces grown in nature's laboratory. However, even here the crystallization processes often cannot go on freely, and imperfect crystals, or perhaps a mass of only a confused crystalline aggregate, may be produced.

The quartz, feldspar, and mica in a granitic rock usually form in such a way that they interfere with each other, and none can build itself into a perfect crystal. However, the student who understands the study of thin sections with a polarizing microscope can prove that each formless grain is crystalline with an orderly internal structure. In a cavity in the granite, there may be crystals of quartz and feldspar, perhaps also of mica, for here each mineral was free to build itself into a perfect crystal.

A familiar example of crystallization is the ice covering a pond, which is as truly crystalline in structure as the perfect snow crystal, but with no crystal faces. Many of us have witnessed the freezing of a little pool of water on a walkway, with the formation of slender crystalline ribs of ice that shoot out from the edge. They form a framework that may soon lose its distinctness as the crystals interfere and the entire surface is frozen. An examination of this ice in polarized light would show it to be composed, as is the granite, of many formless entities.

Many mineral aggregates are composed of particles grown together in a haphazard way and said to be *massive.* How can we tell that this mineral is crystalline rather than amphorous? If cleavage is present, we know the answer. For, as explained later, cleavage is, like crystal faces, an expression of orderly atomic arrangement. Thus, if small flashes of light are reflected from the massive material, cleavage is present and it can be said with certainty that the mineral is crystalline. If cleavage is not present, an examination by x-ray diffraction or in polarized light will help in making a decision. Unfortunately, these methods can only be carried out by a trained person.

Another method, known as *etching,* aids the skillful mineralogist in determining an ordered atomic structure. A liquid (or gas) that has the power of dissolving the substance is allowed to act on a smooth surface for a short time. Then it is removed and the surface is cleaned and examined under a microscope. Often, a multitude of little cavities or pits are found on the surface, the shape of which show clearly how the structure is built. The process of etching is as if the stones of a pyramid were so smooth and closely fitted that no joints were visible, leaving the mason to pull out a number of bricks until one could see the pattern. Figure 3-4 shows the figures etched by hydrofluoric acid on the faces of a crystal of quartz; the variations reveal the complex structure of this mineral.

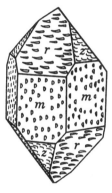

Fig. 3-4. *Etched quartz crystal.*

In general, it can be said that although many minerals are massive, they are usually crystalline, and the amorphous state is the exception. There are, however, certain minerals or varieties of minerals that seem to be intermediate between the distinctly crystalline and the amorphous states; these are called *microcrystalline.* Many fine-grained varieties of quartz such as agate are microcrystalline and resemble amorphous substances. When these minerals are viewed under a high-powered polarizing microscope, one can see they are an aggregate of myriad tiny crystalline quartz particles. Moreover, an x-ray photograph would give the same pattern as that given by a single large quartz crystal ground to a fine powder.

As already mentioned, a small crystal is just as perfect and complete an individual as a similar one of great size. Among the crystals of a given species there is no relation between size and age as there is among the individuals of a species in the animal and plant kingdoms. Some crystals are so minute as to be microscopic; others may be of enormous size, such as the large beryl crystals from Maine and the even greater spodumene crystals found in the Black Hills of South Dakota, some of which measure over 40 feet in length. A cave opened many years ago at Macomb, New York, contained 15 tons of great cubic crystals of fluorite; another cave in Wayne County, Utah, contained a large number of enormous crystals of gypsum, some of them 3 feet or more in length. But the very small crystals and those of enormous size are not essentially different except in the comparatively unimportant respect of magnitude. Figure 3-5 shows a group of gypsum crystals from Naica, Mexico, the largest of which is over 4 feet long.

Nevertheless, there are many interesting points of resemblance between crystals and living plants. Like plants, crystals *grow*—under favorable conditions, so rapidly that the increase in size may be watched not only from day to day but from hour to hour and even from minute to minute. The complex forms that are built up, especially in such cases of rapid growth, are often wonderfully plantlike in aspect. As many will know, the delicate frost figures form quickly on a windowpane in winter; other, more permanent examples are the arborescent or dendritic forms of native gold, silver, or copper. Indeed, the terms used in describing them are given

Fig. 3-5. *Gypsum crystals, Naica, Mexico.*

because of their resemblance to forms of vegetation. Furthermore, as a wounded plant tends to heal itself when, for example, a branch has been broken, so, too, a broken crystal may be more or less healed, but with crystals the material that repairs the injury must be supplied from a surrounding solution. Thus the silica to mend a broken quartz crystal must come from a surrounding solution, and the crystal itself only directs the way in which the atoms from the solution are laid down. It is interesting, however, that the growth takes place more readily on a surface of fracture than on a natural crystal face. In this way the formless grains of quartz in a sandstone often build themselves into complete crystals.

Although a crystal never has an old age in the sense that a plant or an animal does, nevertheless, many crystals change as time goes on if subjected, for example, to the corroding effects of an invading solution. Even beautiful minerals such as the sapphire, emerald, and topaz, which are hard and comparatively insoluble, have this tendency to undergo what is called *chemical alteration,* with the loss of their beauty and a change of chemical substance. This alteration is spoken of again later in the chapter when pseudomorphs are described, but it is worth noting here because it is somewhat analogous to the change that old age brings to a living organism.

CRYSTAL SYMMETRY, CLASS, AXES, AND FORMS

If one is privileged to handle well-formed crystals of a variety of minerals, one will quickly see that there is a wide difference in them. The size, shape, appearance, and number of faces differ from crystal to crystal. On one, all the faces may be similar, whereas on another only two similar faces may be found, and these may be located on opposite sides of the crystal. A little study will show that like faces are arranged symmetrically on the crystal and that this symmetry is the same for all crystals of any given mineral species.

Crystal Symmetry

Three types of symmetry, known as the *elements of symmetry,* are to be found in crystals: (1) symmetry across a plane, (2) symmetry about a point, and (3) symmetry about an axis.

A *plane of symmetry,* also called a *mirror plane,* is an imaginary plane passed through a crystal dividing it in half, so that each half is the mirror image of the other.

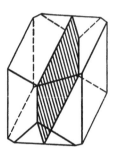

If it were possible then to split a crystal in half along this plane and place one half against a mirror, the image would appear to restore the other half. Similarly, the student's left and right hands are mirrors of each other. Figure 3-6 illustrates a plane of symmetry. Some crystals have as many as nine planes of symmetry, although some have none.

A crystal is said to have a *center of symmetry* if an imaginary line can be passed from any point on its surface through its center to a similar point on the opposite side. Images formed by a pinhole camera are related to the true object in exactly this way

Fig. 3-6.
Symmetry plane.

and to an observer are seen to be inverted relative to the original object. Similarly, simple microscopes and the telescope of the astronomer invert images in the same way. Thus the image of the moon through the astronomer's telescope will appear upside down and reversed.

An *axis of symmetry* is an imaginary line through a crystal about which the crystal may be rotated and repeat itself during a complete revolution. There are 1-, 2-, 3-, 4-, and 6-fold symmetry axes. Notice that while 5-, 7-, 8-, 9-fold, and higher-symmetry axes may occur in solid geometry, they cannot occur in crystals. Figure 3-7 illustrates a 4-fold axis.

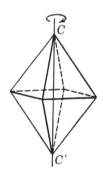

The crystallographer also considers as fundamental certain compound symmetries. These *rotary inversions* consist of a simple rotation combined with an inversion through the center (a center of symmetry). As with the simple rotations, there are five possible axes of rotary inversion: $\bar{1}, \bar{2} = m, \bar{3}, \bar{4},$ and $\bar{6}$. Whereas simple rotations leave a right or left hand unaffected by the rotation, the rotary inversions have the property of changing handedness; and whereas the rotation portion leaves the handedness unchanged, the inversion converts a right hand into a left hand or a left hand into a right hand. This is best seen, as we mentioned above, in the case of a mirror plane. The student should note that a $\bar{1}$ rotary inversion is the symbol used for a symmetry center, m for a mirror plane, and that $3/m$ is commonly substituted for $\bar{6}$, where $3/m$ means a mirror plane perpendicular to a 3-fold axis. Figures 3-6 and 3-7 both have symmetry centers as well as axes and planes; Fig. 3-8 has only a center of sym-

Fig. 3-7.
Symmetry axis.

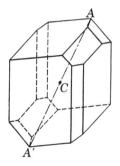

Fig. 3-8.
Symmetry center.

metry. It is possible for certain crystals to lack all elements of symmetry, but only one or two such minerals are known.

The symmetries discussed above are those permitted by the usual internal arrangement of the atoms. In fact, crystals have such a strong tendency to achieve as high as symmetry as possible that two crystals will often join in such a way that a mirror plane, called the *twin plane,* or a 2-fold axis, called the *twin axis,* has been added. The combined crystals then possess a higher level of symmetry than is possible in either of the two crystals individually.

Crystal Classes and Systems

One might think that if these symmetry elements were combined by placing them together at all possible angles, a hopelessly large number of combinations would result. Exploration of all the possibilities leads to a maximum of 32 combinations called the *32 crystal classes.* To the beginner in mineralogy the number of classes may seem large, but that should not discourage one, for only 10 or 12 of the 32 classes are represented in the common minerals.

Because of certain similarities, the 32 crystal classes are grouped together into six units, the *crystal systems:* I, isometric; II, tetragonal; III, hexagonal; IV, orthorhombic; V, monoclinic; VI, triclinic (anorthic). The symmetry similarities of the crystal classes in a given crystal system permit the several classes to be referred to the same lines of reference, the *crystallographic axes.*

Crystallographic Axes

In describing crystals it is helpful to refer them to lines of reference passing through the center of the crystal. The position of these imaginary lines, called *crystallographic axes,* is determined by the crystal's symmetry. For most crystals, the axes either coincide with symmetry axes or are perpendicular to symmetry planes. They are also parallel to the intersection edges of major crystal faces. For some crystals the symmetry permits more than one orientation of crystallographic axes, and the choice is made by morphology.

With the exception of the hexagonal system, crystals are referred to three crystallographic axes lettered *a, b, c.* In the crystal system of lowest symmetry, the triclinic, the axes are all of different length and at oblique angles to each other (Fig. 3-114). In the other crystal systems, there are right angles between some or all of the axes. For a general description of the crystallographic axes, consider those of the orthorhombic system (Fig. 3-9). Here there are three mutually perpendicular axes of different length.

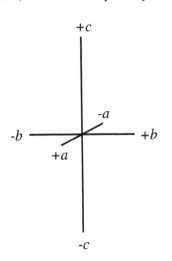

Fig. 3-9. *Orthorhombic crystal axes.*

In the conventional orientation, the *a* axis is horizontal and oriented front to back; the *b* axis is horizontal and left–right, and the *c* axis is vertical. The front of *a,* the right-hand end of *b,* and the upper end of *c* are positive; the opposite ends are negative.

Crystal Form

The term *form* is used here in the special sense of the crystallographer. Although it may be used to express the general outward configuration of the crystal, it is usually reserved to indicate all those crystal faces that have like positions with respect to the elements of symmetry and the crystallographic axes. The names of certain forms appear over and over again, and although they differ in detail in each crystal class and system, the terms *pyramid, prism,* and *pinacoid* are common to several of them.

Pyramid and Dipyramid. When a group of three or more similar faces meet at a point, this group of faces is known as a *pyramid.* If there is a symmetry plane at the base of a pyramid, the number of faces is doubled and the form is called a *dipyramid.* Because the faces of a dipyramid can completely enclose the crystal, it is called a *closed form.* A single pyramid does not enclose the crystal and is called an *open form.* In this sense, the octahedron can be considered a dipyramid of eight faces.

Prism. When a group of three or more similar faces meet such that their lines of intersection are all parallel, this group of faces is known as a *prism.* In contrast to the dipyramid, a prism alone cannot complete a crystal; it thus is an open form and must be in combination with a pyramid or other form. Quartz is a common example of a prism being enclosed by another form, the rhombohedron.

Pinacoid. In contrast to a pyramid or a prism, a *pinacoid* is made up of only two similar faces on opposite ends of the crystal. Yet pinacoids are important in combination with the prism as occurs in beryl, where to complete the crystal, a six-sided prism is combined with a basal pinacoid at the top and bottom. In some cases, as with anhydrite, the crystal may be enclosed by a total of six faces consisting of three pinacoids.

Measurement of Angles

It is helpful in studying minerals to be able to measure the angles between the faces of a crystal, for in this way it is possible to tell a square prism (tetragonal) from a rhombic prism (orthorhombic), a cube from a rhombohedron, or an octahedron from a dipyramid. If the crystal is large enough, the simplest method is to use a contact goniometer, as shown in Fig. 3-10. The crystal is placed between the jaws as shown, so that the two faces whose angle is to be measured are in contact with the jaws, and the edge between these faces is at right angles to them. The angle is then read from the scale.

An inexpensive goniometer can be purchased or one can be made from a protractor. Two arms of thin wood or plastic, shaped like the moveable one of Fig. 3-10, are cut out and then a pivot on which the arms can turn is put through them. It is not necessary or desirable that the arms be permanently attached to the protractor. One pair of edges (the inner edges to the right in the figure) must be exactly in line with the

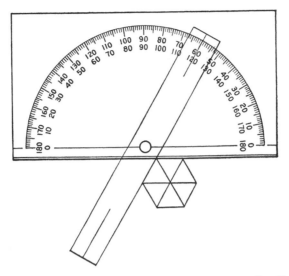

Fig. 3-10. Contact goniometer. (e.g., Ward's, Rochester, New York.)

center or pivot; between them the angle is read off when the arms are placed on the protractor, the pin then passing through its center and one edge through its zero. The other pair of edges (to the left) must be parallel to those mentioned first, so that they give the same angle; the two faces of the crystal whose angle is to be measured are placed between them, as already explained.

For the accurate measurement of angles or for those of a very small crystal, a reflecting goniometer is required. This expensive instrument requires crystal faces of high quality and considerable skill and experience on the part of the operator. For most current work in mineralogy, its use has largely been replaced by the use of single-crystal x-ray diffraction.

Symbols Used for Crystal Classes and Forms

In advanced books the student will find that mineralogists use various symbols for describing the crystal classes and crystal forms. Although a proper explanation of their derivation is beyond the scope of this book, they are mentioned here to alert the student to their existence and the basic way in which they are used.

Crystal Classes. Many systems of indicating the symmetry of these classes have been proposed, but the notations most widely accepted by crystallographers are the *Hermann–Mauguin symbols.* Using them, the symmetry is expressed as follows: axes of rotation are indicated by the numbers 1, 2, 3, 4, 6; axes of rotary inversion by numbers with lines above, as $\bar{1}, \bar{3}, \bar{4}, \bar{6}$ (read "bar one," "bar three," etc.). Symmetry planes are indicated by *m* (mirror plane). An axis of symmetry with a symmetry plane normal to it is indicated by the axis number over *m*, such as 2/*m*, 6/*m*.

It may at first appear that the Hermann–Mauguin symbols do not express the complete symmetry. But it should be pointed out, the symmetry elements operate on one another. For example, consider the mineral zircon, whose crystal class has the symmetry 4/m 2/m 2/m. The 4-fold axis is vertical and at right angles to a horizontal symmetry plane. The 2-fold axes lie at 45° to each other in a horizontal plane with symmetry planes normal to them. Operating with the 4-fold axis generates two more 2-fold axes and two more symmetry planes. The symmetry could be expressed as a 4-fold axis, four 2-fold axes, and five symmetry planes. Similarly, let us consider the symmetry of tourmaline and quartz. The symbol expressing the symmetry of tourmaline is 3m. This indicates that a 3-fold rotation axis lies in a mirror plane and that rotation of the axis would generate two other mirror planes. The symmetry would, therefore, be one 3-fold axis and three mirror planes. The symbol for quartz is 32. This indicates a 3-fold rotation axis with a 2-fold axis at right angles. Operation of the 3-fold axis generates two additional 2-fold axes.

Crystal Form. In advanced books on mineralogy, the forms, faces, and directions in crystals are expressed in simple numbers called *Miller indices* that relate to the intersection of crystal planes with the crystal axes. To differentiate among form, face, and direction, the Miller indices are enclosed as follows: form in braces { }, face in parentheses ( ), direction in brackets [ ].

In using a form symbol, one must specify the crystal system and crystal class, for the same index is used for different forms in the different systems. For example, below are given a few form symbols and the forms they indicate in the highest symmetry class of five of the crystal systems. The Miller indices of the forms in the hexagonal system have four digits instead of three.

Symbol	Isometric	Tetragonal	Orthorhombic	Monoclinic	Triclinic
{001}	Cube	Pinacoid	Pinacoid	Pinacoid	Pinacoid
{111}	Octahedron	Dipyramid	Dipyramid	Prism	Pinacoid
{101}	Dodecahedron	Dipyramid	Prism	Pinacoid	Pinacoid
{110}	Dodecahedron	Prism	Prism	Prism	Pinacoid

Specification of the crystal system, class, and Miller index indicates the number and arrangement of the faces on a crystal. For example, isometric {111} indicates to a mineralogist that this is an octahedron, with eight faces; trigonal {10$\bar{1}$1}, a rhombohedron, with six faces; tetragonal {110}, a prism, with four faces; and monoclinic {001}, a basal pinacoid, with two faces.

The existence of these indices again reflects the regular internal atomic architecture of crystals. For our study the origin of these indices will not be described; should the student encounter them with respect to crystal forms and cleavage, careful comparison of the descriptions in this book with those at a more advanced level will be needed to see the correspondence.

THE CRYSTAL SYSTEMS

The following descriptions consider for each of the six crystal systems the crystal axes, the symmetry of important crystal classes, and the common crystal forms.

Isometric System

All crystals of the isometric system are characterized by having four 3-fold or four $\bar{3}$-fold axes. Of the 32 crystal classes, five are in the isometric system, and although they have different symmetry, they can all be referred to three equal crystallographic axes at right angles to each other (Fig. 3-11). It will be seen that the arrangement of faces about the six ends of the axes of a given crystal is always the same. Only three of these five classes are common enough in minerals to deserve our attention. The name of the crystal class is that of the general form, the form whose faces intersect all three axes at different lengths.

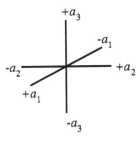

Fig. 3-11.
Isometric crystal axes.

Hexoctahedral Class. The *hexoctahedral* or *galena* class ($4/m\ \bar{3}\ 2/m$) is the most common class in the isometric system. It has the highest crystal symmetry with the following symmetry elements: four $\bar{3}$-fold axes, three 4-fold axes, six 2-fold axes, nine mirror planes, and a center. The principal forms in this class are the cube, octahedron, dodecahedron, trapezohedron, tetrahexahedron, and hexoctahedron. In the following descriptions, the perfect geometrical solid is described, but it should be made clear at the start that most crystals are not model perfect and are frequently malformed, as described on page 48.

Cube. The cube (Fig. 3-12) has six equal faces, each of which is a square, and the angle between any two adjacent faces is a right angle, or 90°. Galena and fluorite often occur in cubes.

Octahedron. A regular octahedron (Fig. 3-13) has eight similar faces, each an equilateral triangle; the angle between any two adjacent faces is 109°28'. Magnetite often occurs as octahedrons.

Dodecahedron. The rhombic dodecahedron (Fig. 3-14) has 12 equal faces, each of which is a rhomb with plane angles of 60° and 120°, the angle between the two adjacent faces being 120°. This is a common form of garnet.

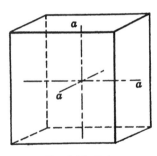

Fig. 3-12. Cube.

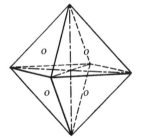

Fig. 3-13. Octahedron.

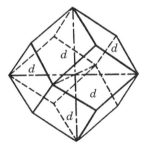

Fig. 3-14. Dodecahedron.

All these three forms may occur on the same crystal. Thus crystals of galena often show the cube and octahedron. Figure 3-15 is generally described as a cube modified by an octahedron, and Fig. 3-16 as an octahedron modified by the cube. If a cube is cut out of a block and the solid angles are sliced away carefully, the new surfaces making equal angles with three cube faces, the result is an octahedron. It is seen that the octahedral faces are little triangles on the solid angles of the cube and are equally inclined to the three cube faces. On the other hand, the cube faces are small squares on the six solid angles of the octahedron. The angle between adjacent faces of a cube and an octahedron is 125°16′.

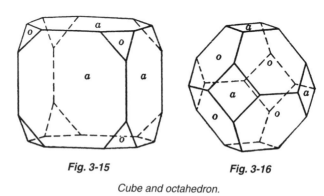

Fig. 3-15   **Fig. 3-16**

Cube and octahedron.

Figure 3-17 shows a combination of the cube and dodecahedron, and Fig. 3-18 shows a combination of the octahedron and dodecahedron. Both the cube and the octahedron have 12 similar edges, and these are cut off equally, or truncated, by the 12 faces of the dodecahedron. In Fig. 3-19 is shown a crystal with the combination of the cube (*a*), octahedron (*o*), and dodecahedron (*d*). The angle between adjacent faces of the cube and the dodecahedron is 135°; between those of the octahedron and the dodecahedron it is 144°44′.

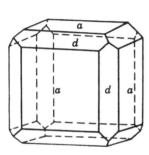
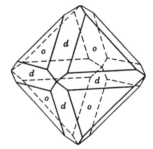
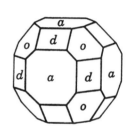

Fig. 3-17. *Cube and dodecahedron.*   **Fig. 3-18.** *Octahedron and dodecahedron.*   **Fig. 3-19.** *Cube, octahedron, and dodecahedron.*

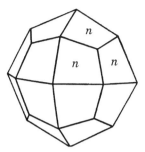

Fig. 3-20. Trapezohedron.

Trapezohedron. A trapezohedron has 24 equal faces, each a four-sided figure or trapezium. Unlike the forms already described, which are always the same, there are several different trapezohedrons, all having the same number of faces but differing in the angles between the faces. It requires a much more extensive study than is possible or necessary for the beginner, to learn how these forms are mathematically distinguished from each other. The trapezohedron in Fig. 3-20 is the most common and is frequently found in garnets.

Figures 3-21 to 3-24 show combinations of the trapezohedron (*n*) with the cube (*a*), octahedron (*o*), and dodecahedron (*d*). The last two are common combinations on garnet.

The trapezohedron is also called a tetragonal trisoctahedron because the form suggests an octahedron in which three faces take the place of a single octahedral face, each face being a four-sided figure or tetragon.

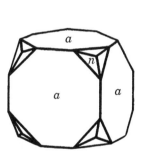

Fig. 3-21. Cube and trapezohedron.

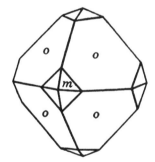

Fig. 3-22. Octahedron and trapezohedron.

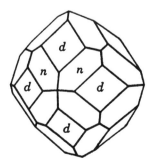

Fig. 3-23

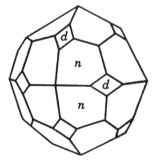

Fig. 3-24

Trapezohedron and dodecahedron.

There is also another trisoctahedron, called a trigonal trisoctahedron, shown in Fig. 3-25, which also has 24 faces, three of these also corresponding to an octahedron face, but each is a three-sided figure (trigon) or an isosceles triangle. This form does not often occur alone but may be seen on complex crystals of galena.

Tetrahexahedron. A tetrahexahedron (Fig. 3-26) has 24 faces, each face being an isosceles triangle and four together having the same position as the face of a cube. Figure 3-27 shows a combination of the cube and a tetrahexahedron; the latter is said to *bevel* the edges of the cube because the two planes are equally inclined to the two adjacent cube faces.

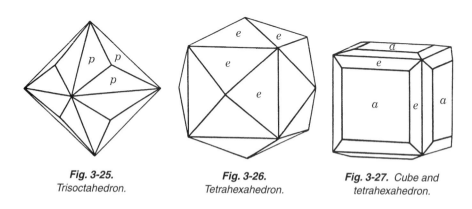

Fig. 3-25.
Trisoctahedron.

Fig. 3-26.
Tetrahexahedron.

Fig. 3-27. *Cube and tetrahexahedron.*

Hexoctahedron. A hexoctahedron (Fig. 3-28) is a 48-faced solid; each face is a scalene triangle, and six faces have the same general position as a face of an octahedron. This is the general form and gives its name to the crystal class.

Figure 3-29 shows a combination of the cube (*a*) with the hexoctahedron (*s*), as found on some fluorite crystals. Figure 3-30 is a common combination of forms in garnet with the hexoctahedron (*s*) beveling the edges of the dodecahedron (*d*).

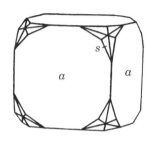
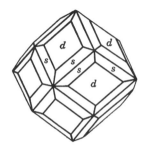

Fig. 3-28.
Hexoctahedron.

Fig. 3-29. *Cube and hexoctahedron.*

Fig. 3-30. *Hexoctahedron and dodecahedron.*

Figure 3-31 (cuprite) and Fig. 3-32 (a rare species, microlite) show some rather complex combinations of the forms described. In Fig. 3-31 the cube (*a*) and the dodecahedron (*d*) predominate; the faces of the octahedron (*o*) are small; *n* and β are faces of two different trapezohedrons. In Fig. 3-32 the octahedron (*o*) predominates, the cube (*a*) is intermediate, and the dodecahedron (*d*) is subordinate; the faces *m* belong to a trapezohedron, and the faces *p* belong to a trigonal trisoctahedron.

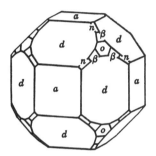

Fig. 3-31. Cuprite.

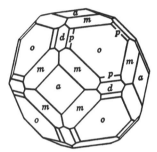

Fig. 3-32. Microlite.

Hextetrahedral Class. The *hextetrahedral* or *tetrahedrite* class ($\overline{4}3m$) has four 3-fold axes, four $\overline{4}$-fold axes, three 2-fold axes, and six mirror planes. This class may show the cube, dodecahedron, and tetrahexahedron but the only characteristic form is the tetrahedron. Some crystals in this class may appear to be malformed octahedrons but actually consist of combined positive and negative tetrahedrons.

Tetrahedron. The tetrahedron (Fig. 3-33) has four faces, each of them an equilateral triangle. It may be considered as the half-form of the octahedron, since half the faces of the octahedron, if every other one is taken, will, if extended, form a tetrahedron (Fig. 3-34). Reducing the symmetry also reduces the number of faces. The tetrahedron is the only common form in this class, but it is frequently found in combination with the cube, as shown in Figs. 3-35 and 3-36. It is seen that the tetrahedron faces (*o*) are

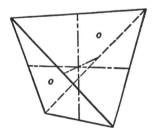

Fig. 3-33.
Positive tetrahedron.

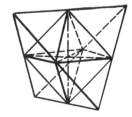

Fig. 3-34. *Relation between octahedron and tetrahedron.*

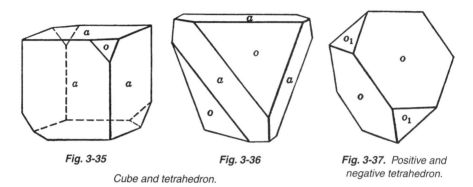

Fig. 3-35	Fig. 3-36	Fig. 3-37. Positive and

Cube and tetrahedron. (under Fig. 3-35/3-36) *negative tetrahedron.* (under Fig. 3-37)

present only on the alternate angles of the cube. Figure 3-36 illustrates a combination of a cube and a tetrahedron in which the latter predominates.

Figure 3-37 shows a combination of the tetrahedron (o) with a similar form (o_1) made up of the four remaining faces of the octahedron. It might be asked why this form cannot be regarded as an octahedron in which four faces are accidentally larger than the others. This is impossible, for it can be shown, by differences of luster, that the eight faces are not *all* alike but make up two forms with four faces each, a combination of the positive and the negative tetrahedron.

Pyritohedral Class. The *pyritohedral* or *pyrite* class ($2/m\ \bar{3}$) has four $\bar{3}$-fold axes, three 2-fold axes, and three mirror planes. This class may show the cube, octahedron, and rhombic dodecahedron, but the only characteristic form in this class is the pyritohedron. The lower symmetry in this class can be observed in the striations found on the cube face, which alternate in different directions, showing that the cube has only 2-fold symmetry instead of 4-fold as in the galena class.

Pyritohedron. The pyritohedron (Fig. 3-38) is a 12-sided solid, or dodecahedron, each face of which is a pentagon; it is thus sometimes called the pentagonal dodecahedron. In crystallography the name *dodecahedron* is usually given only to the rhombic dodecahedron, which has been described (Fig. 3-14), and this form, the pyritohedron, takes its name from the mineral pyrite or iron pyrites, on which it is a common form.

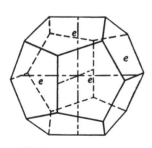

Fig. 3-38. Pyritohedron.

The pyritohedron, which is the half-form of the tetrahexahedron and thus has a lower symmetry, is frequently found in combination with other forms. If in combination with the cube, the solids in Figs. 3-39 and 3-40 result; Fig. 3-41 is a combination of an octahedron and a pyritohedron.

Tetragonal System

All crystals of the tetragonal system are characterized by having a single 4-fold or $\bar{4}$-fold axes. Of the 32 crys-

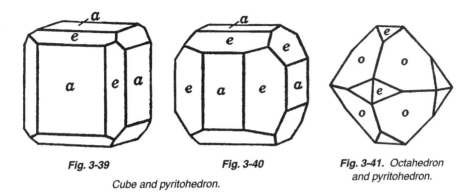

Fig. 3-39

Cube and pyritohedron.

Fig. 3-40

Fig. 3-41. Octahedron and pyritohedron.

tal classes, 7 are in the tetragonal system, and of these only two are important. All the crystals in the tetragonal system are referred to three crystallographic axes at right angles to each other (Fig. 3-42). The two horizontal *a* axes are equal and interchangeable, but the vertical axis, *c,* is of different length. This difference in length reflects a lower symmetry than in the isometric system.

Ditetragonal-Dipyramidal Class. The *ditetragonal-dipyramidal* or *rutile* class ($4/m\ 2/m\ 2/m$) is the symmetry class in which most tetragonal minerals fall. It consists of a 4-fold axis, four 2-fold axes, five planes, and a center, as shown in Figs. 3-43 and 3-44. Common forms include the tetragonal dipyramid, prism, and the basal pinacoid.

An examination of the drawings of tetragonal crystals will show that faces of the same kind are arranged in fours about the vertical axis, indicating that this axis is one of 4-fold symmetry. Intersecting in this axis are four symmetry planes, two of which include the horizontal crystallographic axes, and the other two planes lie between them at 45°. A fifth symmetry plane is at right angles to the others in the plane of the two horizontal crystallographic axes. All four axes of 2-fold symmetry lie in the horizontal symmetry plane; two are the same as the crystallographic axes, and the other two are at 45°.

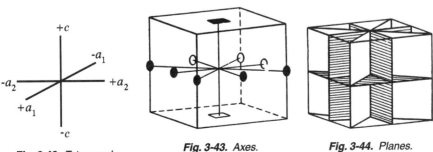

Fig. 3-42. Tetragonal crystal axes.

Fig. 3-43. Axes.

Fig. 3-44. Planes.

Tetragonal symmetry.

Tetragonal Prism and Basal Pinacoid. The tetragonal or square prism (Fig. 3-45) has, like the cube, angles of 90° between the faces, but it differs from the cube in that there are four rather than six faces. The two end faces in Fig. 3-45 do not belong to the square prism but are *basal planes* belonging to the form known as the basal pinacoid and therefore differ from and are not interchangeable with the prism faces. This difference is often shown in the crystal by a difference in smoothness or luster of the two kinds of faces or by a cleavage parallel to faces of one form and not to the other.

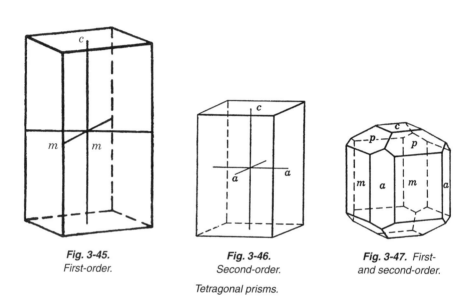

Fig. 3-45.
First-order.

Fig. 3-46.
Second-order.

Fig. 3-47. *First-
and second-order.*

Tetragonal prisms.

In addition to the square prism just described, there is another one, similar but placed at 45°, as shown in Fig. 3-46. Taken alone, these two forms cannot be distinguished from each other. Figure 3-47 shows the two prisms on the same crystal, the faces of one truncating the edges of the other.

On a careful examination of Figs. 3-45 and 3-46 it will be seen that in the prism first mentioned (Fig. 3-45), called the *first-order prism,* the *a* axes join the intersection edges of the prism faces. In Fig. 3-46, the *second-order prism,* the horizontal axes join the midpoints of opposite faces. When only one of the prisms is present, it is customary to set it up as the first-order prism.

Tetragonal Dipyramid. The tetragonal dipyramid, with its eight triangular faces, looks somewhat like an octahedron, but here the faces are isoceles triangles rather than equilateral triangles, and the angle between two faces over a horizontal edge differs from that over the sloping edges. Both angles are characteristic of a given species and differ from one species to another. The crystal form known as a pyramid is found only at one end of the crystal of a lower symmetry class. The term *dipyramid* as used here refers to two pyramids base to base with one pyramid at each end. There may be several tetragonal dipyramids of the same order on one crystal but differing in their angles and consequently, flatter or sharper at the apex.

Corresponding to the first-order tetragonal prism there is a first-order tetragonal dipyramid (Fig. 3-48), and at 45° to this, there is a second-order tetragonal dipyramid (Fig. 3-49), corresponding to the second-order square prism. Here again, it is impossible to distinguish between the two forms when only one is present. However, when both dipyramids are found on the same crystal, the faces of one can be seen to truncate the edges of the other. Figure 3-50 shows the two dipyramids in combination.

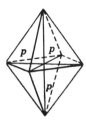 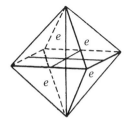 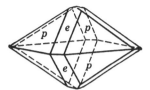

Fig. 3-48.   **Fig. 3-49.**   **Fig. 3-50.** First-
First-order.   Second-order.   and second-order.

Tetragonal dipyramids.

It should be noted that the *a* crystallographic axes in the first-order prism meet the crystal at the point of intersection of four dipyramid faces; in the second-order dipyramid the same axes meet the midpoints of the intersection of a face below with a face above.

In Fig. 3-51 is shown the combination of a first-order prism with a second-order dipyramid; in Fig. 3-52 the reverse combination is shown. Figure 3-52 resembles a cube modified by an octahedron (compare with Fig. 3-15), but it differs from it in that the angles between *p* and *c* are not equal to the angles between *p* and *a*. Figures 3-53

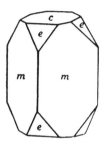 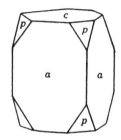

Fig. 3-51.   **Fig. 3-52.**

Combinations of prisms and dipyramids.

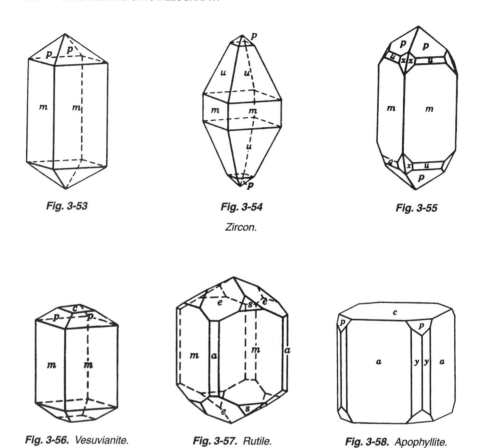

Fig. 3-53   **Fig. 3-54**   **Fig. 3-55**

Zircon.

Fig. 3-56. *Vesuvianite.*   **Fig. 3-57.** *Rutile.*   **Fig. 3-58.** *Apophyllite.*

to 3-55 represent combinations of forms found on crystals of zircon. Figure 3-56 represents vesuvianite; Fig. 3-57, rutile; and Fig. 3-58, apophyllite.

Besides the square prisms, there is the ditetragonal prism made up of eight similar faces. It is shown on the complex crystal represented in Fig. 3-59 with its faces lettered *h*. Corresponding to this prism is the ditetragonal dipyramid, consisting of two eight-sided pyramids base to base, as shown in Fig. 3-60. It is also shown in Fig. 3-59 lettered *z*, in combination with other forms. Figures 3-56 to 3-58 show some of the combinations of forms found on tetragonal crystals.

All the tetragonal forms that have been described belong to the symmetry given on page 27. There are other forms of lower symmetry, but only one, the disphenoid, is common enough to be mentioned.

Tetragonal-Scalenohedral Class. The *tetragonal-scalenohedral* or *chalcopyrite* class ($\overline{4}2m$) consists of a single $\overline{4}$ axis, two 2-fold axes, and two mirror planes. Common forms include the tetragonal prism, pyramid, and basal pinacoid, but the class is characterized only by the disphenoid.

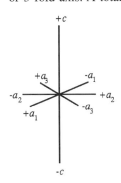

Fig. 3-59.
Scapolite.

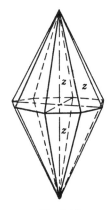

Fig. 3-60.
Ditetragonal dipyramid.

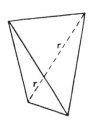

Fig. 3-61.
Disphenoid.

Disphenoid. The disphenoid is a four-faced solid (Fig. 3-61) resembling a tetrahedron but differs from it in that the faces are isosceles rather than equilateral triangles. It can be thought of as the half-form of the tetragonal dipyramid shown in Fig. 3-48. The mineral chalcopyrite commonly crystallizes as disphenoids that are very difficult to distinguish from tetrahedrons.

Hexagonal System

All crystals of the hexagonal system are characterized by having a single 6-, $\bar{6}$-, 3-, or $\bar{3}$-fold axis. A total of 12 of the 32 crystal classes are in the hexagonal system.

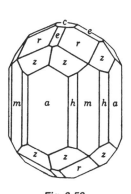

Fig. 3-62. *Hexagonal crystallographic axes.*

Because of that the system is divided into two divisions, a hexagonal division with seven classes characterized by a 6- or $\bar{6}$-fold axis and a trigonal (rhombohedral) division with five classes characterized by a 3- or $\bar{3}$-fold axis. Only the class of highest symmetry is described from each division because most of the common hexagonal minerals are in these classes.

The hexagonal system differs from most of the others in that it has four crystallographic axes of reference (Fig. 3-62) rather than three. The three a axes are of equal length and lie at 120° to each other in the horizontal plane. The fourth axis, c, is vertical and is either shorter or longer than the horizontal axes.

Hexagonal Division, Dihexagonal Dipyramidal Class. The *dihexagonal dipyramidal* or *beryl* class ($6/m\ 2/m\ 2/m$) is the one with the highest symmetry in the hexagonal division. It has one 6-fold axis, six 2-fold axes, seven mirror planes, and a center of symmetry, as shown in Figs. 3-63 and 3-64. Common forms include the dipyramid, hexagonal prism, and the basal pinacoid.

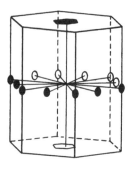

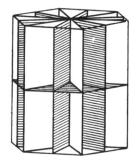

Fig. 3-63. Axes. **Fig. 3-64.** Planes.

Hexagonal symmetry.

Three of the 2-fold axes are coincident with the *a* crystallographic axes, and the other 2-fold axes lie midway between them in the horizontal plane. The 6-fold axis is coincident with the vertical crystallographic axis. Six of the symmetry planes are vertical, and each includes one of the 2-fold axes; the seventh plane is horizontal. On examination of the crystal drawings, one will see that this symmetry is shown by the 6-fold arrangement of faces about the end of the vertical axis and the 2-fold arrangement of faces about the ends of the horizontal symmetry axes.

Hexagonal Prism and Basal Pinacoid. The hexagonal prism shown in Fig. 3-65 is made up of six similar vertical faces, with angles of 120° between them. This is an illustration of a *first-order prism,* in which the crystallographic axes may be seen meeting the intersection edges of prism faces. If we rotate the crystal 30° about the vertical axis, we bring it into another position, which satisfies the symmetry requirements equally well. Here the *a* axes meet the center of each face at right angles, and in this position it is known as a *second-order prism* (Fig. 3-66). If only one of these prisms is present on a given crystal, it is normally oriented as a first-order prism with a face toward the observer. However, if both are present, as in Fig. 3-69, the best-developed one is set as the first-order and the other as the second-order prism.

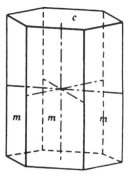

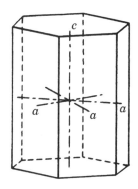

Fig. 3-65. First-order. **Fig. 3-66.** Second-order.

Hexagonal prisms.

It is obvious that since the hexagonal prism is made up of only vertical faces, it is an open form and another form must be in combination with it. In Figs. 3-65 and 3-66 the *basal pinacoid* is shown at the top and bottom of the crystals. The basal pinacoid is then made up of only two similar faces at opposite ends of the *c* axis.

Hexagonal Dipyramid. There are two hexagonal dipyramids, corresponding to the two hexagonal prisms. These dipyramids are made up of 12 faces, each an isosceles triangle, six above and six below. The angles that the faces make with each other or with the prism faces are characteristic of a given species.

Figure 3-67 illustrates the dipyramid of the first order and Fig. 3-68 the dipyramid of the second order. If these figures are compared with Figs. 3-65 and 3-66, it will be seen that the faces of each of these dipyramids lie directly above and below the faces of the corresponding prism. If a dipyramid and a prism of different orders appear on the same crystal, it is customary to set the dipyramid as first order. Figure 3-69, a beryl crystal, shows a combination of first-order (*m*) and second-order (*a*) prisms, first-order (*p*) and second-order (*s*) dipyramids, and the basal pinacoid, *c*.

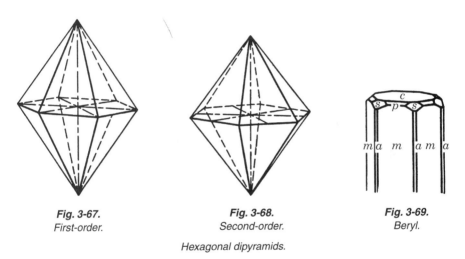

| Fig. 3-67. | Fig. 3-68. | Fig. 3-69. |
| *First-order.* | *Second-order.* | *Beryl.* |

Hexagonal dipyramids.

In addition to the prisms mentioned, there is a 12-sided prism known as the dihexagonal prism. This form is made up of 12 similar vertical faces grouped in pairs at the ends of the three horizontal crystallographic axes (Fig. 3-70). Although the faces are similar in appearance, the alternate angles are different, and there are thus only the six vertical planes of symmetry.

Corresponding to the dihexagonal prism there is a *dihexagonal dipyramid,* bounded by 12 similar triangular faces, six above and six below. This form, shown in Fig. 3-71, is usually present only as small faces in combination with other forms. Figure 3-72 is a drawing of a beryl crystal in which the faces lettered *n* and *v* are dihexagonal dipyramids. Figure 3-73 is an enlarged map and shows the faces as they would appear on looking directly down on a beryl crystal much like the one in Fig. 3-72. Note also in Fig. 3-73 the first-order prism (*m*) and dipyramids (*u* and *p*), the second-order prism (*a*) and dipyramid (*s*), and the basal pinacoid (*c*).

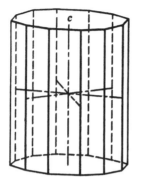

Fig. 3-70. *Dihexagonal prism.*

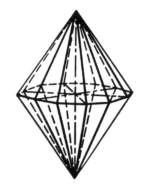

Fig. 3-71. *Dihexagonal dipyramid.*

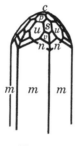

Fig. 3-72

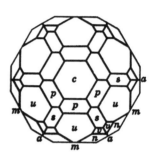

Fig. 3-73

Beryl.

Trigonal Division, Trigonal-Scalenohedral Class. The *trigonal-scalenohedral* or *calcite* class ($\overline{3}2/m$) is the class in the trigonal division that has the highest symmetry. Although important minerals such as quartz, tourmaline, and dolomite occur in other classes of the trigonal division, this is the only class that we describe in detail. As shown in Figs. 3-74 and 3-75, there are three vertical planes of symmetry at 60° to each other,

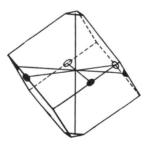

Fig. 3-74. *Axes.*

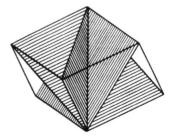

Fig. 3-75. *Planes.*

Symmetry: trigonal-scalenohedral class.

three 2-fold symmetry axes coinciding with the three *a* axes, one $\bar{3}$-fold symmetry axis coinciding with the *c* axis, and a center of symmetry. In the forms belonging to this class, the faces are in threes or groups of three about the extremities of the vertical axis.

Rhombohedron. The rhombohedron is a six-sided solid, each face of which is a rhomb, as shown in Figs. 3-76 to 3-78. There are a great many rhombohedrons differing in the angles between the faces, and thus the general appearance may be flattened and obtuse (Fig. 3-76) or elongated and acute (Fig. 3-78). The rhombohedron looks somewhat like a cube, with the line joining two opposite angles placed in a vertical position. In fact, the cube may be thought of as a special case of the rhombohedron with face angles of 90° and as coming between the obtuse and the acute rhombohedrons.

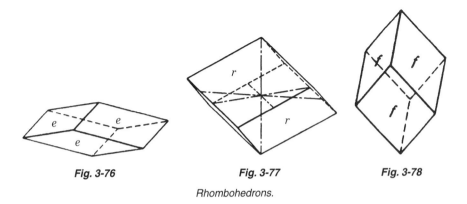

| Fig. 3-76 | Fig. 3-77 | Fig. 3-78 |

Rhombohedrons.

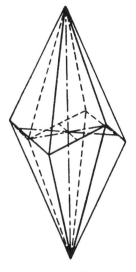

Fig. 3-79.
Scalenohedron.

Scalenohedron. The scalenohedron is a 12-sided solid (Fig. 3-79), looking a little like a double six-sided pyramid, but the faces are scalene triangles and the edge between top and bottom faces is zigzag, up and down, instead of horizontal as in the dipyramid. Moreover, only the alternate angles between the faces over the edges that meet in the vertex are alike; in other words, there are two sets of three each, those of one set being more obtuse than those of the other.

The number of species crystallizing in the trigonal-scalenohedral class is very large, and the crystals of some of them, as, for example, calcite, are highly complex. In the calcite shown here (Figs. 3-80 to 3-84) the faces *r, f,* and *e* belong to different rhombohedrons; *v,* to a scalenohedron; *m* is the first-order hexagonal prism; *c,* the basal plane.

Figure 3-85 represents a more complex crystal, also of calcite, and Fig. 3-86 gives a basal projection of it. Here there are several rhombohedrons, *r, e,* and *f;* the scalenohedrons, *v* and *t;* and the prism, *m.* These figures show well the symmetry about three planes meeting at angles of 60°.

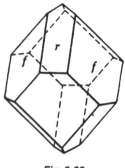

Fig. 3-80

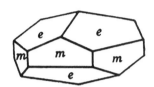

Fig. 3-81

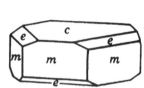

Fig. 3-82

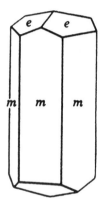

Fig. 3-83

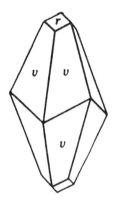

Fig. 3-84

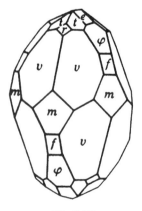

Fig. 3-85

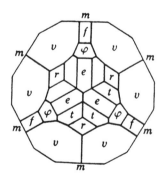

Fig. 3-86

Calcite.

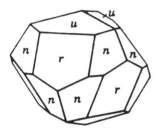

Fig. 3-87. Hematite.

Figure 3-87 shows a crystal of hematite; *u* and *r* are faces of two rhombohedrons, and *n*, the face of a scalenohedron.

Orthorhombic System

Crystals of the orthorhombic system are characterized by the presence of either two 2-fold axes or two mirror planes and the absence of any 3-, 4-, or 6-fold symmetry axes. The forms of the crystal classes in the orthorhombic system are referred to three crystallographic axes of unequal length at right angles to each other (Fig. 3-88). There is no unique *c* axis. The axes are interchangeable and which axis is chosen as *c*, the vertical axis, has been determined largely on the habit of the crystal. If crystals are elongate in one direction, this direction is usually chosen as *c*. If, however, the crystals are tabular with a prominent pinacoid, the *c* axis may be chosen at right angles to it. In some cases the intersection of two equivalent cleavage directions, as in barite, has been taken as the *c* axis. There are three crystal classes in the orthorhombic system, but only the one with highest symmetry, the rhombic-dipyramidal, is described.

Rhombic-Dipyramidal Class. The *rhombic-dipyramidal* or *barite* class (2/m 2/m 2/m) has three mutually perpendicular symmetry planes, three axes of 2-fold symmetry coincident with the crystallographic axes at the intersections of the symmetry planes, and a symmetry center (Figs. 3-89 and 3-90). The principal forms in this class are the prism (Fig. 3-91), dipyramid (Fig. 3-92), and the pinacoid (Fig. 3-93).

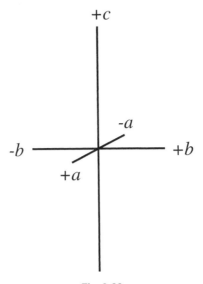

Fig. 3-88.
Orthorhombic crystal axes.

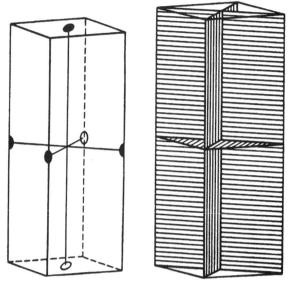

Fig. 3-89. *Axes.*   **Fig. 3-90.** *Planes.*

Orthorhombic symmetry.

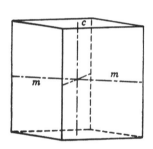

Fig. 3-91.
Prism and base.

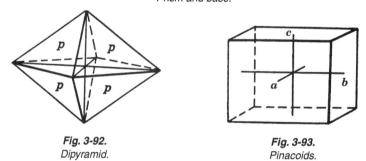

Fig. 3-92.
Dipyramid.

Fig. 3-93.
Pinacoids.

Orthorhombic crystal forms.

Prism. The orthorhombic prism is made up of four similar faces, as shown in Figs. 3-91, 3-94, and 3-95. Unlike the tetragonal system, in which the prism is vertical, here there are three prisms of four faces each, one parallel to each of the three crystal axes. They are designated as follows:

First-order prism: faces parallel to *a;* intersects *b* and *c*

Second-order prism: faces parallel to *b;* intersects *a* and *b*

Third-order prism: faces parallel to *c;* intersects *a* and *b*

In the mineral descriptions in Chapter 7, when it is stated that crystals are prismatic or have a prismatic cleavage, unless otherwise stated, the third-order prism is implied.

In no case will the angles between adjacent faces of a prism be 90°; angles measured over alternate edges will be alternately acute and obtuse. The orthorhombic prism is an open form and therefore must be accompanied on a crystal with another form or forms. In Figures 3-91, 3-94, and 3-95, the enclosing form is a pinacoid.

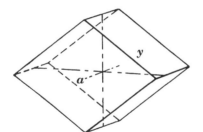

Fig. 3-94. *First-order prism and front pinacoid.*

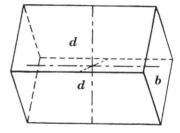

Fig. 3-95. *Second-order prism and side pinacoid.*

Dipyramid. The orthorhombic dipyramid is an eight-faced form with four faces above and four below. It resembles somewhat the tetragonal dipyramid, but because of the difference in length of the horizontal axes, its cross section is a rhomb, not a square. Although the eight faces are similar, the edges between them belong in three sets, with different angles for each. There may be a variety of orthorhombic dipyramids, differing in their angles, and each corresponding to a given prism. In Fig. 3-99, *m* and *s* are third-order prisms, and *e* and *f* are the corresponding dipyramids.

Pinacoids. Unlike the tetragonal and hexagonal systems in which there is a single pinacoid that intersects the vertical axis, *c,* there are now three possible pinacoids, each of which intersects one crystallographic axis and parallels the other two:

Front or *a* pinacoid intersects *a;* parallels *b* and *c.*

Side or *b* pinacoid intersects *b;* parallels *a* and *c.*

Basal or *c* pinacoid intersects *c;* parallels *a* and *b.*

In Figure 3-93 the three pinacoids are shown in combination. This resembles a cube since the angles between the faces are 90°, but differs from it in that the faces belong in three sets, which are not similar to each other. Some crystals show the difference well, in variation, in luster, or in etching, or a cleavage may be parallel to one set and not to the others.

Figures 3-96 to 3-98 show orthorhombic crystals with various combinations of forms. Figure 3-99 illustrates well the various prisms; here faces lettered *h* and *k* are first-order; *d,* second-order; *m* and *s,* third order. Figure 3-100 is a projection of the same crystal on the base.

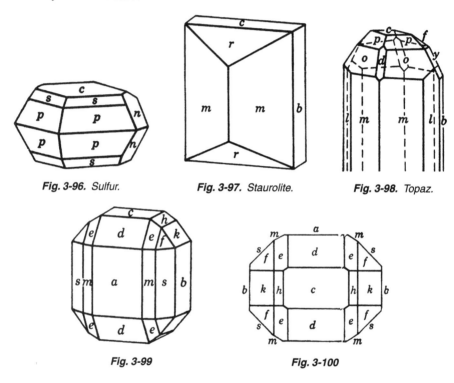

Fig. 3-96. *Sulfur.*   **Fig. 3-97.** *Staurolite.*   **Fig. 3-98.** *Topaz.*

Fig. 3-99   **Fig. 3-100**

Monoclinic System

Fig. 3-101.
Monoclinic crystal axes.

The monoclinic system is characterized by having a single 2-fold axis coincident with the *b* axis, a single mirror plane that lies in the *a*–*c* plane, or both a symmetry plane and a 2-fold axis. The crystal forms of the monoclinic system are referred to three crystal axes of different length: the vertical axis, *c;* the right–left axis, *b;* and the front–back axis, *a.* Axis *b* is at right angles to *a* and *c,* but *a* is inclined at an angle β to *c* and to the *b*–*c* plane (Fig. 3-101). The system is called *monoclinic* because of this one inclined axis. Of the three crystal classes in the monoclinic system, we consider only one, the prismatic class.

Prismatic Class. The *prismatic* or *gypsum* class (2/m) has a single mirror plane, at right angles to a 2-fold symmetry axis and a symmetry center (Figs. 3-102 and 3-103). There are only two types of forms, the prism and the pinacoid. It is called the prismatic class because the fourth-order prism is the general form (i.e., its faces intersect all three axes).

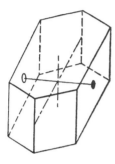 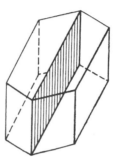

Fig. 3-102. *Axis.*   **Fig. 3-103.** *Plane.*

Monoclinic symmetry.

Prism. Four different four-faced prisms are present in the monoclinic system (Figs. 3-106 to 3-109). Two of them are similar to corresponding prisms in the orthorhombic system. These are the first-order, parallel to the *a* axis, and the third-order (vertical prism), parallel to the *c* axis. Because of the inclined *a* axis the second-order prism does not exist; neither does the dipyramid. The eight faces of the orthorhombic dipyramid become two independent prism forms, the faces of which intersect all three axes. These are fourth-order prisms, one positive (Fig. 3-108), the other negative (Fig. 3-109). The prism is positive if the faces at the upper end of the crystal intersect the positive end of the *a* axis; if they intersect the negative end of *a,* the prism is negative.

Pinacoids. In the monoclinic system as in the orthorhombic there are three pinacoids, each of which is parallel to two crystallographic axes and intersects the third. They are the front pinacoid, *a;* the side pinacoid, *b;* and the basal pinacoid, *c.* Because of the inclination of the *a* axis, the basal pinacoid slopes to the front, so that the angle between pinacoids *a* and *c* is obtuse. However, pinacoids *b* and *c* and *a* and *b* are at right angles to each other (Fig. 3-104).

We have seen that because of the inclination of the *a* axis the second-order prism is not present in the monoclinic system. Instead, there are two pinacoids called second-order. The positive form lies over the obtuse edge between the front and basal pinacoids, and the negative form lies over the acute edge. It should be emphasized that these are independent forms and that the presence of one does not necessitate the presence of the other.

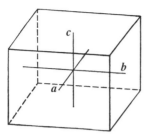

Fig. 3-104.
Monoclinic pinacoids.

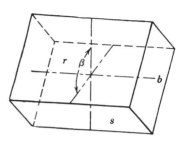

Fig. 3-105.
Second-order.

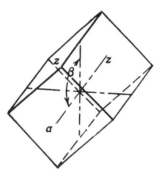

Fig. 3-106.
First-order prism.

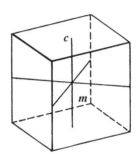

Fig. 3-107. *Monoclinic
third-order prism and base.*

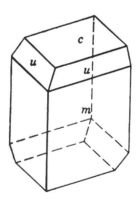

Fig. 3-108. *Positive.*

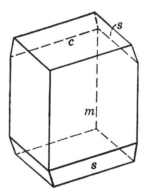

Fig. 3-109. *Negative.*

Fourth-order prisms.

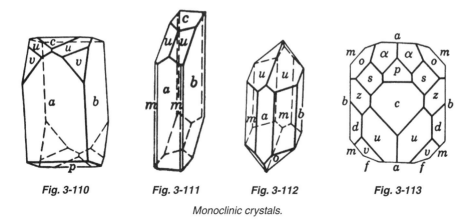

| Fig. 3-110 | Fig. 3-111 | Fig. 3-112 | Fig. 3-113 |

Monoclinic crystals.

Figures 3-110 to 3-113 are drawings of various monoclinic crystals with the forms lettered as follows: *m* and *f,* third-order prisms; *u* and *v,* positive fourth-order prisms; *o* and *s,* negative fourth-order prisms; *a, b,* and *c,* pinacoids. Figure 3-113 shows at right angles to the *c* axis a projection of a complex crystal; here the symmetry parallel to the *b* faces is clearly exhibited.

Triclinic System (Anorthic)

The triclinic system is characterized by the absence of all symmetry except a 1-fold or a $\bar{1}$-fold axis of symmetry. The crystal forms are referred to three unequal axes, all oblique to each other (Fig. 3-114). The system is named because of the three oblique axes. Of the two possible crystal classes, only one, the pinacoidal, is described here.

Pinacoidal Class. The *pinacoidal* or *albite* class ($\bar{1}$) possesses only a center of symmetry. As a consequence, the only crystal form permitted is the pinacoid, a two-face form, with faces on opposite sides of the crystals—hence the name of the class.

Figure 3-115, axinite, and Fig. 3-116, albite feldspar, show two triclinic crystals. Here it is seen that the like planes are in sets of two each: one in front, the other

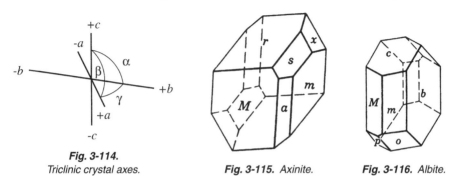

Fig. 3-114.
Triclinic crystal axes.

Fig. 3-115. Axinite.

Fig. 3-116. Albite.

Triclinic crystals.

behind, represented in dashed lines. In Fig. 3-116 there is some resemblance to a monoclinic crystal, but the angle between the faces *b* and *c* is not 90°, as it must be there; moreover, the angles *bm* and *bM* are different, as are also the angles *bo* and *bp*. Hence *m* and *M* are different forms, as are *o* and *p*. The subject of triclinic crystals will not be carried further, because of its great complexity. Fortunately, only a few common minerals crystallize in the triclinic system, so that the beginner is not often confronted with study of them and the problem of their orientation.

Twins

When one finds that two or more individual crystals of the same mineral have grown together such that certain crystallographic directions in them are parallel whereas others are in reverse position, the intergrowth is called a *twin* or *twin crystal.* By forming twins, the combined crystals achieve a higher symmetry than that allowed by the internal symmetry of the mineral.

The different orientations of the two individuals are related to each other in one of two ways: (1) as though one crystal has been rotated 180° about an axis common to both called the *twin axis* (frequently a 3-fold symmetry axis but never a 2-, 4-, or 6-fold axis), or (2) as though one part had been generated from the other by reflection over a plane called the *twin plane,* which is usually parallel to a face of a common crystal form but is never parallel to a symmetry plane. Rotation on an even-fold symmetry axis or a reflection across a symmetry plane would bring all parts of both crystals into parallel position. A twin is designated by its twin plane or twin axis, known as its *twin law.* In addition, it is designated as a *contact twin* or a *penetration twin.* In a contact twin the two individuals are in contact on the composition plane, usually the twin plane (Fig. 3-117), whereas in the penetration twin, the two crystals interpenetrate each other (Fig. 3-118). Contact twins can be thought of as though one part were derived from the other by reflection across a plane (Fig. 3-119), as in the illustration of the gypsum twin (Fig. 7-115). Figure 3-120 shows a contact-twin of columbite in which the difference in direction of the striations of the two halves also shows that the crystal is twinned. Figure 3-121 is a penetration twin of staurolite.

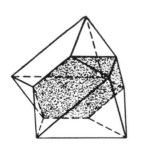

Fig. 3-117. *Twinned octahedron
(spinel twin).*

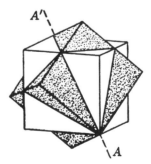

Fig. 3-118.
Fluorite penetration twin.

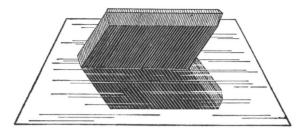

Fig. 3-119. Twinning by reflection.

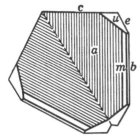

Fig. 3-120. Columbite.

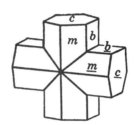

Fig. 3-121. Staurolite.

If more than two individuals are twinned according to the same law, the aggregate is called *repeated* or *multiple twinning.* If in a multiple twin the twin planes are not all parallel but are parallel to other faces of the same form, it is called a *cyclic twin,* a twin commonly found in rutile (Fig. 3-122). Figures 3-123 to 3-125 show repeated twins, which are often very regular. The twin aggregate may be made up of perhaps three, five, six, or even eight parts of crystals or complete crystals, arranged symmetrically to resemble a star in many instances. These twins must be examined closely, as they may on first examination resemble a hexagonal crystal (Fig. 7-87).

Fig. 3-122. Rutile twin.

If the twin planes are closely spaced and all parallel, it is called *polysynthetic twinning.* Figure 3-145 shows polysynthetic twinning lamellae in a crystal of magnetite. However, it is best illustrated by a piece of a triclinic feldspar showing fine lines or striations on a surface of basal cleavage; these lines are simply the edges of the thin successive parallel plates. If the specimen is held so as to catch the reflection of light from a distant window and is turned through a very small angle, first one set of edges reflects and then the other set. This type of twinning is illustrated by albite in Fig. 7-140.

It must be understood that in most specimens showing twinning, the revolution spoken of has not actually taken place. The rule is simply given in this form to show best the geometrical relations of the two parts. Still it is most interesting to note that

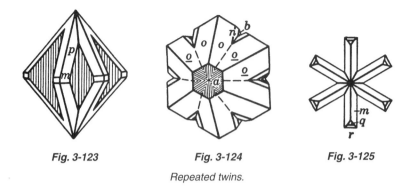

| *Fig. 3-123* | *Fig. 3-124* | *Fig. 3-125* |

Repeated twins.

in a few instances it is possible to cause the atoms of part of a crystal to change their position and produce twinning artificially, as, for example, by pressure in the proper direction.

Thus Fig. 3-126 represents a cleavage piece of calcite placed with an obtuse edge on a firm surface and then pressed by a knife (not too sharp) at *a*. Steady uniform pressure serves to reverse the position of the atoms in the part lying to the right so that the whole is pushed to the side and assumes a twinned position with reference to the rest. If the pressure is applied skillfully, no change in the transparency of this part takes place, and the new surface *gce* is perfectly smooth. It should be noted that to perform this experiment successfully, clear calcite known as *Iceland spar* should be used. In nature, pressure may have produced twinning after the formation of the crystal; it is then called *secondary twinning*. The twinning layers or lamellae observed in most cleavage masses of otherwise clear calcite may often be explained in this way. The similar layers that are common on large crystals of pyroxene and cause a separation or "parting" parallel to the basal plane, appearing much like the easy fracture called *cleavage,* are caused by twinning.

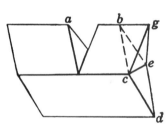

Fig. 3-126.
Calcite: artificial twinning.

It is evident that since the crystal faces on a single specimen of a species may be grouped in a great variety of ways, it is not always easy to decide whether or not a given intergrowth is a twin; the decision often requires careful study, exact measurement of angles, and optical examination. For example, it is common to find quartz crystals crossing each other at a great variety of angles, but real twins like that in Fig. 3-128, known as a *Japanese twin,* are rare. Minerals in each crystal class have characteristic twins that may help identify them.

Isometric System. The isometric system has an abundance of even-fold axes and mirror planes. The only common form not a mirror plane is the octahedron. Twins are either contact twins, in which two crystals are joined on an octahedron face as is common in spinel and diamond (Fig. 3-117), or penetration twins as in fluorite (Fig. 3-118).

Tetragonal System. In the tetragonal system there are no mirror planes parallel to either tetragonal dipyramid. The faces of these forms can thus be twin planes. In cassiterite (Fig. 3-127) and rutile (Fig. 3-122), the second-order tetragonal dipyramid is a twin plane. Repeated twinning can produce a hexagonal shape called a cyclic twin (see Plate II-2).

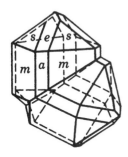

Fig. 3-127.
Cassiterite.

Hexagonal System. Twins are relatively rare within the hexagonal division largely because twin axes cannot be 6- or 2-fold nor twin planes parallel to symmetry planes. However, within the trigonal division a number of twins are important. In calcite, twinning on the basal pinacoid, the positive rhombohedron, and secondary twinning on a negative rhombohedron are most common. The latter is the type shown in Fig. 3-126. Two or three types of twinning are common in quartz, and it is difficult to distinguish twinned from untwinned crystals. However, interrupted striations or paired vicinal faces may help. In most cases quartz is twinned according to one or both of two twin laws: the *Brazil twins,* with the second-order prism as the twin plane, and the *Dauphine twins,* with *c* as the twin axis. These are penetration twins and rarely show reentrant angles or any morphological evidence of twinning. The *Japanese twin* (Fig. 3-128) is the only common quartz twin showing a reentrant angle.

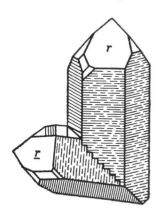

Fig. 3-128.
Quartz, Japanese twin.

Orthorhombic System. The most common twins in the orthorhombic system are the aragonite contact twins (Figs. 7-86 and 7-87). In this case the twin planes are third-order prism faces. Repeated twinning, as in chrysoberyl, often produces a pseudohexagonal form (Fig. 3-124). However, careful inspection will show that along one junction, the separate members join along an irregular surface.

Monoclinic System. Because of the much lower symmetry in this system, several types of twinning are not only possible but are common. Twinning on the front pinacoid is common in gypsum (Fig. 7-115), pyroxene, and hornblende. In orthoclase the most common twinning, known as *Carlsbad* twin, is a penetration twin with *c* as the twin axis and joined on an irregular surface essentially parallel to the side pinacoid (Fig. 3-129).

Triclinic System. Since there is a complete absence of either mirror planes or 2-fold axes, twinning may occur in any orientation. In albite feldspar there is repeated

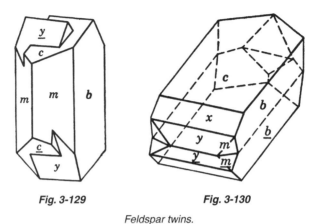

Fig. 3-129   Fig. 3-130

Feldspar twins.

twinning with the side pinacoid as both the twin and the composition plane; this is called *albite twinning* (Fig. 7-140). In microcline twinning occurs not only on the side pinacoid but it is also twinned according to the *pericline law,* with a twin axis perpendicular to the side pinacoid. In addition, twins similar to those of the monoclinic system may be present.

Crystal Habit

The *habit* of a crystal is its characteristic and common form. For many minerals the crystal habit is so characteristic that it serves as a means of identification. Such properties can be discerned and described without specific knowledge of crystallography.

A number of terms are used to describe the general shape of single crystals, which in many cases is the habit. Thus a mineral may be described as *acicular* if the mineral occurs in the form of needles such as natrolite; *capillary or filiform,* as in millerite, when it occurs in hairlike masses; or *columnar,* as in tourmaline, if the elongate crystals are thicker than 1 mm. When the crystals are flat as in a board, they are said to be *tabular* as in barite, but if elongate and flattened like a knife blade as in kyanite, they are called *bladed.* Twinning is so characteristic of some minerals that it can be considered as a habit, particularly in the feldspars (Figs. 3-129 and 3-130).

Malformed Crystals

Most of the crystals of minerals would give a poor impression of nature's workmanship to one who always expected to see them exactly like carefully made models, or like the figures shown earlier. The cubes of galena that we find are often flattened or drawn out. An octahedron (Fig. 3-131) may be flattened to look like Fig. 3-132; a dodecahedron (Fig. 3-133) may take the forms shown in Figs. 3-134 and 3-135. These forms are not poorly made, like a poor model; on the contrary, the size of the similar faces on a crystal may vary, and therefore the shape of the solid as a whole may vary, but the *angles* between them remain the same. Moreover, when we study a

Fig. 3-131.
Octahedron.

Fig. 3-132.
Malformed octahedron.

Fig. 3-133.
Dodecahedron.

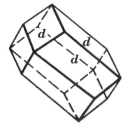

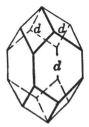

Fig. 3-134

Fig. 3-135

Malformed dodecahedrons.

crystal more carefully, we find that what is really essential is not the size or shape of each face but the way in which the atoms are arranged. For example, in a cube the fact that the structure is the same in the three directions parallel to the crystal axes is an essential point. It follows that in the cube not only are the angles between two adjacent faces always 90°, but the six cube faces are all similar. Therefore, if there is easy fracture, called *cleavage,* parallel to one cube face, there will be the same cleavage parallel to the others. Moreover, the etching and luster on all the cube faces will be the same, and the actual size of the faces is a matter of no importance. In fact, in one species the cubes are sometimes lengthened so that they are like fine hairs.

Similar remarks can be made in regard to the malformed octahedron and dodecahedron already illustrated, and indeed, about any malformed crystal. The symmetry in the atomic structure, and hence the angles between the faces, remain unchanged, although the external geometrical symmetry is not that of the ideal model.

Another good example of what is possible in a malformed crystal can be explained by referring to Fig. 3-136, a cube with octahedral faces on its solid angles. Instead of this ideal form, it is common to find in natural crystals no two of the triangular octahedral faces of the same size; some of them may even be absent; the cube faces also vary (Fig. 3-137). But such a crystal is not essentially different from Fig. 3-136, for every octahedral face is identical with each of the others and is equally inclined to the three adjacent cube faces, that is, to the three crystallographic axes, even if all the faces differ in size. In other words, unless the sizes of the faces vary systematically (Fig. 3-37), it is always the position of the face, not its size, which is essential.

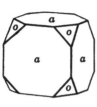
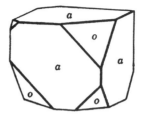

Fig. 3-136 **Fig. 3-137**

Cube and octahedron.

In the same way, a cube may in nature look like a square tetragonal prism, for all the angles between the faces are right angles in both cases, and the goniometer will not tell the difference between them, as has already been explained, but the atomic structure of the two is not to be confused. In the tetragonal prism there is the same arrangement in the horizontal directions but a different arrangement along the vertical direction. Hence the square top of the crystal appears different from the four vertical faces, and we may find cleavage parallel to one set and not to the other. For example, the mineral apophyllite forms pseudocubes in combination with a dipyramid that resembles a cube and octahedron (Fig. 3-138), but there is a cleavage parallel to the base (*c*) which imparts a pearly luster, while a high glossy luster occurs on the faces of the square prism (*a*).

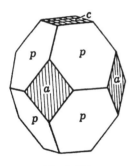

Fig. 3-138.
Apophyllite.

Because so much variation is possible in the size of the similar faces on a crystal, and hence in the shape of the whole, the practical study of natural crystals is much more difficult than the study of models that give the ideal geometrical symmetry. A careful comparison of the well-developed crystals in Fig. 3-139 with the wooden models (Fig. 3-140) that represent them will show that it is the exceptional crystal that is a perfect geometrical solid. Most crystals are so implanted on the rocks or embedded in them that only part of the crystal has been developed. Thus a quartz crystal is often attached at one extremity, and only the other end has had a chance to grow freely. Or crystals may be implanted on a surface of rock so that only a series of minute faces and angles is visible. In that event the study of the crystal is a difficult matter requiring much skill and experience, and the beginner should not be discouraged because one cannot at once tell the form of a crystal. Even here, however, some conclusion can often be drawn from the shape of the faces; thus if they are equilateral triangles, they probably belong to an octahedron; if rhombs, to a rhombic dodecahedron; and so on.

Besides the malformed crystals just considered, which although they look irregular are really perfect in regard to the position of the faces and the angles between them, there are others that are really deformed. Some unusual conditions attending

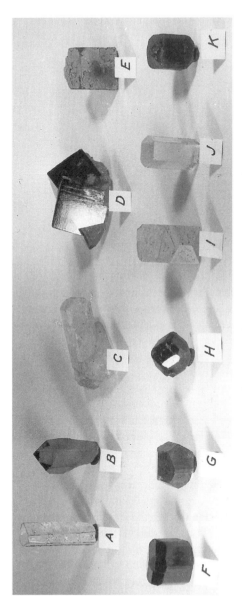

Fig. 3-139. Crystals: A, calcite; B, quartz; C, barite; D, fluorite; E, galena; F, tourmaline; G, pyrite; H, garnet; I, orthoclase; J, beryl; K, vesuvianite.

Fig. 3-140. Models.

the growth of the crystal or perhaps some force that has acted upon it since it was formed may have bent or twisted it out of its normal shape. Such crystals vary not only in the size and shape of the faces but also in the angles between the faces. Thus the faces may be curved, as in the barrel-shaped crystals of pyromorphite, or convex as in common in crystals of the diamond; or the whole crystal may be bent, like some crystals of quartz or stibnite, or some kinds of chlorite and gypsum (Fig. 3-141).

Fig. 3-141. *Gypsum.*

Aside from this curving and twisting, a crystal may have had its shape more or less changed by a force exerted in the rock since it formed. It may even have been broken and later cemented together, so that many irregularities may result. Figure 3-142 shows a crystal of beryl that has been broken into many pieces; these have been slightly displaced from each other, and the whole has been cemented together by quartz.

Fig. 3-142. *Beryl crystal broken and cemented by quartz.*

Other irregularities of crystals occur besides those mentioned. The faces of crystals, instead of being perfectly smooth, are often rough, perhaps because they are made up of a multitude of points. Or they may be covered with fine lines, or *striations,* like those on the cubic faces of pyrite (Fig. 3-143), which are explained by the successive combination of faces of another form (the pyritohedron) in narrow lines with the cube face. In the magnetite crystal (Fig. 3-144) the fine lines represent striations on a dodecahedral face due to the presence of the octahedron. This *oscillatory combination,* as it is called, may even make the crystal nearly round, like some prismatic crystals of tourmaline. Striations may be due to twinning, as is common with the triclinic feldspars (see p. 45; also Fig. 7-140). Figure 3-145 shows an octahedral crystal of magnetite with twinning lamellae appearing as striations on an octahedral face.

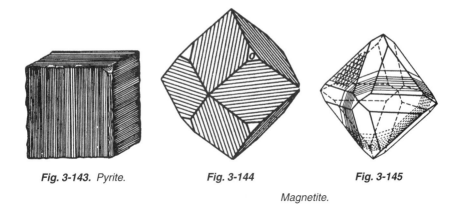

Fig. 3-143. Pyrite.   *Fig. 3-144*   *Fig. 3-145*

Magnetite.

Other crystals may have faces with a multitude of little elevations or depressions, the latter like the pits spoken of on page 14 as having been produced by etching; in fact, they can sometimes be explained as etching by nature. The same cause—the action of some partial solvent after the formation of the crystal—often explains the rough faces to which we have alluded. The careful examination of some crystals may show the replacement of a face by two or more others varying a little from it in angular position. The four slightly raised faces that take the place of the cube face on some English fluorite crystals are good examples. Such planes are often called *vicinal* planes (from the Latin *vicinus,* neighboring).

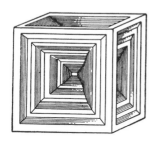

Fig. 3-146. Halite.

Crystals that have formed rapidly may have only a more or less regular skeleton shape, like the crystal of salt represented in Fig. 3-146 and the drawing (Fig. 7-60) on p. 192. Some salt crystals show one face distinctly but with a depression in the center, so that they are called hopper-shaped crystals. The cavernous crystals of pyromorphite and vanadinite are other examples. Crystals often enclose foreign substances of both solid and liquid material. Some quartz crystals contain water, occasionally with a movable bubble of gas. In others the liquid may be carbon dioxide, with a bubble of the same substance in the form of gas. Such a crystal must have been formed under great pressure, sufficient to keep the gas in the liquid form. Fragments of such crystals heated to high temperature fly to pieces with great violence, because of the sudden release of gas formed by heating the liquid.

More commonly, crystals contain foreign solid substances of many kinds; quartz crystals enclose clay, particles of carbon, chlorite, rutile, tourmaline, and others. The famous groups of calcite crystals from Fontainebleau and from the Bad Lands of South Dakota (Fig. 3-147) contain some 60% quartz sand. It is most remarkable that the force of crystallization was powerful enough under such circumstances to marshal into place the atoms forming the calcite. In some crystals impurities are regularly arranged, and a curious effect is obtained in a cross section if it is cut and

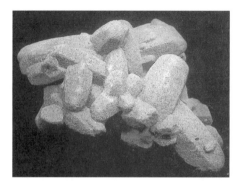

Fig. 3-147. *Sand crystals, South Dakota.*

Fig. 3-148.
Garnet enclosing quartz.

Fig. 3-149. *Chiastolite.*

polished. Figure 3-148, garnet enclosing quartz, shows such an effect. The variety of andalusite called chiastolite with carbonaceous inclusions affords another interesting example (Fig. 3-149).

Pseudomorphs. The word *pseudomorph* means "false form," and the name is applied to a specimen having the form characteristic of one species and the chemical composition of another. This seeming contradiction is easily explained. Most chemical compounds are liable to undergo a change or alteration when subjected to certain conditions, such as moisture, the action of alkaline waters, or acid vapors. Thus the mineral cuprite (Cu_2O) is rather easily changed chemically to malachite, the carbonate of copper. Anhydrite ($CaSO_4$) assumes water and changes to the hydrated sulfate gypsum ($CaSO_4·2H_2O$); pyrite (FeS_2) changes to the hydrated oxide limonite [$FeO(OH)·nH_2O$].

Now, in these and similar minerals, if the original specimen was in crystals, the external form may be perfectly preserved, although the chemical composition and the atomic structure have changed. Hence we describe the false forms mentioned as pseudomorphs of *malachite* after *cuprite, gypsum* after *anhydrite,* and *limonite* after *pyrite.*

Other examples are pseudomorphs of chlorite after garnet, pyromorphite after galena, and kaolin after feldspar. In a few rare instances, where the same chemical compound occurs in nature in two distinct crystalline forms, each with its own atomic structure, a change may take place in the structure without alteration of the chemical substance. Thus the rare mineral brookite (TiO_2) may be changed to rutile (also TiO_2), or marcasite (FeS_2) to pyrite (FeS_2). Such pseudomorphs have the special name *paramorph.*

The cases in which the original substance has entirely disappeared and some other has come to take its place are also called *pseudomorphs*. Thus we occasionally find quartz in the form of calcite, or of fluorite, or of barite, that is, a pseudomorph after one of these; also replacement of crocidolite by silica, producing tiger's eye; replacement of shells and even animal remains by opal; cassiterite in the form of orthoclase feldspar; and native copper in the form of aragonite. Even fossil wood may be said to be a pseudomorph of quartz or opal after the original wood, the structure of which it sometimes preserves with wonderful perfection. If instead of replacing the mineral, it has been leached away, leaving holes that retain the original shape of the mineral, it is called a *boxwork*. As with pseudomorphs, it may be possible from these features to identify the original mineral.

Crystal Aggregates

When crystals occur alone, the forms are usually developed on all sides with some of the regularity of the ideal model. Thus perfect garnets are found in mica schist or granite, and gypsum crystals both as individuals and as twins are found in clay. But it is still more common to find crystals grouped together either irregularly, as in the majority of cases, or perhaps in parallel position.

Parallel Grouping. One common type of crystal grouping, one that the beginner is likely to confuse with twinning, is parallel growth. In such aggregates, crystals or parts of crystals are parallel to each other, so that the axes of all have the same directions and are not inclined as in most twins. This is illustrated by a pile of cubes with faces parallel and having reentrant angles between them. Some crystals of many species are arranged in this way, but in every case it will be found that if the group is held so that it reflects the light, the faces on adjoining crystals that reflect at the same time are always similar faces. An octahedron of fluorite, built up of a multitude of little cubes in parallel position, is a common example. Figure 3-151 shows a complex crystal of analcime, formed of a number of single crystals all of which are parallel; it is hence *not* a twin. Figure 3-150, quartz, illustrates well this parallel grouping.

Parallel grouping is most interesting when the result is an aggregate built up with branching and rebranching parts like the limbs of a shrub or tree, forming what are called *arborescent* or *dendritic groups*. Here all the crystals or parts of crystals have like axes in the same direction. This is shown in Fig. 3-152, a crystal of native copper. In Fig.

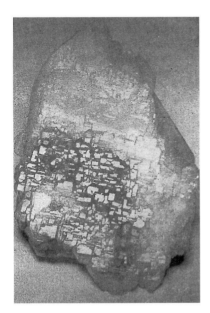

Fig. 3-150. *Quartz, parallel growth, Greenwood, Maine.*

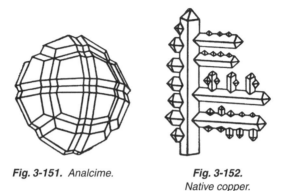

Fig. 3-151. Analcime.

Fig. 3-152.
Native copper.

Fig. 3-153.
Hematite (Eisenrose), St. Gothard, Switzerland.

3-153 the little plates of hematite are grouped together with such variation in their positions that the top of the aggregate has the shape of a rosette. Such a crystal is called by the German word *Eisenrose,* iron rose. If the crystal plates are arranged in the form of latticework as sometimes occurs in rutile or cerussite, the aggregate is called *reticulated.*

Another interesting type of aggregate is formed when a number of crystals are implanted on the surface of another which has obviously so influenced their growth that they are in parallel groups and in a definite position relative to it. Many examples have been noted, such as chalcopyrite crystals on sphalerite and rutile crystals on a tabular crystal of hematite, as shown in Fig. 3-154. Figure 3-155 is a related example; it consists now of rutile alone and has been described as a pseudomorph (see p. 54) of rutile after hematite. In the natural healing of the broken surface of a crystal, such as quartz (alluded to on p. 15), it follows, almost as a matter of course, that the new growths are oriented in directions parallel to the old ones. A surface covered with a layer of small crystals, although not a parallel growth, is called *drusy;* quartz is commonly found this way lining cavities.

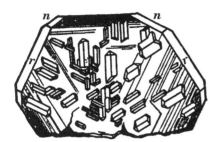

Fig. 3-154. Rutile on hematite.

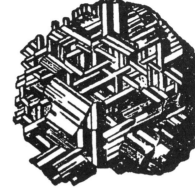

Fig. 3-155. Rutile after hematite.

Mineral Aggregates. In addition to the parallel growths already mentioned, mineral aggregates are common in which the individuals that make them up are not crystals as we have learned to regard them. In fact, in most mineral collections at least half of the specimens will show no crystal faces and are said to be *massive.* There are, however, important distinctions of habit among massive minerals.

It has been pointed out that the presence of cleavage in a mineral is proof of its crystallinity. The cleavage may be in the same direction in all parts of the specimen as if it were a fragment of one large crystal, but more commonly cleavage direction changes, showing that the specimen is made up of many grains with different orientations. In such a specimen the mass is obviously crystalline, and the aggregate is said to be *cleavable* and *granular.* Such a mass of galena is really made up of a multitude of little grains, each of which has its own directions of cleavage and presents to the eye its own edges. If the individual grains are large, the aggregate is said to be *coarse-granular;* if small, it is *fine-granular.* From the latter we pass to the closely *compact* kinds in which the state of aggregation may not be at all obvious to the eye; they may then be said to be *impalpable.* But this extreme is rare; for example, a piece of white marble, even if it is so fine-grained that the particles cannot be seen by the eye, usually sparkles in a strong light from the reflection of the multitude of minute cleavage faces.

This granular texture may belong also to other minerals which do not have cleavage as evidence of crystalline structure, and then the grain boundaries are difficult to see without the aid of polarized light. In some granular kinds of pyroxene, the small grains are found to be imperfect crystals.

Lamellar is a term applied to a mass made up of layers, whether separable or not. If the mass is in thin leaves or plates that can be separated from each other, it is called *foliated,* as with graphite. When the separation takes place as readily as in a piece of mica, it is called *micaceous.* If the scales are divergent or featherlike, it is said to be *plumose.*

If the mass is made up of little columns, it is called *columnar,* as in some specimens of calcite. When there are distinct fibers, the mass is said to be *fibrous,* as in asbestos, in which the fibers are easy to pull apart, or *separable*. There are many intermediate kinds between *fine-fibrous* and *coarse-columnar*.

If all the fibers, or little columns, or leaves go out from a center like the spokes of a wheel, the aggregate is said to be *radiated* (e.g., wavellite, Fig. 3-156), or perhaps *stellate* when it is star-shaped. If the layers are arranged in parallel position about one or more centers, the aggregate is said to be *concentric* (e.g., malachite, Fig. 3-157).

Fig. 3-156. *Wavellite, Hot Springs, Arkansas.*

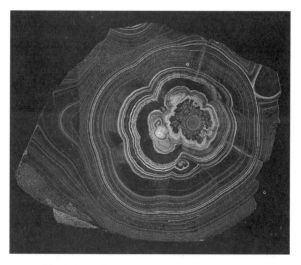

Fig. 3-157. *Malachite, Ural Mountains, Russia.*

All these terms—granular, foliated, lamellar, columnar, fibrous, radiated, and concentric—ordinarily refer to the aggregate when it is more or less distinctly crystalline.

When the external surface of the mineral is in the form of rather large rounded prominences, it is called *mammillary* (e.g., hematite, Fig. 3-158); if the prominences are smaller, somewhat resembling a bunch of grapes, it is called *botryoidal* (e.g., prehnite, smithsonite, and chalcedony, Fig. 3-159). If the surface is made up of little spheres or globules, it is called *globular* (e.g., hyalite or prehnite). If the surface resembles that of a kidney, it is called *reniform* (e.g., hematite, Fig. 3-160). It should be understood that there is no sharp line dividing these different surfaces, and thus the term *colloform* is used to include all of them.

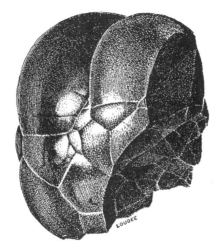

Fig. 3-158. Mammillary hematite.

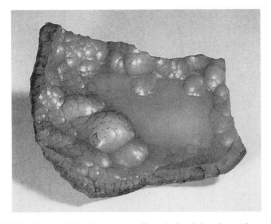

Fig. 3-159. Botryoidal chalcedony, Desolation Island, southern Chile.

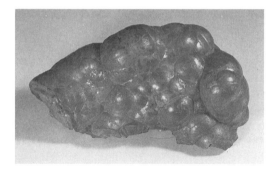

Fig. 3-160. Reniform hematite, England.

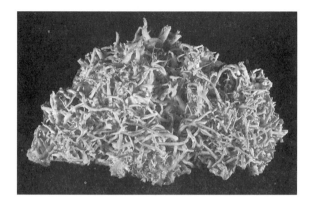

Fig. 3-161. Aragonite, flos ferri, *Styria, Austria.*

Fig. 3-162. Stalactitic limonite, Ural Mountains, Russia.

Minerals that take the form of a delicate, branching coral are called *coralloidal,* such as certain varieties of aragonite known as *flos ferri* (Fig. 3-161). If they are made up of forms like small stalactites, they are said to be *stalactitic* (e.g., limonite, Fig. 3-162).

If the material has clustered about a center like those curious forms of impure calcium carbonate called concretions, common in clay (Fig. 3-163), it is said to be *concretionary. Dendrites,* or dendritic forms, are those that have more or less the shape of a branching shrub or tree, like the forms of manganese oxide (see Fig. 3-164) often seen on surfaces of smooth limestone or enclosed in moss agates. Some dendritic forms are made up of little crystals grouped together in parallel position (see p. 55).

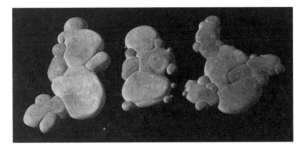

Fig. 3-163. *Concretions.*

Fig. 3-164. *Dendrites of manganese oxide.*

If the mineral occurs in narrow bands of different color or texture, it is said to be *banded.* When in rounded masses the size of a pea, *pisolitic* (Fig. 7-59), as masses of spheres the size of fish roe, *oölitic,* and as masses the size of almonds usually in basalt, *amygdaloidal.* If a round rock cavity is partially filled with minerals, it is called a *geode* (Fig. 7-130).

4

Physical Properties of Minerals

In addition to the external form of minerals as shown in crystals and crystalline aggregates, there are many other physical properties that characterize minerals and aid in their identification. Most of these properties can be divided into three groups based on (1) cohesive forces, (2) density, and (3) the action of light. Other properties less frequently used in identification depend on heat, magnetism, electricity, taste, and odor.

PROPERTIES DEPENDING ON COHESION

The properties of minerals that depend on the atomic forces of cohesion include cleavage, parting, fracture, hardness, and tenacity.

Cleavage

As stated previously, the internal orderly arrangement of the atoms is revealed not only by the crystal form but also by the *cleavage*. Mineral cleavage may be defined as *the smooth easy fracture that takes place between atomic planes across which there is a weak bonding or large interplaner spacing.*

There are three types of cleavage recognizable in the field: *blocky,* in which broken fragments are completely bounded by cleavage; (2) *prismatic,* two cleavage directions yielding fragments usually elongate to the *c* crystallographic axis; and (3) *platy,* with only one cleavage direction. A description of cleavage requires giving its quality such as perfect, good, fair, difficult, and so on, and also the crystallographic form to which it is parallel. For example, galena has perfect cubic cleavage, diopside good prismatic cleavage, and beryl poor basal pinacoidal cleavage. Like crystal forms, cleavage must be consistent with the symmetry. Thus if cleavage is present parallel to one face of a cubic crystal, it must be present parallel to the other two cube faces as well.

Since all crystal forms in the isometric crystal system are closed forms, every isometric cleavage should be blocky. Because of its perfect cubic cleavage, galena breaks into a multitude of little cubes (Fig. 4-1). If an individual cleavage face is carefully examined with a hand lens, one can observe rectangular outlines from the other two cleavage directions. Fluorite has good octahedral cleavage and, with care, from a cube of the mineral, one can cleave out a nearly perfect octahedron. This cleavage octahedron is readily distinguished from the octahedral crystal because its uneven splintery faces are not as uniform as those of the crystal. Careful inspection of an individual cleavage face will also reveal the presence of little triangles formed by the intersection of the other three cleavage directions. Fine cubic crystals of fluorite should be handled with care, for the solid angles are easily broken, giving the appearance of Fig. 3-15, a cube modified by an octahedron. Sphalerite, another isometric mineral, has perfect dodecahedral cleavage (six directions) and from a crystal it is possible to cleave out a small dodecahedron. Careful inspection of an individual cleavage surface will reveal a rhombic pattern with angles of 60° and 120°.

The rhombohedron is also a closed form, and thus calcite's perfect rhombohedral cleavage (Fig. 4-2), as well as cleavages in the isometric system, can be said to be blocky. However, a mineral with two or more cleavages of unequal quality can also yield blocky fragments. Thus anhydrite breaks into rectangular blocks resulting from unequal cleavages parallel to three pinacoids; and barite breaks into blocky fragments with cleavage parallel to a prism and a pinacoid.

Prismatic cleavage consists of two similar cleavage directions (three in the hexagonal system) whose intersection is parallel to a crystal axis, usually *c,* as is well illustrated by any of the amphiboles and pyroxenes, where it is an important feature in their identification. Careful observation of an individual cleavage face will usually reveal straight cleavage lines running in the elongate direction of the cleavage fragment; the ends of the fragment are usually rough and uneven. In some cases a mineral with two directions of platy cleavage of unequal quality may yield pseudoprismatic cleavage.

Fig. 4-1. *Cubic cleavage, galena.*

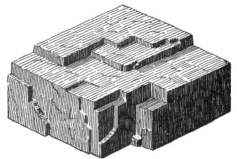

Fig. 4-2. *Rhombohedral cleavage, calcite.*

The feldspars have two pinacoidal cleavages of almost equal quality that intersect at 90° or nearly 90°, yielding elongate pseudoprismatic fragments.

Platy cleavage is a single cleavage direction parallel to a pinacoid. This is frequently parallel to the basal pinacoid and is called *basal cleavage*. When the mineral can be separated into very thin sheets, as in the micas, it is called *micaceous cleavage*.

A mineral can have several types of cleavage, as shown by barite (Fig. 4-3) and gypsum. Gypsum (Fig. 7-116) has perfect cleavage parallel to the side pinacoid, yielding plates almost as thin as those of mica. The plates show two other cleavages on their edges, the "snap" cleavage parallel to the front pinacoid and the "bend" cleavage (bends without breaking) parallel to a prism.

Cleavage occurs in varying degrees of perfection. In some minerals, as in the micas, it is easy and highly perfect; in others it is obtained with difficulty and is said to be imperfect, interrupted, or difficult. In some minerals cleavage may exist but is so difficult to obtain that it is described as being without cleavage. When struck with a hammer, a quartz crystal shows only a concoidal fracture which is characteristic of the mineral. However, a quartz crystal heated and then plunged into cold water may develop cleavage parallel to rhombohedral planes.

It should be pointed out that even if a crystal does not actually show broken surfaces, a cleavage is often clearly indicated by a fine pearly luster on the face of the crystal to which it is parallel. This is seen on the basal plane of apophyllite and on the side pinacoid of stilbite. This pearly luster is due to the presence of cleavage rifts, although the crystal has not actually parted, just as many layers of closely packed plastic film show a pearly luster

Fig. 4-3.
Cleavage in barite.

because of the repeated reflections. Cleavage rifts can often be seen in a transparent crystal; a flat clear crystal of barite or of celestite (Fig. 4-3) often shows on the basal pinacoid traces of the prismatic cleavage in two directions, making an angle of about 104° with each other.

A massive specimen of a mineral may show cleavage as a multitude of little smooth faces changing position with that of the grains to which they belong; if these grains are very small, the cleavage may appear only as a fine spangling of the surface, as mentioned on page 57.

In all advanced books in mineralogy, the cleavage and also partings are specified without comment in terms of their Miller indices. Although these numbers are exceeding simple, a thorough understanding actually requires a rather advanced level of study in crystallography. However, since certain Miller indices are used frequently, representative Miller indices for some of the more common cleavages are listed below.

Miller Indices of Some Common Cleavages

Blocky Cleavages		Prismatic Cleavages		Platy Cleavages	
Cubic	{001}	Prism parallel to *c*	{110}	Basal pinacoid	{001}
Octahedral	{111}	Prism parallel to *b*	{101}	Side pinacoid	{010}
Dodecahedral	{011}	Prism parallel to *a*	{011}	Front pinacoid	{100}
Rhombohedral	{10$\bar{1}$1}	Hexagonal prism	{10$\bar{1}$0}	Hexagonal basal pinacoid	{0001}

Parting

Parting is a breaking of a mineral along planes of structural weakness. The weakness results from imperfections during crystal growth or from twinning. During crystal growth interruptions may occur in which the mineral stops growing and the growth surface becomes coated or contaminated with impurities before resuming. Similarly, polysynthetic or lamellar twinning may occur during crystal formation or may be induced later by pressure. Because the breaking is along crystallographic planes, it resembles cleavage and is sometimes mistaken for it. However, cleavage is present in all specimens of a mineral and theoretically can be produced between any two atomic planes parallel to the cleavage direction. Parting, on the other hand, is not shown by all specimens, only those that have been subjected to the proper growth or pressure. Moreover, the number of planes is limited, and between those planes there is an irregular fracture. For example, in twinned crystals parting will take place only along the composition plane. Examples of parting are the rhombohedral parting of corundum and hematite, the basal parting of pyroxene, and the octahedral parting of magnetite.

Fracture

The nature of the surface given by fracture, when not the smooth surface of cleavage, is often an important property and may aid in distinguishing mineral species. Thus glass, as well as quartz and many other minerals, shows a shell-like fracture surface that is called *conchoidal* (Fig. 4-4) or, if less distinct, subconchoidal.

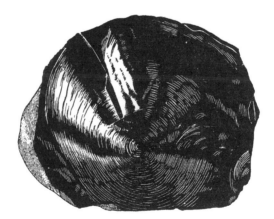

Fig. 4-4. *Conchoidal fracture in obsidian.*

More commonly, the fracture is simply said to be *uneven,* when the surface is rough and irregular. Occasionally, it is *hackly,* like a piece of fractured iron. *Earthy* and *splintery* are other terms sometimes used and easily understood.

Hardness and Tenacity

The resistance that the smooth surface of a mineral offers to a point or edge, tending to scratch it, is its *hardness* (designated H). A diamond easily makes a scratch on a smooth topaz crystal; the topaz scratches a quartz crystal; the quartz scratches a glass surface; and the glass in turn scratches calcite. Each substance named is harder than that which it scratches, or, in other words, softer than the one by which it is scratched.

In 1824, an Austrian mineralogist, F. Mohs, selected 10 common minerals as a scale to determine relative hardness. These are designated by the numbers from 1 to 10 in order of increasing hardness as given in the accompanying list. Crystallized varieties should be taken in each case, that is, a crystal with even surfaces or a smooth cleavage fragment.

Mohs Scale of Hardness

1. Talc	6. Orthoclase
2. Gypsum	7. Quartz
3. Calcite	8. Topaz
4. Fluorite	9. Corundum
5. Apatite	10. Diamond

When a mineral is said to have a hardness of 4, it means that it is scratched as easily as fluorite. If the hardness of a mineral is given as 5½, the mineral is a little harder than apatite and a little less hard than orthoclase.

The student should practice with the minerals in this scale up to corundum, and then experiment with them on some other known minerals until it is learned to what degree of hardness each number corresponds. The student should soon become so proficient as to be able to determine the lower grades of hardness by scratching with the knife (a hardness of about 5.5) without the use of the reference minerals in the scale.

One will find at once the following general distinctions between minerals of the several grades.

No. 1 has a soft, greasy feel like talc and graphite, and flakes of it will be left on the fingers.

No. 2 can be scratched easily by the fingernail, as can a cleavage piece of gypsum.

No. 3 can be cut easily by a knife and just scratched by a copper coin but is not scratched by a fingernail.

No. 4 is scratched by a knife without difficulty but is not so easily cut as calcite.

No. 5 is scratched with a knife with difficulty.

No. 6 is not scratched by a knife but is scratched with a file and will scratch ordinary glass.

No. 7 scratches glass easily but is scratched by topaz and a few other minerals.

Minerals that are as hard as, or harder than, quartz are few and include most of the highly prized gems.

The beginner will need a word of advice in regard to testing for hardness. In the first place, treat a mineral, especially a crystal, with as much consideration as possible. A scratch on a piece of plate glass, like a daub of colored paint on a white wall, is a little thing, but it may have a sadly disfiguring effect. Similarly, a scratch on a crystal disfigures it and destroys its value in large measure in the eyes of one who is a true mineralogist. Hence make as minute a scratch as possible, no longer than —, and if possible, put the scratch where it will show least. Treat the crystal as if it had feelings that would be hurt by the cut, and never scratch its smooth surface wantonly. A fine crystal should always be identified by tests which will mar it in no way.

In making the hardness test, it is necessary to be sure to distinguish between a real scratch on a smooth surface and the crushing of a rough surface or the separation of grains by the knife-edge; a very hard mineral may often appear to be scratched in this way. The danger of making a mistake of this kind is less if, besides the useful knife-point test, the mineral is rubbed on a piece of glass. Have a small piece at hand (do not disfigure a windowpane). Do not make the opposite mistake of calling a white mark left by a soft mineral on the glass a scratch. The mark of a soft mineral can be rubbed off; a true scratch is permanent.

It is necessary to remember that minerals are often altered; that is, they may have undergone some chemical change, particularly on the surface, which has rendered them soft when the original mineral was really hard. Thus it is often easy to make a scratch on a crystal of corundum because of a surface change to mica, although the mass within is very hard. Make sure that you are scratching a fresh, unaltered surface.

There are other properties that also depend on the strength of bonding of atoms of a mineral. They include the following, which are grouped under the general head of *tenacity.*

Malleability. A mineral is malleable if it can be flattened under the blow of a hammer without breaking or crumbling into fragments. Malleability is conspicuously true of gold and silver and makes it possible to beat out gold into extremely thin leaves. The property of malleability belongs only to the native metals and, in an inferior degree, to a few compounds of silver.

Ductility. A substance that can be changed in shape by pressure, especially if it is capable of being drawn out into the form of wire, is ductile. Gold, silver, and copper are ductile. Among minerals only the native metals possess this property in a high degree.

Sectility. A mineral is sectile if it can be cut by a knife like cold wax, so that a shaving may be turned up with care, yet the mineral breaks with a blow and is not completely malleable. Chlorargyrite is eminently sectile. Gypsum and chalcocite are imperfectly sectile. No sharp line separates the minerals that show this property from the truly brittle minerals.

Brittleness. A mineral is brittle if it separates into fragments with a blow of the hammer or with a cut by a knife. Brittleness is observed, in varying degrees, in most minerals.

Elasticity. When a mineral is capable of being bent or pulled out of shape and then returns to its original form when relieved, it is elastic. Elasticity is shown well by a plate of mica.

Flexibility. A mineral that bends easily and stays bent after the pressure is removed is flexible. Flexibility is shown by talc.

SPECIFIC GRAVITY OR RELATIVE DENSITY

It has been shown that much can be learned about a mineral by visual inspection, for if crystals are present, their symmetry may enable one to place it at once in a crystal system. If it is not in crystals, at least the habit of the aggregate and the presence or absence of cleavage can be observed. By a simple trial its hardness may be determined. While holding it, one should also note whether it seems distinctly heavy or light* compared with common substances of similar appearance. In this way one can obtain the first approximation of its density.

Comparative weight is one of the important properties of minerals; it is called specific gravity (designated G). It is defined as the ratio of the weight of a volume of a substance to the weight of the same volume of some other material taken as a standard. For solids the standard is water. Thus if a mineral weighs 3 pounds and an equal volume of water weighs 1 pound, the specific gravity is 3. For accurate determinations the temperature of the water must be at 4°C (39.2°F), the temperature at which water has its maximum density. However, for normal work, water at room temperature gives satisfactory results. *Density* and specific gravity are frequently used interchangeably. But density, defined as mass per unit volume, requires the citation of units: for example, grams per cubic centimeter (g/cm^3) or pounds per cubic foot; whereas specific gravity is a number expressing a unitless ratio. Whereas the density of diamond must be given as 3.52 g/cm^3, its specific gravity is given without units as 3.52.

Most of us from everyday experience have unconsciously developed an idea of specific gravity, that is, what a nonmetallic object of a given size should weigh. If you

---

*In these statements, as in some similar cases, the word *heavy* is used instead of *dense,* that is, of high specific gravity, and also the word *light* to express the opposite character; the meaning will be clear even if these terms are not employed quite scientifically.

were picking up colored pebbles at the seashore, you would probably gather many without thinking whether they were heavy or light. But if you were to pick up a pebble of barite, which would look like the other stones, you would probably examine it carefully. It seems too heavy, G = 4.5. The common stones we are used to handling are mostly quartz (G = 2.65), feldspar (G = 2.60–2.75), and calcite (G = 2.72). Thus the specific gravity for nonmetallic minerals and most rocks is within this narrow range, and minerals with much greater or lesser specific gravity we consider unusual. With a little experience, by hefting a specimen one can judge whether it is heavy, light, or about average for its size.

Although barite with specific gravity 4.5 seems heavy for a nonmetallic mineral, one would not consider a metallic object of the same size with this specific gravity unusually heavy. Unconsciously, we have developed a different standard for metallic objects—we expect them to be heavier. In general, metallic minerals have higher specific gravities than those of nonmetallic minerals, and G = 5.0, that of the common mineral pyrite, is average for them. Thus graphite (G = 2.2) seems light, whereas silver (G = 10.5) seems heavy.

Nonmetallic minerals can be roughly divided into three classes based on their specific gravities: (1) those that have a noticeably low specific gravity, such as gypsum; (2) those of average specific gravity, such as quartz; and (3) those with a high specific gravity, such as barite.

1. Minerals of relatively low density with specific gravity not higher than 2.5. Examples are:

Mineral	G	Mineral	G
Borax	1.7	Stilbite	2.2
Sulfur	2.05	Gypsum	2.3
Halite	2.1	Sodalite	2.3
Serpentine	2.2	Apophyllite	2.4

The zeolites (as stilbite above) mostly fall between G = 2.0 and G = 2.3.

2. Minerals of average density; specific gravity 2.5–3. Common examples are:

Mineral	G	Mineral	G
Nepheline	2.6	Feldspar	2.6–2.75
Quartz	2.66	Talc	2.8
Calcite	2.7	Muscovite	2.9

The common minerals tourmaline (G = 3.0–3.2), apatite (G = 3.2), vesuvianite (G = 3.4), amphibole (G = 2.9–3.4), pyroxene (G = 3.2–3.6), and epidote (G = 3.25–3.5) fall between this and the following group. Some varieties of garnet also belong here; others have a specific gravity up to 4.3.

3. Nonmetallic minerals of high density: specific gravity 3.5 or above; those for which a fragment seems heavy in the hand are known as *heavy minerals*. Examples are:

Mineral	G	Mineral	G
Topaz	3.5	Corundum	4.0
Diamond	3.5	Rutile	4.2
Staurolite	3.7	Witherite	4.3
Strontianite	3.7	Barite	4.5
Celestite	3.9	Zircon	4.7

Several minerals with an adamantine luster also belong here. These include compounds of lead, the commonest of which are cerussite, the carbonate; anglesite, the sulfate; and pyromorphite, the phosphate. They all have a specific gravity between 6 and 7. Copper minerals such as cuprite (G = 6.0) and silver minerals such as proustite (G = 5.5) are included in this group. Also here are scheelite, cassiterite, and cinnabar, the minerals of the heavy metals tungsten, tin, and mercury, respectively.

Minerals with metallic luster have an average specific gravity of about 5, that of pyrite and hematite. If the density of a metallic mineral is less than 4, it seems light as graphite (G = 2.2). If the density is 7 or above, it seems notably heavy, as galena (G = 7.5). Uraninite, which has a submetallic luster, has a remarkably high specific gravity of 9.7. All of the native metals have relatively high specific gravities, which vary from arsenic (G = 5.7), the lowest, to platinum, the highest (G = 21.5).

The quick judgment of specific gravity that comes with practice is always of value, but it should be applied with discretion, for the mineralogist must be continually on guard against being misled. In the first place, the size of the mass is an important factor, for a big lump of quartz seems heavy, of course, though its specific gravity is not relatively high. Also, we may get a wrong impression in handling a specimen if the mineral we are interested in forms only a small part of it; a little galena in a large mass of quartz will not make it heavy. If the mineral is open and porous and is made up of interlacing fibers, like some specimens of cerussite, it may appear light, even if the specific gravity is actually high. The eye may thus be deceived by the appearance of bulk if the solid mass present is not great.

A major factor influencing the specific gravity of a mineral is the mass of the atoms composing it. Thus for two isostructural minerals, the one composed of atoms with the higher atomic weight will usually have the higher specific gravity. As an illustration consider the minerals rutile, pyrolusite, and cassiterite listed below, all having the same structure and 1:2 ratio of cation to anion (metal to oxygen).

Change in Specific Gravity with Change in Cation

Mineral	Chemical Formula	Cation Atomic Weight	G
Rutile	TiO_2	47.9	4.2
Pyrolusite	MnO_2	54.9	4.8
Cassiterite	SnO_2	118.7	7.9

The type and strength of the bonding forces holding the atoms of a mineral together is another factor influencing specific gravity. The stronger the bonds and the more atoms that can be packed into a given volume, the higher the specific gravity of the mineral. This is illustrated dramatically in the polymorphs of carbon, diamond and graphite. Diamond is the hardest of minerals, with G = 3.52; whereas graphite, one of the softest, has G = 2.2. In diamond a single type of covalent bond holds the carbon atoms closely together. In graphite, which has two types of bonds, one bond holds the carbon atoms closely together in the form of sheets, while a second bond between the sheets holds them together very loosely, giving the mineral its low specific gravity and low hardness.

Measurement of Specific Gravity

As mentioned above, a first approximation of the specific gravity can be made by hefting a mineral. By doing this one can say that a mineral is heavy, light, or of average weight for its size. But classification into such a broad category may be of little help in determination of a specific mineral; a more precise measurement is necessary. However, it is possible in both laboratory and field camp for the beginning mineralogist to make a simple but relatively precise measurement of specific gravity. Its determination is most satisfying, for unlike many other tests, it results in a number that can be looked up in a table where the mineral or minerals to which it conforms are listed. See Appendix III, in which minerals are listed in order of increasing specific gravity.

We have seen that specific gravity is a ratio of the weight of a given volume of mineral to the weight of the same volume of water. But to determine specific gravity it is not practicable to compare the weights of equal volumes directly. Although it is a simple matter to weigh, for example, a piece of calcite, it is essentially impossible to determine its volume accurately. Hence it is necessary to make use of a well-known principle in hydrostatics, that when a body is immersed in water it is buoyed up and weighs less than it does in the air; this loss of weight is equal to the weight of the water it displaces. Hence if we first find the weight of the fragment on the pan of a delicate balance and then its weight while immersed in water and subtract the second weighing from the first, the difference is the weight of the equal volume of water. Some modern balances are designed with a hook underneath so that a pan suspended by a fine wire can be accommodated for weighing a specimen immersed in water.

For example, the weight of a little quartz crystal is 3.455 grams in air; in water it is 2.156; the loss of weight, or weight of a volume of water exactly equal to it, is therefore 1.299; hence the specific gravity is

$$G = \frac{3.455}{3.455 - 2.156} = \frac{3.455}{1.299} = 2.66$$

Jolly Balance. Because specific gravity is a ratio, it is not necessary to weigh the specimen but merely to obtain values that are proportional to its weight in both air and water. Such values are obtained with the Jolly balance by the stretching of a

spring. The instrument (Fig. 4-5) consists of a spiral spring, one end of which is attached to the top of a movable scale; from the other end hang two pans, the lower one immersed in water. When the spring and pans are in a "balanced" position, the scale is adjusted to zero. A small fragment whose specific gravity is to be determined is placed in the upper pan. By means of a rack-and-pinon gear, the spring is stretched to bring it to the balanced position and the reading on the scale (n_1) noted. The mineral is then placed in the lower pan and the scale reading (n_2) noted at this new balanced position.

The specific gravity is then given by the expression

$$G = \frac{n_1}{n_1 - n_2}$$

For example, consider the readings taken using a small quartz crystal: If $n_1 = 14.5$ centimeters and $n_2 = 9.02$ centimeters, then

$$G = \frac{14.5}{14.5 - 9.02} = \frac{14.5}{5.45} = 2.65$$

Fig. 4.5. *Jolly balance. (Eberbach, Ann Arbor, Michigan.)*

Beam Balance. The beam balance (Fig. 4-6) is a simple device for determining specific gravity which, like the Jolly balance, does not require definite weights. It can be constructed easily and inexpensively in any size, large enough to handle 1-kilogram specimen or small enough to give accurate readings on a 1-gram specimen. After the frame is made with two uprights, a thin piece of wood is cut in the form of the beam, *abc.* This should be graduated in inches and tenths of inches or in centimeters and millimeters from *b*, the axis, to the end *c*. A fine wire or needle passed through the beam at *b* serves as the axis. At *a*, the end of the short arm, two pans or

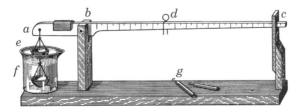

Fig. 4-6. *Beam balance.*

wire baskets are arranged so that one is in air and the other, suspended by a single wire, is in water, *f.* When the pans are empty, the beam is balanced by a small piece of lead between *a* and *c* on the short arm. A final adjustment to the zero position is made by moving a little rider to some point, as *d,* where it serves to make *bc* exactly horizontal, as shown by a mark on the upright at end *c.* A number of counterbalances of different weights are needed, but it is not necessary to know their exact weights. They can be hooks as at *g,* coiled wire, wire hooks carrying washers, or fish sinkers.

After the beam is balance to the "zero" position, the mineral fragment is placed in the upper pan and a counterweight chosen that will balance the beam when placed near the end of the long arm. When balanced, the position of the counterweight (n_1) is recorded; it is proportional to the weight of the mineral in air. The specimen is then transferred to the lower pan and the beam again balanced by moving the *same* counterweight closer to the fulcrum. The reading of this position (n_2) is proportional to the weight of the mineral in water. The specific gravity can then be determined by using the same expression as that above for the Jolly balance.

PROPERTIES DEPENDING ON LIGHT

Of the properties observed by the eye, several have not been mentioned yet in detail. These are the properties that depend on the reflection or absorption of light: (1) *luster* or the appearance of the surface independent of the color, due to the way the light is reflected; (2) *color;* and (3) degree of *transparency.*

Luster

Luster is the quality of the light observed reflected from the surface of a mineral. The differences are not always easy to describe, but the eye notes it at once and, after a little training, seldom makes a mistake. The types of luster distinguished are as follows.

1. *Metallic luster:* the bright reflectance of a metallic surface such as steel, lead, tin, copper, or gold. In substances having bonds where the electrons are shared, as in the metallic bond and in certain covalent bonds (e.g., sulfides), free electrons (conductors) or easily detached electrons (semiconductors) may occur. It is the presence of these free or mobile electrons that produces the bright reflectance of the metallic luster. These free electrons also absorbed any light transmitted, rendering the mineral quite opaque even on very thin edges. The luster of some minerals, such as columbite, is said to be submetallic when they lacks the full luster of the metals. A few minerals have varieties with metallic and others with nonmetallic luster; hematite is an example.

2. *Nonmetallic luster:* the duller reflectance observed when most of the light passes into the mineral and only a small portion of the incident light is reflected from the surface. Upon entering a mineral, this light is bent or refracted in varying degree. As with specific gravity, the refracting power of a mineral can be measured accurately and used as a confirmatory test in mineral identification. A number of types of

nonmetallic luster related to the nature of the refraction can be distinguished and are described by the following terms:

3. *Vitreous, or glassy luster:* that of a piece of broken glass. This is the luster of most quartz and of a large part of the nonmetallic minerals.

4. *Adamantine, or the luster of the diamond:* the brilliant, almost oily luster shown by some very hard minerals, such as diamond and corundum, and also by some compounds of the heavy metals, such as the carbonate (cerussite) and the sulfate (anglesite) of lead. All these refract the light strongly and have a high refractive index.

5. *Resinous or waxy:* the luster of a piece of resin, as shown by most kinds of sphalerite; near this, but often quite distinct, is *greasy luster,* shown by some specimens of milky quartz and nepheline.

6. *Pearly, or the luster of mother-of-pearl:* common when a mineral has very perfect cleavage and hence has partially separated into thin plates. The basal plane of crystals of apophyllite shows pearly luster.

7. *Silky, the luster of a skein of silk or a piece of satin:* characteristic of some minerals in fibrous aggregates, such as the variety of gypsum called satin spar and of most asbestos.

The luster of minerals is also described according to the brightness of the surface; it is called *splendent* in freshly fractured galena, but *dull* in jasper. Other surfaces may be described in self-explanatory terms such as *glistening* or *glimmering,* according to the nature of the surface.

Color

When white light strikes the surface of a mineral, part of it is reflected and part refracted into the mineral. If none of the reflected or refracted light is absorbed, the mineral is colorless. If the mineral is filled with minute bubbles, the refracted light is scattered in all directions and the mineral will appear white. If certain wavelengths (colors) are absorbed (selective absorption), color results from a combination of wavelengths that reach the eye. For example, if red has been absorbed, the mineral will be blue, the complementary color.

The color of most minerals with metallic luster is constant, such as the brass-yellow of pyrite, the brownish bronze of bornite, and the copper-red of nickeline. This is because light shining on a mineral with a metallic luster elevates electrons from the valence band into the conduction band, whose absorption properties are determined by the nature of the particular metallic or covalent bonds of the individual mineral. Because surface alterations produce a tarnish on these and other metallic minerals, it is important that a fresh surface showing the true color of the mineral be examined. A striking example is given by the copper mineral bornite, which is a reddish-bronze color on a fresh surface. The mineral is called "peacock copper ore" and "purple copper ore" because even on short exposure, its bronze color is covered by a blue-violet tarnish. Although less striking, another copper mineral, chalcopyrite, which is brass-yellow when fresh, quickly becomes coated with an iridescent film.

The color of nonmetallic minerals includes all colors of the spectrum. It varies not only from one species to another, but there may be a great variation within a single species. For example, corundum occurs in all colors and tourmaline may be colorless or various shades of red, blue, green, brown, and black. Moreover, tourmaline may show several colors within a single crystal. For some nonmetallic minerals color is directly related to one of the major constituent elements, and since it is constant and characteristic, it is in those cases an important means of identification. The manganese minerals rhodonite and rhodochrosite are always red; the copper minerals malachite and dioptase are always green, but azurite, also a copper mineral, is always blue. All three copper minerals contain Cu^{2+}, but the structure of azurite affects the absorption differently from the others. It is important for the student to learn in which of the nonmetallic minerals color is a constant property that can be relied upon as a distinguishing criterion.

Minerals with ionic bonds where electrons have been transferred from the cation to the anion, in general, have a nonmetallic luster and, when pure, are colorless. However, many of these have color varieties that are as common as the colorless. The color is the result of small or even trace amounts of impurity ions of one of the transition elements: V, Cr, Mn, Fe, Co, Ni, Cu. These ions, called *chromophores,* are a major factor in selective absorption and hence in color. In the discussion of the atom (see Chapter 5) we learned that electrons are in different shells or orbitals. The transition elements are characterized by having *d* orbitals that are only partially filled. When these empty *d* orbits are surrounded by other atoms in the crystal, the empty orbits have a higher energy than the filled orbits. If the radiant energy of visible light shining on a mineral is sufficiently great, it will elevate an electron from a low-energy *d* state to an empty higher-energy *d* state and be absorbed in the process. Thus a very minor amount of a transition element becomes a major factor in the coloration of minerals, as in the red color of ruby (corundum) or the green color of emerald (beryl).

The color of some nonmetallic minerals with no chromophoric ions is attributed to imperfections in the crystal structure, called *color centers.* The imperfection may be a structural void, a deformation of the lattice, or a deficiency or excess of ions of a constituent element. The color results from the excitation of unpaired electrons to a higher energy level, causing selective absorption and the transmission of other wavelengths, as in purple fluorite, smokey quartz, or blue halite.

Other minerals have mixtures of multivalent ions, such as $Fe^{2+}-Fe^{3+}$, $Mn^{2+}-Mn^{4+}$, or $Fe^{2+}-Ti^{4+}$. In this case the incident light may cause an electron to jump from one ion to another and light will be absorbed. In very small quantities, these may also absorb light to produce a color, such as the light green color in aquamarine or the blue in kyanite and sapphire. In higher quantities, the combination commonly results in a black mineral. Thus the submetallic opaque black of magnetite is from mixed-valency iron, while that in pyrolusite is from mixed-valency manganese. If the mixed-valency state is less abundant, the mineral (except in small fragments) may also appear black, as in augite, hornblende, and tourmaline. It is thus often important to determine if a black mineral is truly opaque.

Streak. It is often important, especially with a mineral having metallic luster, to test the color of the fine powder or the color of the streak. The slight scratch that is given to test the hardness will often show the streak, but a better way is to have at hand a piece of unglazed white porcelain, or ground glass, upon which the mineral can be rubbed. This method shows, for example, that an iron-black hematite crystal with a bright metallic luster has a red streak. For many minerals this is a highly diagnostic property. Minerals of nonmetallic luster usually have a streak that differs but little from white even if the mineral itself is dark-colored or even black, as, for example, the several varieties of tourmaline.

Play of Colors. When describing the color of a mineral, some peculiarities noted in its distribution may receive special names. A mineral is said to show a play of colors when, like precious opal, it exhibits internally the various prismatic colors when the mineral is turned.

Pleochroism. The appearance of different colors when a crystal is viewed in transmitted light in different directions is known as *pleochroism*. If only two directions have distinct colors, the property is called *dichroism*. It is commonly found in minerals colored by multivalent ions. This is a common property that can be greatly enhanced by the use of plane-polarized light as when viewed with a polarizing microscope. Pleochroism is often seen in transparent crystals of epidote, tourmaline, and in the pink gem variety of spodumene, kunzite.

Opalescence. A pearly reflection from the interior of a mineral, like the effect of a glass of water to which a few drops of milk have been added, is called opalescence because such an effect is common with opal. Opalescence is also seen in moonstone (feldspar).

Iridescence. A mineral is iridescent when it shows a series of colors due to light undergoing reflective interference with itself either on the surface or in the interior. External iridescence is caused by the presence of a thin surface film or coating; internal iridescence is often caused by the presence of thin twinning lamellae or minute air spaces, along cleavage planes.

Chatoyancy and Asterism. Some minerals have a silky appearance in reflected light, resulting from closely packed fibers, parallel inclusions, or cavities. When a cabochon gem is cut from such a mineral, a band of light crosses the stone at right angles to the inclusions. As the stone is turned, the band moves from side to side as in a cat's eye, and the name *chatoyancy* is appropriately given to this phenomenon. Fibrous gypsum and tiger's-eye quartz are examples of chatoyant minerals. In some minerals, inclusions or cavities may be oriented by the crystal structure in three identical directions, as in hexagonal crystals. Light reflected from the base of such a crystal, or the surface of a cabochon cut from the mineral, appears as a six-pointed star, formed by a beam of light at right angles to each set of inclusions. This triple chatoyancy, called *asterism,* is best known in the two varieties of corundum known as star ruby and star sapphire. Asterism is also seen in some rose quartz when viewed parallel to the 3-fold symmetry axis. A similar six-pointed star is seen when a point source of light is viewed through certain types of phlogopite mica (Fig. 4-7). In

Fig. 4-7. *Asterism in phlogopite.*

corundum the inclusions are oriented by the hexagonal structure; in phlogopite by the pseudohexagonal structure.

Fluorescence and Phosphorescence. One of the most fascinating properties of certain minerals, and one that first interests many in the study of minerals, is the phenomenon of *fluorescence.* We are all familiar with the spectral colors from red to violet. Of these, red has the longest wavelength and violet the shortest. Beyond the violet that we can see are still shorter wavelengths known as ultraviolet. If, on exposure to ultraviolet light in a dark room, a mineral emits visible light, it is said to fluoresce. If a mineral collection is examined under such conditions, some remarkable discoveries may be made. A drab piece of calcite may become a vivid pink, and a piece of chalky white scapolite may turn a bright yellow, whereas the most colorful crystal in the collection as seen in white light may have no color at all.

Unlike most other properties of minerals, fluorescence is difficult to predict. If, for example, we are sure that a specimen is calcite, we can say that it will have a hardness of 3, a specific gravity of about 2.72, and a good rhombohedral cleavage, but the only way in which we can be sure that it will fluoresce is to try it. Two specimens of calcite may look identical, yet one will fluoresce and the other will not. The name of this property derives from *fluorite,* because it was first observed in this mineral. However, only a few fluorite specimens will fluoresce. As with color, most mineral fluorescence is caused by the presence of small amounts of an impurity element, which in this case is usually a lanthanide element such as cerium, gadolinium, or samarium, which has substituted for some of the calcium. Fluorescence therefore cannot be used as a tool in determinative mineralogy except in rare cases. One such exception is willemite from Franklin, New Jersey, which fluoresces a bright green. Another is the mineral scheelite, which usually fluoresces a pale blue. Because the uranyl ion causes fluorescence, many minerals with small amounts of uranium and many uranium minerals such as autunite will fluoresce.

Just as the wavelength of visible light varies from place to place in the spectrum, so does the wavelength of ultraviolet light vary. Common source lamps usually produce either a long-wave or a short-wave ultraviolet. One source of ultraviolet may have no effect on a specimen, whereas a light of a different source may produce a vivid fluorescence. If possible, therefore, one should have more than one source. One should exercise caution not to look directly at a black light or at too long exposure, as it can burn the retina. Viewing specimens through a piece of windowglass (do *not* use plastic) is a worthwhile precaution.

Some fluorescent minerals will continue to glow after the ultraviolet light has been turned off. This property is known as *phosphorescence*. Only a few of the fluorescent minerals, such as certain specimens of willemite and calcite, show this property to a marked degree. Others may phosphoresce but for so short a period that the eyes cannot become adjusted to the darkness quickly enough to observe it.

Thermoluminescence and Triboluminescence. Some minerals when heated below red heat will emit visible light and are said to be *thermoluminescent*. It is now thought that such thermoluminescence is the result of defects in the crystal lattice either from strain or from damage produced during the radioactive decay of uranium. As with fluorescence, the color of the light emitted is quite independent of the color of the mineral. It is an interesting experiment to heat a thermoluminescent mineral in a darkened room. As the specimen begins to get warm, it will give off a faint glow that becomes brighter with continued heating. Some specimens will continue to glow for many minutes after removal from the flame. This property is most commonly observed in fluorite, and the variety known as *chlorophane* was so named because of the green light emitted. Other thermoluminescent minerals are scapolite, apatite, calcite, and feldspar; but again it should be pointed out that only a few specimens of a given mineral will have this property.

Some minerals when rubbed or struck with a hammer will emit light and are *triboluminescent*. This property is less spectacular than fluorescence and thermoluminescence and is usually observed only momentarily. However, it is easily seen in a dark room if two pieces of milky quartz are rubbed briskly together. Other minerals that commonly show this property are fluorite, sphalerite, and lepidolite.

Transparency and Optical Properties

A mineral is said to be *transparent* when it is so clear that an object can be seen through it with perfect distinctness, as through a piece of windowglass, a plate of selenite, or a thin sheet of mica (Fig. 4-8). A mineral is *translucent* when it transmits light, as does a piece of thin porcelain or jade, and allows only the outline of an object to be seen through it (Fig. 4-9). It is *subtranslucent* when light is transmitted only on

Fig. 4-8. *Transparent mica.*   **Fig. 4-9.** *Translucent gypsum.*

the edges. If no light is transmitted even on thin edges, the mineral is said to be *opaque*. Most metallic and submetallic minerals are opaque.

The determination of the optical properties of the nonopaque minerals plays an important part in their identification. Advanced identification methods requires a detailed understanding of refraction, polarization, and interference of light. Here we briefly describe refraction and simple polarization.

Refractive Index. Of the optical properties, the refractive index is the most important. The refractive index (N) of a material is the ratio between the velocity of light in air (V_1) and its lesser velocity in the denser medium (V_2), or $N = V_1/V_2$. Fortunately, since the refractive index is a ratio an absolute measurement of the speed of light is not required, and we need only to consider the velocity of light in air to be 1 and thus $N = 1/V_2$. When light passes from one medium into another of greater refractive index, it is refracted, that is, bent toward the normal to the surface (Fig. 4-10). The

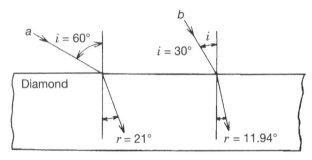

Fig. 4-10. *Refraction of light.*

greater the bending, the higher the refractive index. Most of us have seen this phenomenon by plunging a stick diagonally into water and observing that it appears to be bent at the air–water surface.

If grains of quartz ($N = 1.55$) and zircon ($N = 1.95$) are placed together in a beaker of water ($N = 1.33$) they are both visible, but the zircon stands in higher relief. This is because the light as it passes from water to zircon is bent more than in passing from water to quartz. If, however, grains of cryolite ($N = 1.34$) are placed in the water, they will virtually disappear, for the refractive index of the cryolite essentially equals that of the water and there is little or no refraction at the interface. The refractive index is a characterizing property of each nonopaque mineral. Therefore, its determination is a major contributing factor in mineral identification. We shall see that one way the mineralogist determines refractive index is, in a manner, similar to the above. It is called the *immersion method* and is carried out using a polarizing microscope.

Isotropic Versus Anisotropic. Optically transparent substances can be divided into two groups: isotropic and anisotropic. The *isotropic* includes noncrystalline substances such as gases, liquids, and glass; it also includes isometric crystals. In them, light moves in all directions with equal velocity, and thus each isotropic substance has a single refractive index. All crystals except those of the isometric system are *anisotropic*. In them, the velocity of light varies with crystallographic direction and, except for special orientations, two refractive indices can be measured in any crystal section.

Polarized Light and Double Refraction. Ordinary light is considered a wave motion vibrating in all directions at right angles to the direction of propagation. If the wave motion is constrained to vibrate in a single plane, the light is said to be *plane polarized.* There are several ways in which light can be polarized. One way is by double refraction. Light passing through an anisotropic crystal, in all but a few special directions, is resolved into two polarized rays vibrating at right angles to each other. This is called *double refraction.* The only common mineral in which the phenomenon can be seen readily is calcite, especially in the transparent variety, Iceland spar.

Figure 4-11 illustrates this property well; here the single cross on the paper beneath appears double; one cross (to the eye looking perpendicularly down on the surface) has its arms in the continuation of the lines beneath; the other is pushed to one side. Neither cross appears quite black except at the two points where they intersect. Each image of the cross has its own direction of polarization. To determine the direction of polarization, the images are viewed through a polarizing filter, *polaroid,* in which the direction of polarization is known. On turning the filter, a position is reached at which one of the images disappears. The polarizing directions of the remaining image and that of the filter then coincide. On turning the filter 90°, the other image will disappear. Before the discovery of polaroid, prisms were made from Iceland spar, which eliminated one of the two rays, permitting the other to emerge as plane-polarized light. The device, known as the Nicol prism, was used in the construction of the first polarizing microscopes. Today, polaroid is used instead of Nicol prisms.

Fig. 4-11. Double refraction in calcite.

A polarizing microscope differs from a biological type in that two polarizers, with vibration directions at right angles to each other, are used. One polarizes the light in which the subject material on the stage is examined, the other is mounted in the tube above the stage. With the advent of the Nicol prism in the nineteenth century, the polarizing microscope became an effective tool for mineral identification and for the study of the optical properties of crystals. Its use requires considerable skill and knowledge of crystal optics and an understanding of the complex interference properties of polarized light.

Immersion Method. In preparation for the determination of the refractive index by the immersion method, the mineral is crushed to a coarse powder, and a few grains placed on a glass slide are immersed in a liquid of known refractive index. By a simple test while observing the magnified grains under the microscope, one can tell whether the mineral has a higher or lower refractive index than the liquid. If the mineral has a higher refractive index, other grains are successively immersed in liquids of higher refractive index until a match is reached and the grains become essentially invisible. The refractive index of the mineral, N, is then known, for it is the same as the calibrated liquid. Standard tables can then be used to assist in identification of the mineral.

For anisotropic minerals, the determination of refractive indices is more complicated. Let us examine a tiny cleavage fragment of calcite in an immersion liquid under the polarizing microscope in which the light illuminating the stage is polarized in a known direction. It is possible to turn the calcite grain so that one of its directions of polarization coincides with that of the microscope light. In this position one index of refraction of the grain can be compared with that of the liquid. The other index can be compared with the liquid on turning the grain 90°. Thus by trial and error, the two indices of refraction of the calcite fragment lying on a cleavage face can be determined.

Two refractive indices can be determined on every section of an anisotropic mineral, but the values obtain depend on the crystallographic orientation and thus are different for each section. In principle, if we measure the refractive indices of a large number of such sections from a series of random grains, we would find that the observed mineral has a minimum refractive index and a maximum refractive index. These maximum and minimum refractive indices are a diagnostic property of each mineral that can be used in their identification. Because there are now on the market both used and reasonably priced new polarizing microscopes, it is feasible for the amateur to actually make refractive index measurements, but the details of doing this are beyond the scope of the present work.

Properties Depending on Heat

The fusibility of minerals or the ease with which they can be melted is an important property of many minerals. This is discussed on page 107 with the description and the use of blowpipe tests. The thermal conductivity of crystals is another property that has been measured for many crystals. Crystals react to the conductivity of heat in a manner similar to their transmission of light. Heat conduction in an isotropic crystal is the same in all crystallographic directions. But in anisotropic crystals, it varies with crystallographic direction. Recently, meters have been developed to separate diamonds from diamond simulates. These depend on thermal conduction. Diamond is the best heat conductor of any material, including all the metals. Because of this, when a diamond is touched with a hot point, it conducts the heat away so fast that the meter can quickly separate diamonds from simulates.

Magnetic Properties

A few minerals have the property of being attracted by an ordinary horseshoe magnet. As with color, this property requires the presence of a transition metal, but instead of trace quantities, the element must be a major constituent and must be near the middle of the series. Minerals that lack the presence of a transition metal or other magnetic ions are called *diamagnetic*. In our discussion of the atom in Chapter 5 we will learn that electrons are in different shells or orbitals, and the transition elements are characterized by having *d* orbitals, which are only partially filled and have spins that are, where possible, unpaired. These unpaired spins cause each transition-metal ion to act as if it is a miniature bar magnet. This effect is greatest in iron and manganese, so these are the only naturally important magnetic ions.

If the magnetic ions in a mineral have a completely random orientation, as in garnet, pyroxene, and amphibole, the mineral is called a *paramagnetic* mineral. In some minerals there is a natural tendency for pairs of magnetic ions to align in opposite directions so that there is spin pairing between adjacent magnetic ions. This is called *antiferromagnetism* and is the case in the common minerals siderite, rhodochrosite, hematite, and goethite. Noticeable magnetism is relatively rare and occurs only in special cases known as ferrimagnetism, where there is an excess of magnetic ions aligned in one particular direction. The most important case occurs in the mineral magnetite. In this particular case, all of the iron ions in the tetrahedral site are aligned in one direction and all of the magnetic iron ions in the octahedral site are aligned in the opposite direction. Since there are twice as many iron atoms in the octahedral site as in the tetrahedral site, there is a strong net magnetic moment. The ferrimagnetic properties of magnetite and related spinels are so important industrially in the manufacture of transformers and magnetic recording tapes that this group is known as the ferrites to the physicists. Ferrimagnetism is also found in the minerals pyrrhotite and maghemite. In both of these cases, the pure stoichometic compound is antiferromagnetic, but in both pyrrhotite and maghemite, there is a systematic absence of iron atoms as indicated in the formula $Fe_{1-x}S$ for pyrrhotite, which results in a net magnetism. True ferromagnetism, in which all of the magnetic atoms are aligned in a single direction, is found in the rare minerals kamacite and ferroplatinium.

Some specimens of magnetite are themselves magnets and have the power of attracting little particles of magnetite or iron (Fig. 4-12). They have north and south

Fig. 4-12. *Lodestone showing magnetism.*

poles and, if hung by a thread, will swing around until the poles come into the magnetic meridian, that is, the direction assumed by a compass needle. This kind of magnetite is called *lodestone.* Pyrrhotite is much less magnetic than magnetite, and the magnetic varieties of platinum are not common; both may have polarity as the lodestone. A few minerals, such as hematite, ilmenite, and franklinite, are sometimes slightly magnetic either because of imperfect antiferromagnetism or because they contain a little admixed magnetite.

In the presence of a high-intensity magnetic field, it is possible to separate para-magnetic minerals which are attracted by the magnetic field, such as garnets, pyroxenes, and amphiboles, from diamagnetic minerals such as quartz, calcite, and fluorite, which are slightly repulsed by the field. By adjusting the strength of the field, various paramagnetic minerals may be separated from one another. Surprisingly, pyrite and marcasite will separate with the diamagnetic minerals. This is because of the special covalent bond formed between iron and sulfur in these two minerals.

ELECTRICAL PROPERTIES

There are a number of electrical properties of minerals, but they belong to a more extensive study of minerals and are mentioned only briefly here. A number of minerals, including diamond, sulfur, and topaz, take on a static electric charge when rubbed with a piece of silk or fur and show this by their power of attracting light substances such as bits of paper or straw. This property was observed in amber in about 600 B.C., and our word *electricity* is taken from the Greek word for *amber.*

Piezoelectricity

Another kind of electrical charge that can be developed in certain minerals was discovered in quartz by Pierre and Jacques Curie. They found that if pressure was exerted along an *a* axis of quartz, a positive electrical charge is set up at one end of the axis and a negative charge at the other end. This is known as *piezoelectricity* (pressure electricity). It was reasoned by the Curies that if quartz developed an electric current when squeezed, might not the converse be true? That is, would not the quartz be deformed if subjected to an electric field? Experiments proved this to be true, and a great industry resulted. During World War II, 55 million quartz oscillator plates were used for controlling the frequency of radios used for military purposes. Today, many millions of tiny quartz plates, cut to exacting crystallographic angles, control the time of watches and other timepieces.

It was found that piezoelectricity could be generated in other minerals. At first, there seemed to be no correlation between the property and the type of crystal, for representatives were found in each crystal system. Then it was determined that the property was confined to minerals that crystallized in a symmetry class that lacked a center of symmetry. Furthermore, the electric charge developed only at the ends of polar axes: that is, those axes that had different forms at opposite ends. Tourmaline is such a mineral but is less effective than quartz as a radio oscillator. However, small amounts of it are used in the manufacture of piezoelectric pressure gauges. Tourmaline is hexagonal with *c* as the polar axis, and plates cut normal to the *c* axis will generate an electric current when subjected to a transient pressure. Tourmaline gauges were developed in 1945 and were used by the United States to record the blast pressure for each test atomic explosion.

Pyroelectricity

Electric charges similar to piezoelectricity are induced by heating crystals lacking a symmetry center. This is known as *pyroelectricity*. This property was noted long ago in tourmaline. When crystals that had been heated on an open hearth were cooled, ashes would collect and cling to one end and not the other. Whereas in the past, natural minerals were used as radio-oscillator plates or pressure transducers, these have largely been replaced by artificially grown crystals or by crystals of manufactured chemical compounds.

TASTE AND ODOR

Taste belongs only to the few minerals that dissolve to some extent in water. The terms employed are familiar and hardly need explanation. *Saline* means the taste of common salt; *alkaline,* of soda; *bitter,* of epsom salts; *sour,* of an acid; *astringent,* of iron vitriol; *sweetish astringent,* of alum; *cooling,* of saltpeter. Halite, ordinary salt, and sylvite can usually be identified by their taste. Although the tasting of natural minerals presents no undue hazard, the tasting of soils should generally be avoided because of the fungi present.

An odor is characteristic of only a few minerals. Some varieties of limestone, barite, or quartz have a fetid odor, or odor of rotten eggs, especially when they are freshly broken or rubbed sharply; this is usually due to the presence of a sulfur compound. Moistened clay and some claylike minerals when breathed upon give off a peculiar argillaceous odor. Bitumen and some allied substances have a bituminous odor.

A sharp blow across the surface of a piece of arsenopyrite often produces a peculiar garlic odor like that obtained by heating the same mineral with a torch; it is, in fact, due to the same cause. Similarly, a blow on a mass of pyrite may yield a sulfurous odor.

5

Mineral Chemistry

Early in a study of minerals it seems appropriate to consider the materials of which minerals are made. An important part of our definition of a mineral is that it is a *chemical element* or *compound*. At the start, then, it is necessary to understand what chemical elements are.

In the laboratory the chemist analyzes various substances, that is, separates them into different kinds of matter of which they are composed. However, the process of analysis or the chemical separation can be carried only so far, for the chemist soon obtains substances that cannot be further decomposed. For example, common salt can be separated into two kinds of matter, the metal sodium and the gas chlorine, but neither of these can be separated further. They are thus called elementary substances or *elements*.

Although there is no limit to the number of combinations of chemical elements that can be combined into chemical compounds, the number of elements is relatively small. There are 88 elements that occur naturally and hence are the materials of which minerals are made. Although the elements extend up to atomic number 92, atomic numbers 43, 61, 85, and 87, technetium, promethium, astatine, and francium, respectively, must be produced synthetically by nuclear reactions. Elements up to atomic number 83, bismuth, are stable, but elements with higher atomic numbers are unstable and are the radioactive decay products of the very long-lived radioactive elements 90, thorium, and 92, uranium. Most of the natural elements are relatively rare, for the minerals that make up over 99% of the earth's crust are composed of only 10 elements.

The units that compose the elements are atoms. Although atoms themselves are made up of even smaller particles, the atom is the smallest subdivision that retains the characteristics of the element. In 1912, a Danish physicist, Niels Bohr, proposed a model that is still helpful in visualizing the atom. According to this model, the atom resembles a minute solar system with a central nucleus orbited by electrons at different distances or energy levels. The atom differs from the solar system in that the electrons do not orbit the nucleus in a single plane but in a spherical shell. The nucleous of the hydrogen atom is a single proton that is circled by a single electron. The nuclei of atoms of other elements are composed of both protons and neutrons. Protons and neutrons have essentially the same mass but the proton carries a positive electric charge, whereas the neutron is electrically neutral and keeps the protons in the nucleous from flying apart. Each orbiting electron carries a negative electrical charge, and since the atom is electrically neutral, there are as many electrons as there are protons. The mass of the atom is concentrated in the nucleus, for an electron is 1857 times lighter than a proton. Atoms are very small, far too small to be seen with the highest magnification of a microscope. Yet because of the rapidly moving electrons in successive shells, atoms have a large effective diameter, as much as 100,000 times the diameter of the nucleus. Nevertheless, diameters of atoms have been determined and are expressed in angstrom units (Å). An angstrom unit is one ten-millionth of a millimeter. For example, the diameter of an atom of gold is 2.88×10^{-8} centimeters but is usually given as the radius, 1.44 Å.

The atoms of the elements vary in weight as well as in size, and each has an *atomic weight*. The atomic weight of an element is a number expressing its relative weight in terms of an isotope of carbon, ^{12}C, which has been assigned an atomic weight of 12.000. The natural elements and their atomic weights are given in Table 5-1. Following the name of each element is its shorthand designation, the chemical symbol: C, carbon; Al, aluminum; Si, silicon; O, oxygen; and so on. In many cases the initial letters of the Latin name for a metal is used, as Fe from the latin *ferrum,* for iron; Ag, from *argentum,* silver; Au, from *aurum,* gold.

The atoms of the 88 natural elements differ from one another in the number of protons; 1 in hydrogen, the lightest, to 92 in uranium, the heaviest. The number of protons, which equals the number of electrons, is called the *atomic number.* When the elements are arranged according to atomic number, they show a periodicity of properties, that is, elements with like chemical properties are repeated periodically. This idea was put forth by the Russian chemist Dimitri Mendeleev in 1869, who arranged the elements according to their atomic weight rather than atomic number. The order of the two is just about the same except for a few minor reversals in the order of atomic weights. The *periodic table* showing the natural elements is shown in Fig. 5-1.

According to the Bohr model of the atom, the electrons are pictured as orbiting the nucleous at definite energy levels or shells concentric to the nucleus. These shells were originally assigned the letters *K, L, M, N, O, P, Q,* or by the numbers 1 to 7 with *K* equal to 1, closest to the nucleus, and with *Q* equal to 7, farthest from the nucleus. Each shell, with the exception of the *K* shell, has various subshells designated: *s, p, d, f.* When filled, the subshell share definite numbers of electrons occupying them as follows:

Shell	K(1)	L(2)	M(3)	N(4)
Subshell	s	s, p	s, p, d	s, p, d, f
Electrons	2	2, 6	2, 6, 10	2, 6, 10, 14

TABLE 5-1 Atomic Weights, 1993[a]

Element	Symbol	No.	Atomic Weight	Element	Symbol	No.	Atomic Weight
Actinium	Ac	89	227.029	Neodymium	Nd	60	144.24
Aluminum	Al	13	26.982	Neon	Ne	10	20.180
Antimony	Sb	51	121.760	Nickel	Ni	28	58.693
Argon	Ar	18	39.948	Niobium	Nb	41	92.906
Arsenic	As	33	74.922	Nitrogen	N	7	14.007
Barium	Ba	56	137.327	Osmium	Os	76	190.23
Beryllium	Be	4	9.012	Oxygen	O	8	15.999
Bismuth	Bi	83	208.980	Palladium	Pd	46	106.42
Boron	B	5	10.811	Phosphorus	P	15	30.974
Bromine	Br	35	79.904	Platinum	Pt	78	195.08
Cadmium	Cd	48	112.411	Polonium	Po	84	[209][b]
Calcium	Ca	20	40.078	Potassium	K	19	39.098
Carbon	C	6	12.011	Praseodymium	Pr	59	140.908
Cerium	Ce	58	140.115	Protactinium	Pa	91	231.036
Cesium	Cs	55	132.905	Radium	Ra	88	226.025
Chlorine	Cl	17	35.453	Radon	Rn	86	[222][b]
Chromium	Cr	24	51.996	Rhenium	Re	75	186.207
Cobalt	Co	27	58.933	Rhodium	Rh	45	102.906
Copper	Cu	29	63.546	Rubidium	Rb	37	85.468
Dysprosium	Dy	66	162.50	Ruthenium	Ru	44	101.07
Erbium	Er	68	167.26	Samarium	Sm	62	150.36
Europium	Eu	63	151.965	Scandium	Sc	21	44.956
Fluorine	F	9	18.998	Selenium	Se	34	78.96
Gadolinium	Gd	64	157.25	Silicon	Si	14	28.086
Gallium	Ga	31	69.723	Silver	Ag	47	107.868
Germanium	Ge	32	72.61	Sodium	Na	11	22.990
Gold	Au	79	196.967	Strontium	Sr	38	87.62
Hafnium	Hf	72	178.49	Sulfur	S	16	32.066
Helium	He	2	4.003	Tantalum	Ta	73	180.948
Holmium	Ho	67	164.930	Tellurium	Te	52	127.60
Hydrogen	H	1	1.008	Terbium	Tb	65	158.925
Indium	In	49	114.818	Thallium	Tl	81	204.383
Iodine	I	53	126.904	Thorium	Th	90	232.038
Iridium	Ir	77	192.217	Thulium	Tm	69	168.934
Iron	Fe	26	55.845	Tin	Sn	50	118.710
Krypton	Kr	36	83.80	Titanium	Ti	22	47.867
Lanthanum	La	57	138.906	Tungsten	W	74	183.84
Lead	Pb	82	207.2	Uranium	U	92	238.029
Lithium	Li	3	6.941	Vanadium	V	23	50.942
Lutetium	Lu	71	174.967	Xenon	Xe	54	131.29
Magnesium	Mg	12	24.305	Ytterbium	Yb	70	173.04
Manganese	Mn	25	54.938	Yttrium	Y	39	88.906
Mercury	Hg	80	200.59	Zinc	Zn	30	65.39
Molybdenum	Mo	42	95.94	Zirconium	Zr	40	91.224

[a]Atomic weight of $^{12}C = 12.000$.
[b][ ], atomic weight of most abundant isotope.

Group Ia (s¹)	IIa (s²)	IIIb (d¹)	IVb (d²)	Vb (d³)	VIb (d⁴)	VIIb (d⁵)	VIIIb (d⁶)	VIIIb (d⁷)	VIIIb (d⁸)	Ib (d⁹)	IIb (d¹⁰)	IIIa (p¹)	IVa (p²)	Va (p³)	VIa (p⁴)	VIIa (p⁵)	VIIIa (p⁶)
1 H 1.008																	2 He 4.003
3 Li 6.941	4 Be 9.012											5 B 10.811	6 C 12.011	7 N 14.007	8 O 15.999	9 F 18.998	10 Ne 20.180
11 Na 22.990	12 Mg 24.305											13 Al 26.982	14 Si 28.086	15 P 30.974	16 S 32.066	17 Cl 35.453	18 Ar 39.948
19 K 39.098	20 Ca 40.078	21 Sc 44.956	22 Ti 47.867	23 V 50.942	24 Cr 51.996	25 Mn 54.938	26 Fe 55.845	27 Co 58.933	28 Ni 58.693	29 Cu 63.546	30 Zn 65.39	31 Ga 69.723	32 Ge 72.61	33 As 74.922	34 Se 78.96	35 Br 79.904	36 Kr 83.80
37 Rb 85.468	38 Sr 87.62	39 Y 88.906	40 Zr 91.224	41 Nb 92.906	42 Mo 95.94	43	44 Ru 101.07	45 Rh 102.906	46 Pd 106.42	47 Ag 107.868	48 Cd 112.411	49 In 114.818	50 Sn 118.710	51 Sb 121.760	52 Te 127.60	53 I 126.904	54 Xe 131.29
55 Cs 132.905	56 Ba 137.327	57 * La 138.906	72 Hf 178.49	73 Ta 180.948	74 W 183.84	75 Re 186.207	76 Os 190.23	77 Ir 192.217	78 Pt 195.08	79 Au 196.967	80 Hg 200.59	81 Tl 204.383	82 Pb 207.2	83 Bi 208.980	84 Po [209]	85	86 Rn [222]
87	88 Ra 226.025	89 ** Ac 227.029															

*Lanthanide Series

f¹	f²	f³	f⁴	f⁵	f⁶	f⁷	f⁸	f⁹	f¹⁰	f¹¹	f¹²	f¹³	f¹⁴
58 Ce 140.115	59 Pr 140.908	60 Nd 144.24	61	62 Sm 150.36	63 Eu 151.965	64 Gd 157.25	65 Tb 158.925	66 Dy 162.50	67 Ho 164.930	68 Er 167.26	69 Tm 168.934	70 Yb 173.04	71 Lu 174.967

**Actinide Series

90 Th 232.038	91 Pa 231.036	92 U 238.029

Atomic Number — 26 Fe 55.845 — Symbol / Atomic Weight

Fig. 5-1. *Periodic table of the natural elements.*

In the lighter elements between hydrogen and argon (atomic numbers 1 to 18) electrons are added in regular succession to the *K, L,* and *M* shells until all *s* and *p* subshells, including *M-p* (*3p*), are filled, yielding a total of 18 electrons. These 18 elements constitute the first three "short" periods of the periodic table. In addition to possessing an electric charge, electrons also carry a magnetic moment called the spin (Fig. 5-2). Each subshell fills initially with all the spins in the same direction until the subshell is half filled. Then as additional electrons are added to complete the subshell, the spins of the added electrons are aligned in the opposite direction so that electrons with opposing spins become paired. Unpaired spins are important for proper understanding of the covalent bond, color, and magnetism.

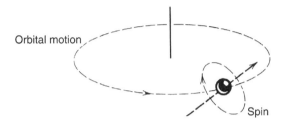

Fig. 5-2. *Orbital and spin motions of the electron.*

Addition of one electron to the argon superstructure results in an atom of potassium, and the addition of two electrons, in an atom of calcium. However, these two electrons enter the *N-s* (*4s*) shell, not the *M-d* (*3d*), which remains vacant. With the *N-s* (*4s*) shell now filled, additional electrons enter the *M-d* (*3d*) shell. A sequence of 10 elements intervenes before the *M-d* (*3d*) is filled and further addition to the *N* shell may continue. These 10 elements, scandium through zinc, are the transition metals and do not resemble the elements of the short periods. Starting with gallium, electrons again enter the *N* (*4p*) shell, forming a sequence from gallium through bromine, which do resemble the elements of the three short periods. The fourth period, the first "long" period, ends with the noble gas krypton. The fifth and sixth periods are similar to the fourth, with each containing transition elements, 39 to 48 in the fifth period and 57 to 80 in the sixth period. The sixth and seventh periods follow the pattern of the fourth and fifth. However, inserted at elements 57 and 89 are an additional set of 14 possible elements related to the infilling of the *4f* and *5f* subshells.

A close relation exists between the electronic structure of an atom, the chemical properties of the element, and its place in the periodic table. In groups Ia through VIIa, the number of electrons in the outermost orbit is the same for all elements in the group and is equal to the group number. For example, the atoms of elements in group Ia have one electron, whereas elements in group VIIa have seven electrons. It is these outer orbital electrons called *valence* electrons, which largely determine the chemical properties of an element. The elements in group VIIIa are the noble gases, which, with the exception of helium, which has only two electrons in the *K* shell, have eight electrons in their outermost shells. This is the most stable configuration.

CHEMICAL BONDING

There is a strong tendency for all atoms to achieve stable, completely filled outer shells with a noble-gas configuration. This can be achieved by gaining or losing electrons or by sharing electrons. If an atom gains or loses electrons, it is no longer neutral and is called an *ion*. A positive ion is one in which electrons have been lost and is called a *cation*. If electrons have been gained by adding electrons to an unfilled *p* subshell, the ion has a negative charge and is called an *anion*. The gain or loss of a single electron gives a unit charge. Thus if magnesium were to lose its two valence electrons, the chemical symbol would be Mg^{2+}. If oxygen were to complete its *p* shell by gaining two electrons, the chemical symbol would be O^{2-}. As each electron has a magnetic moment called the spin, a single electron from one atom can pair itself with a single electron from another atom in such that their spins are aligned antiparallel in much the same way as two small bar magnets adhere to one another when opposite poles are brought together. Thus atoms are held together either by electrostatic attraction between oppositely charged ions, the *ionic bond,* or by spin pairing their electrons, the *covalent bond.*

Ionic Bond

Sodium (in group I) has a single valence electron that is easily lost, leaving the atom with a single positive charge and with a noble-gas configuration of neon. On the other hand, chlorine (group VII), with seven electrons in its valence shell, can easily attain the stable configuration of the noble gas argon by adding an electron and thus assume a single negative charge. As shown by the fact that it conducts electricity, a water solution of halite (common salt) contains positive ions of sodium (Na^+) and negative ions of chlorine (Cl^-). If the volume of the solution is sufficiently reduced by evaporation, the attractive forces of the opposing electrical charges cause the ions to join together as a crystal nucleus (Fig. 5-3*a*). On continued evaporation more and more ions attach themselves to the growing nucleus to form a crystal that settles out of solution. This *ionic bond* is the principal type of chemical bonding in minerals.

Covalent Bond

Another way that atoms can achieve the noble-gas configuration is by sharing electrons. The outer regions of the electron clouds for two atoms interpenetrate so that some electrons do double duty in the joined atoms. This sharing by electron spin pairing, called the *covalent bond,* is the strongest of the chemical bonds and commonly forms molecules. A simple example is illustrated by the element chlorine, whose atoms, as we have seen, have seven electrons in the valence shell. Two chlorine atoms join to form a molecule of chlorine gas by one electron from each completing the outer shell of the other (Fig. 5-3*b*) and such that the spin pairing locks them firmly together so that both achieve the noble-gas configuration of argon. It is this same type of bonding that holds one carbon atom to another in diamond. Carbon (in group IV of the periodic table) has four electrons available. In the diamond structure every carbon atom is joined to four others by sharing electrons, as shown in

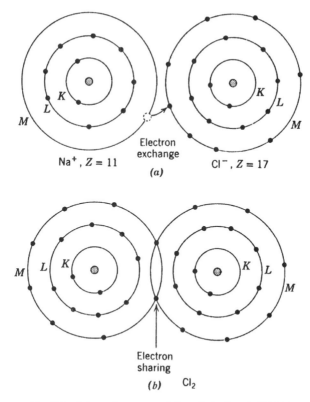

Na$^+$, $Z = 11$ Electron exchange   Cl$^-$, $Z = 17$

(a)

Electron sharing

(b)   Cl$_2$

Fig. 5-3. *Schematic representation of electron exchange.*

Fig. 5-4. The sulfides and the nonmetal native elements are minerals that possess covalent bonding.

Metallic Bond

The metallic bond is the force to which the metals owe their cohesiveness. As in the covalent bond, the atoms share their electrons; however, in the metallic bond the

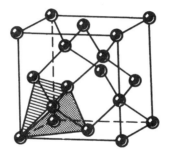

Fig. 5-4. *Diamond structure.*

outer electrons of the atoms are delocalized and weakly tied to individual atoms so that most of the electrons owe no allegiance to any nucleous and are free to drift through the structure without disrupting the bonding. It is to the presence of these free electrons that metals owe their high reflectivity, electrical conductivity, malleability, and ductility. The native metals are the only minerals that have metallic bonding.

Van der Waals Bond

The van der Waals bond is the weakest bond, one that ties essentially neutral groups of atoms into a lattice by virtue of a small residual electrical charge on the molecular surfaces. Although it is common among organic crystals, it is not common in minerals. The best example is the case of ice, in which atoms of hydrogen and oxygen are strongly held together by covalent bonds to form molecules of water. However, the distribution of electrical charge around the molecule is not uniform, so some parts of the molecule are more positive and other parts more negative. When water freezes to form ice, the weak electrical charges on the molecules are attracted to each other and form a van der Waals bond between the molecules of water. Similar bonding occurs in native sulfur.

Observations have shown that the relatively simple concept of the Bohr model with electrons moving in well-defined orbitals was too simplistic. A model based on *quantum mechanics* pictures the motion of the electrons in terms of the probability of finding an electron within a certain volume. The quantum theory is mentioned merely to let the student know that this theory is required for the advanced structural analysis of crystals. However, an adequate discussion of it would be long and complicated and would not contribute materially to our understanding of basic mineral chemistry.

CHEMICAL ANALYSES AND FORMULAS

As has already been explained, minerals are chemical elements or compounds, and the chemical composition is the most important of its many properties. Thus, when all other methods of identification fail, a quantitative chemical analysis must be made. Such an analysis gives the percentage of each element in a mineral and, knowing this, a chemical formula can be written.

The traditional way of making a quantitative analysis was by the *wet method,* in which the mineral was dissolved in an appropriate solvent and the percentage of the elements in solution determined by one of several different techniques. This method was time consuming and could be performed only by a skilled chemist. Today most analyses are made by instrumental methods. The most widely used are *x-ray fluorescence analysis, atomic absorption, optical emission spectroscopy,* and *electron microprobe analysis.* The results of these analyses are given in the weight percentages of the elements or of the oxides present in the mineral analyzed. The equipment for each method is very expensive and must be operated by a trained technician. It is important for the student to know that such methods exist even though they are beyond one's reach.

Except for the native elements, minerals are compounds of two or more elements. Quantitative analyses, as reported by the chemist, give the weight percent of the metals or of the oxides of the elements present. From the weight percent a formula can be calculated indicating the atomic proportions of the elements. These analyses usually add up to slightly less or slightly more than 100%. Consider the analysis for enargite, Cu_3AsS_4.

	1 Weight Percent	2 Normalized Percent	3 Atomic Weight	4 Atomic Proportion	9 Atomic Ratio
Cu	48.17	48.30	63.55	0.7600	2.9827
As	19.05	19.10	74.97	0.2548	1.0000
S	32.52	32.60	32.06	1.0186	3.9976
	99.74	100.00			

The percentages in column 1 as reported are recalculated in column 2 to 100% by dividing each by the total, 99.74, and multiplying by 100. These percentages do not represent the ratios of the various atoms. To determine the atomic proportions (given in column 4), the weight of each element is divided by its atomic weight (given in column 3). From these numbers the ratios of the various elements can be determined. By dividing the atomic proportions by the smallest proportion and rounding to the nearest whole number (column 5) we see that the ratios of elements as determined by the chemical analyses are Cu/As/S = 3:1:4. The chemical formula of enargite is thus Cu_3AsS_4.

Most minerals are oxidized compounds and the chemical analyses for these is conventionally reported as percentages of oxides. By calculations similar to those reported above for enargite, the ratios of the oxides and thus the chemical formulas can be determined. However, the weight percentages of the oxides must be divided by the formula weight for the oxides (i.e., the sum of the atomic weights of the elements in the oxide). Consider the analysis of zicon, zirconium silicate, as an example.

	1 Weight Percent	2 Normalized Percent	3 Formula Weight	4 Proportion	5 Ratio
ZrO_2	67.6	67.2	123.22	0.5454	1
SiO_2	33.0	32.8	60.09	0.5458	1
	100.6	100.0			

The ratios of oxide formulas in column 5 are $ZrO_2/SiO_2 = 1:1$, and thus the formula for zircon is $ZrO_2 + SiO_2 = ZrSiO_4$.

Few minerals are pure chemical compounds, for *impurities* are usually present. They exist from barely detectable traces to appreciable amounts. In many minerals several different ions may occupy a given site in the crystal structure. As mentioned earlier, olivine $(Mg,Fe)_2SiO_4$ is such a mineral. In it, iron and magnesium substitute for each other in all proportions and a solid solution series exists between forsterite, Mg_2SiO_4, and fayalite, Fe_2SiO_4. Small amounts of other elements may also be present. Consider the following analysis of an olivine crystal from Nevada.

	1 Weight Percent	2 Normalized Percent	3 Formula Weight	4 Formula Proportion
SiO_2	40.05	40.29	60.08	0.6705
MgO	47.34	47.63	40.30	1.1819
FeO	11.79	11.86	71.85	0.1651
CaO	0.22	0.22	56.08	0.0039
	99.40	100.00		

	5 *Atomic Proportion* Cations	6 Oxygens	7 Oxygens, 4 Basis	8 Cations	9 Atomic Ratio
Si	0.6705	1.3410	1.9926	0.9962	1.0000
Mg	1.1819	1.1819	1.7562	1.7562	1.7629
Fe	0.1651	0.1651	0.2453	0.2451	0.2456
Ca	0.0039	0.0039	0.0058	0.0058	0.0058
		2.6916	4.0000		

The analysis above illustrates the calculation of the formula proportions of the oxides and cations in olivine. Columns 1 to 4 are similar to those in the zircon analysis, but column 5 lists the proportions of the metals. These are the same values as the formula proportions because one "formula" of the oxide contributes one metal atom. The proportions of oxygen are not the same. Each atomic proportion of the divalent metals contributes one oxygen atom, whereas each atomic proportion of silicon contributes two. This is evident in column 6.

We know that olivine, with the formula $(Mg,Fe)_2SiO_4$, has four oxygen atoms per formula unit. To determine the cation proportions in terms of four oxygens, we divide 4 by 2.6916 (= 1.4861) and multiply the cation numbers in column 5 by the result. Assuming one SiO_2 and two for the total divalent cations, the formula, neglecting the minor amount of CaO, is $(Mg_{1.75}Fe_{0.25})SiO_4$. The composition of olivine is frequently given in terms of percentages of the end members of the series, forsterite (Fo) and fayalite (Fa), which are given by the formula proportions of MgO and FeO. Thus $Fo_{87.5}$ and $Fa_{12.5}$ would express the composition.

It is a simple procedure to reverse the process, that is, to determine the theoretical composition of a mineral from its formula. Take enargite, Cu_3AsS_4, as an example. The atomic weights, divided by their sum, give the weight percentages as follows:

	1 Atomic Ratio	2 Atomic Weight	3 Atomic Weight Proportion	4 Weight Percent
Cu	3 ×	63.55 =	190.65	48.4
As	1 ×	74.97 =	74.92	19.0
S	4 ×	32.06 =	128.24	32.6
			393.71	100.0

CRYSTAL CHEMISTRY

The properties of minerals depend not only on the chemistry but also on the geometrical arrangement of the constituent atoms, molecules, or ions into crystal structures. Although the idea of the atom and the building of crystal structures by arrays of atoms was understood by the nineteenth century, it was not until the introduction of x-ray diffration early in the twentieth century that it actually became possible to measure the distance between planes of atoms and determine the location of atoms relative to one another. There are many different techniques in which x-rays are used in the study of a crystal. But in all of them an x-ray beam enters a crystal and is diffracted by the atomic planes of the crystal lattice. The diffracted rays fall on a photographic film and darken the film as would a ray of light. From the positions of the dark spots, the distances between atomic planes and lattice dimensions can be calculated, and from their intensities, the positions of various atoms can be determined. Complex modern instruments count the diffracted x-rays with scintillation counters, which take the place of the x-ray film.

Ionic Radii

In ionic crystal structures we think of ions as spheres packed together in an orderly three-dimensional array. However, they do not possess finite surfaces and must be considered as very dense nuclei surrounded by space occupied by clouds of electrons whose density diminishes with increased distance from the nucleus. The size of an ion, usually given in terms of the radius, depends on its interaction with other ions.

Between a pair of oppositely charged ions there is an attractive force directly proportional to the product of their charges and inversely proportional to the square of the distance between the nuclei (Coulomb's law). As they approach each other a repulsive force arises from the negatively charged electron clouds and the positively charged nuclei. The distance at which the repulsive forces equal the attractive forces is the interatomic distance of the pair of ions. These interatomic distance can be determined accurately from x-ray diffraction data on the interplanar spacings of crystals. Generally, the radii of individual ions cannot be determined directly, but if the radius of one ion can be determined, the radii of the other ions can be calculated, and from the vast collection of known interplanar spacing, the radii of all other ions can be calculated. Fortunately, there are certain rare substances such as MgSe and LiI where the effect of the Mg and Li on the interplanar spacings can be ignored. From the interplanar distances of these substances, the ionic radii of Se^{2-} and I^- have been determined directly. Starting with these determined radii, the ionic radii of most of the elements have been obtained by calculation.

In general, cations are smaller, more rigid, and display less variation in size than anions and crystal structures can be thought of as closely packed anions with the smaller cations arranged between them. In Table 5-2 are given the ionic radii of the elements that make up most of the common minerals. Note in the table that for elements with more than one valence state, the higher the valence, the smaller the ionic radius.

TABLE 5-2   Charges and Radii of the Common Ions

Atomic Number	Element	Symbol and Ionic Charge	Ionic Radius (Å)
3	Lithium	Li^+	0.68
4	Beryllium	Be^{2+}	0.35
5	Boron	B^{3+}	0.23
6	Carbon	C^{4+}	0.15
8	Oxygen	O^{2-}	1.40
9	Fluorine	F^-	1.33
11	Sodium	Na^+	0.97
12	Magnesium	Mg^{2+}	0.66
13	Aluminum	Al^{3+}	0.51
14	Silicon	Si^{4+}	0.42
15	Phosphorus	P^{5+}	0.35
16	Sulfur	S^{6+}	0.30
		S^{2-}	1.84
17	Chlorine	Cl^-	1.81
19	Potassium	K^+	1.33
20	Calcium	Ca^{2+}	1.00
22	Titanium	Ti^{4+}	0.68
23	Vanadium	V^{3+}	0.74
		V^{5+}	0.59
24	Chromium	Cr^{3+}	0.62
25	Manganese	Mn^{2+}	0.80
		Mn^{4+}	0.60
26	Iron	Fe^{2+}	0.74
		Fe^{3+}	0.64
27	Cobalt	Co^{2+}	0.72
28	Nickel	Ni^{2+}	0.69
29	Copper	Cu^+	0.96
		Cu^{2+}	0.73
30	Zinc	Zn^{2+}	0.74
38	Strontium	Sr^{2+}	1.12
40	Zirconium	Zr^{4+}	0.79
47	Silver	Ag^+	1.26
50	Tin	Sn^{4+}	0.71
52	Tellurium	Te^{2-}	2.21
56	Barium	Ba^{2+}	1.34
74	Tungsten	W^{6+}	0.62
80	Mercury	Hg^{2+}	1.10
82	Lead	Pb^{2+}	1.20
92	Uranium	U^{4+}	0.97

Coordination and Radius Ratio

When cations and anions join in a crystal structure, they tend to coordinate, that is, to surround themselves with as many ions of opposite charge as size permits. When joined together in ionic bonds, they can be considered as spheres in contact, and the geometry is simple. The coordinated ions are arranged around the central coordinating ion in such as way that lines joining their centers outline a geometric solid known as a *coordination polyhedron.* The number of anions that can cluster about a central cation is the coordination number of the cation with respect to the given anion. Figure

5-5 is a model of the halite (sodium chloride) structure. In this model each sodium ion coordinates six larger chlorine ions, and thus chlorine is in 6-coordination with respect to sodium (Fig. 5-7b).

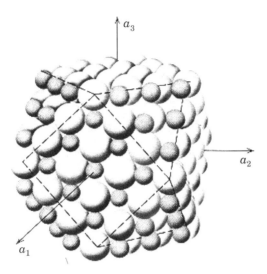

Fig. 5-5. Halite packing model.

Cations may also be regarded as the coordinating ion. In halite each chlorine ion has six sodium neighbors and thus is in 6-coordination with respect to sodium. Since sodium and chlorine are both in 6-coordination, there must be equal numbers of each, which is in agreement with the formula NaCl. On the other hand, examination of the structure of fluorite, calcium fluoride (Fig. 5-6), reveals that each fluorine ion has four closest calcium neighbors and is thus in 4-coordination with respect to calcium. But each calcium ion coordinates eight fluorine ions. There are thus twice as many fluorine as calcium ions, which agrees with the formula, CaF_2.

The type of coordination polyhedron and hence the coordination number depend on the relative size of the coordinated ions. The relative size is expressed by a *radius ratio* R_A/R_X, where R_A is the radius of the cation and R_X the radius of the anion in angstrom units. Take, for example, halite, NaCl. From Table 5-2 we find that the radius of chlorine (Cl^-) is 1.81 Å and the radius of sodium (Na^+) is 0.97 Å. Thus the radius ratio of Na/Cl is $R_{Na}/R_{Cl} = 0.97/1.81 = 0.54$.

If the radius ratio is less than 1 but greater than 0.732, the coordination polyhedron is a cube with anions at the eight corners. This is 8, or *cubic coordination* (Fig. 5-7a). At the critical ratio of 0.732, the anions not only touch the coordinating cation but also touch each other. For lesser values of radius ratio, 8-coordination is not as stable as 6-coordination, in which the coordinated ions lie at the apices of an octahedron (Fig. 5-7b). This is called *octahedral coordination* and is the stable coordination between 0.732 and its lower limit, 0.414.

For radius ratios less that 0.414, 4-coordination, in which the coordinated ions lie at the apices of a tetrahedron, is stable. Thus 4-coordination is called *tetrahedral*

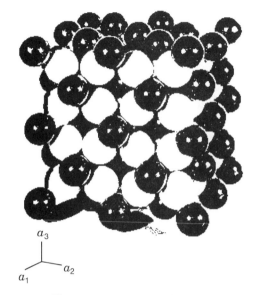

a_3

a_2

a_1

Fig. 5-6. Packing model of fluorite.

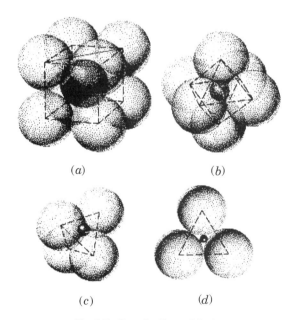

(a)   (b)

(c)   (d)

Fig. 5-7. Coordination polyhedra.

coordination. The limiting radius ratio in which the coordinated ions touch the central ion and each other is found to be 0.225 (Fig. 5-7c). Thus between radius ratios of 0.414 and 0.225, tetrahedral coordination is stable. This is the coordination not only of the phosphates, $(PO_4)^{3-}$, and sulfates, $(SO_4)^{2-}$, but also that of the $(SiO_4)^{4-}$ group that typifies the silicate minerals. Between radius ratios of 0.225 and 0.158, triangular or 3-coordination is stable (Fig. 5-7d). This coordination is found in the borates and is typified by the $(CO_3)^{2-}$ group of the carbonates. Factors other than those given influence the coordination pattern in crystals, but for most minerals, the radius ratio principle is valid.

Polymorphism

X-ray structure investigations have revealed why a specific chemical element or compound occurs in two or more structural forms. Consider diamond and graphite. The crystal structure of these two forms of carbon shows that diamond is harder and denser than graphite because the carbon atoms are more closely packed and held together by stronger bonds. Similarly, calcium carbonate $(CaCO_3)$ is found as two different minerals, calcite and aragonite, with differences in their structures reflected in their physical properties. Although the minerals kyanite, andalusite, and sillimanite share the same chemical composition (Al_2SiO_5), their structural differences give each its own distinctive physical properties.

The phenomenon of a chemical element or compound existing in two distinct physical forms is called *dimorphism;* in three forms, *trimorphism.* However, more generally used is the term *polymorphism* ("many forms") for an element or compound that exists in two or more structural modifications.

Crystal structure not only accounts for polymorphic forms but also resolves other apparent inconsistencies. For example, it explains the fact that ruby, composed of the light elements aluminum and oxygen, has nearly as high a density as sphalerite, made up of the heavier elements zinc and sulfur.

Isostructure

There appears to be little similarity between halite (sodium chloride, NaCl) and the rare mineral periclase (magnesium oxide, MgO). Halite is water soluble with low density, hardness, and melting point; periclase is insoluble in water with high density, hardness, and melting point. Yet structure analysis shows a marked similarity. The magnesium ions in periclase are in 6-coordination with respect to oxygen, and in halite each sodium ion coordinates six chlorine ions. The halite model (Fig. 5-5) could serve equally well for periclase. Minerals such as these with identical geometric arrangements in their structures are said to be *isostructural.*

Minerals that contain the same anion or anionic group often belong to *isostructural groups.* For example, the garnet group and the spinel group of minerals constitute two major isostructural groups, with large ranges in chemical composition within each group. The variation of the properties of group members result from various cations occupying a given structural site in the crystal.

Ionic Substitution and Solid Solution

Because in nature the solutions and melts from which minerals crystallize contain many elements, it is the rule rather than the exception for minerals to contain elements not required for their stable formation. Such additional elements are often present in minute amounts and may act as coloring agents, such as chromium or vanadium in emerald and iron in aquamarine. These metals are not present as tiny bits of metallic chromium or iron but as ions (e.g., Fe^{2+}, Fe^{3+}, or Cr^{3+}) substituting for a major element in the mineral. For example, in halite small percentages of potassium (K^+) may randomly substitute for some of the sodium (Na^+) in the structure; similarly, some fluorine (F^-) may replace some chlorine (Cl^-) anions.

There are many mineral examples in which ionic substitution is extensive. Substitution of one ion for another takes place readily if the ions differ in radius by less than 15% provided that the overall neutral charge of the mineral is maintained. For example, in the olivine group are the minerals forsterite (magnesium silicate, Mg_2SiO_4) and fayalite (iron silicate, Fe_2SiO_4). Iron (Fe^{2+}) and magnesium (Mg^{2+}) answer the requirements of similar ionic radii and have identical charges and thus can substitute for each other in all proportions. Such substitution is called *solid solution* or by some, *isomorphous substitution*. The properties of the intermediate members of the solid-solution series change with composition. With no iron, forsterite is colorless, but with increasing iron the mineral darkens, going from light-toned olive-green to dark green to black in fayalite. The specific gravity also increases with a progressive increase in iron. In peridot, the gem variety of olivine, iron substitutes for about 10% of the magnesium of forsterite.

In the garnet group of isostructural minerals, ionic substitution is the rule rather than the exception. An excellent example is the solid-solution series that exists between pyrope ($Mg_3Al_2Si_3O_{12}$) and almandine ($Fe_3Al_2Si_3O_{12}$). As in the olivine series, magnesium and iron substitute for each other in all proportions. With increasing iron there is a corresponding increase in density and refractive index and a change in color. The purplish red rhodolite is an intermediate variety in which about one-third of the magnesium of pyrope is replaced by iron.

Exsolution

When minerals crystallize at high temperatures, their high internal thermal energy allows for less stringent space requirements on the structure than would be permissible at lower temperatures, and ionic substitution may be extensive. An example is found in the feldspars. Potassium feldspar (orthoclase) formed at relative high temperature can tolerate considerable amounts of sodium ions in place of potassium ions. When such an orthoclase cools, the poor fit of the sodium ions in the potassium position results in stresses, forcing the sodium to migrate through the structure to form domains (small localized areas) of sodium feldspar (albite). This phenomenon is called *exsolution* and, in the example given, is responsible for *perthite*.

Corundum offers another excellent example of exsolution of interest to the gemologist. Corundum is a compound of only aluminum and oxygen, Al_2O_3, but at the high temperature at which it crystallizes, its structure can accommodate some titanium in solid solution substituting for aluminum. Relatively rapid cooling yields a homogeneous crystal, but on slow cooling, the titanium combines with oxygen and is exsolved as acicular (needlelike) crystals of rutile. Because of the constraining influence of the hexagonal crystal structure of the corumdum, the rutile needles are oriented at 60° to each other. If enough of these rutile crystals are present, a gemstone cut with a convex top, *cabochon,* will show a star effect, or asterism.

Powder Method

Because the position and intensity of the diffracted x-rays are characteristic of each mineral, x-ray diffraction may be used in mineral identification (page 97). The most common method for this determination is called the *powder method.* In this method the mineral is powdered and the crystalline particles are picked up and bonded together into a needlelike spindle. The spindle, containing many randomly oriented particles, is mounted at the center of the cylindrical camera (Fig. 5-8a). On striking the spindel-shaped mount, the x-ray beam is diffracted by the many atomic planes of

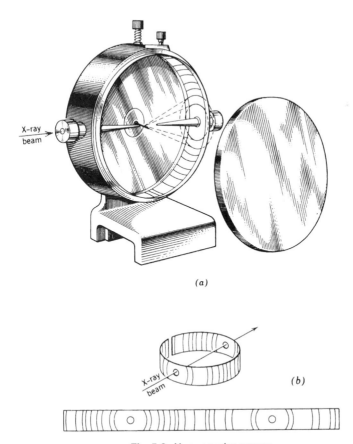

(a)

(b)

Fig. 5-8. *X-ray powder camera.*

the crystalline particles. The diffracted rays are intercepted by and recorded as curved lines on a photographic film (Fig. 5-8*b*) snugly wrapped around the inside wall of the camera. Since for each mineral the spacing and intensity of these lines is characteristic, the powder method has become a major means of mineral identification and has been called the mineral "fingerprint." Because intensities of the diffracted x-rays are more readily measured directly with scintillation counters, the powder cameras using film have now largely been replaced by large machines.

CHEMICAL TESTS

Unfortunately, the equipment and accessories needed for modern chemical determinations are expensive and not available to the amateur or beginning student. Yet because this sophisticated equipment exists, most modern textbooks of mineralogy no longer include the simple time-honored chemical tests for mineral identification. Moreover, only a small amount of inexpensive equipment is necessary for making these tests. Chief among these are the *blowpipe tests* long used by the mineralogy student. We believe these tests, which are highly diagnostic for specific elements, still serve a useful purpose for the amateur collector and in the field camp and are introduced here in a simplified form. A more extensive treatment may be found in the earlier edition of this book or in any mineralogy book published prior to 1970.

Before attempting any chemical tests, the student should have exhausted all possible physical tests, including luster, hardness, cleavage, color, and habit as well as a determination of specific gravity. The chemical test should then be used only to distinguish among a small number of possibilities for which a particular confirmative test can be selected. As all the tests described require only small amounts of chemicals or mineral, the principal precaution while performing the tests is to wear safety glasses or goggles. In the rare circumstance that any hot material or chemical comes in contact with the skin, the area of skin affected should be rinsed with water immediately to minimize the area of skin affected.

Blowpipe Tests

Blowpipe tests determine the presence or absence of elements in a mineral, not the amounts; they are thus qualitative, not quantitative, chemical tests. But to know that a given element is present in an unknown mineral is a major step, and perhaps a definitive one, in its identification. For example, assume that by its physical properties an unknown has been narrowed down to one of two minerals, barite or celestite. When a fragment of the mineral is introduced into a blowpipe flame, the flame becomes green, indicating barium, and the mineral in thus barite, barium sulfate. Had the mineral been celestite, strontium sulfate, the flame would have been red, the test for strontium.

A blowpipe (Fig. 5-9) is a tapering tube, usually made of brass and often with a bulb for collecting moisture, curved at right angles near the narrow end with a heavy tip that has a very fine orifice. The blowpipe is operated continuously by puffing up

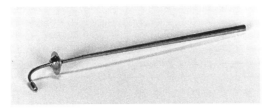

Fig. 5-9. *Blowpipe. (Miners, Riggins, Idaho.)*

the cheeks and blowing air through the mouth while breathing normally through the nose but allowing sufficient air to enter the mouth to keep the cheeks inflated. Air entering the larger end of the blowpipe emerges as a thin, high-velocity stream from the other end. When this airstream is directed into a luminous source flame (gas or candle), rapid combustion takes place, producing a very hot flame. Although not very satisfactory, a temporary blowpipe can be made from a piece of glass tubing. Today, the blowpipe is usually replaced by a small butane torch (Fig. 5-10), which gives a similar flame and relieves the operator from blowing. We will continue, however, to call it a blowpipe flame.

Fig. 5-10.
Butane torch. (Microflame, Minnetonka, Minnesota.)

Equipment for Blowpiping

1. Lamp	4. Charcoal blocks
2. Blowpipe or butane torch	5. Platinum wire
3. Nichrome-tipped forceps	6. Glass tubing
	7. Small magnet

Blowpipe Flame. An important distinction must be made between a *reducing flame* and an *oxidizing flame.* The flame (Fig. 5-11) should consist of three parts: an

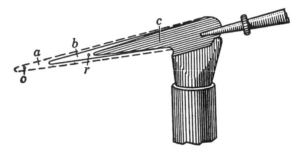

Fig. 5-11. Blowpipe flame.

inner blue cone (c), a yellow luminous cone (b), and the outer, almost invisible enve-lope (a) extending far beyond. In the inner cone and the yellow cone the gas is only partially burned, and there is a deficiency of oxygen. A mineral fragment (called the *assay*) placed at (r) will, if possible, lose oxygen and be *reduced.* For example, a small fragment of ferric oxide, Fe_2O_3, held in this part of the flame will lose oxygen and be reduced to ferrous oxide, FeO. In the outer part of the flame, on the other hand, there is an excess of oxygen from the surrounding air and the tendency is to impart oxygen to the assay placed at its tip (o). Here manganese oxide, MnO, for example, is changed to a higher oxide, Mn_2O_3. The oxidizing flame can be enhanced by placing the blowpipe inside the source flame while the reducing flame is enhanced by placing the blowpipe just at the edge of the source flame. Students can test their proficiency by placing a small fragment of tin on a charcoal block. In a reducing flame the tin bead will remain bright and shiny, whereas in the oxidizing flame the bead will turn white from a coating of tin oxide.

Fusion by Blowpipe Flame. The hottest part of a blowpipe flame is just beyond the visible portion at (a) in Fig. 5-11 and may reach a temperature of 1500°C. The ease with which a mineral fuses may be diagnostic in its identification. In making this test the assay held in forceps (Fig. 5-12) is placed in the hottest portion of the flame. It

Fig. 5-12. Forceps.

may melt completely or there may be only a rounding of the thin edges. In both cases the mineral is said to be *fusible.*

The method of experiment described gives in the first place an approximate determination of the melting point or degree of fusibility. The following *scale of fusibility* is used to define the fusibility of the different minerals.

1. *Stibnite:* fusible in the ordinary yellow gas flame or candle flame, even in large fragments.
2. *Natrolite:* fusible in fine needles in the ordinary gas flame, or in larger fragments in the blowpipe flame.
3. *Almandite, or iron-alumina garnet:* fusible to a globule without difficulty with the blowpipe, if it is in thin splinters.
4. *Actinolite:* fusible to a globule in thin splinters.
5. *Orthoclase:* thin edges can be rounded without great difficulty.
6. *Bronzite:* fusible with difficulty on the finest edges.

The following list gives the name of some minerals, most of them common, with the degree of fusibility of each according to this scale.

Stibnite, galena	1
Cryolite, apophyllite, pyromorphite	1½
Amblygonite, witherite, prehnite, arsenopyrite	2
Rhodonite, analcime	2½
Gypsum, barite, celestite, fluorite, epidote	3
Oligoclase	3½
Albite	4
Apatite, hematite, magnetite	5
Bronzite	6
Quartz, calcite, topaz, sphalerite, graphite	Infusible

At the same time that one tests the fusibility and flame color of a mineral, the student must be on the watch for attendant phenomena. For the fragment, instead of fusing quietly, may:

1. *Swell up,* that is, throw out little globules or curling processes (as stilbite)
2. *Intumesce,* that is, bubble up and then fuse (as scapolite and most zeolites)
3. *Exfoliate,* that is, swell up and open out in leaves (as apophyllite and, even more, vermiculite)
4. *Glow brightly,* without melting (as calcite)
5. *Decrepitate,* that is, break violently into fragments (as in barite)

After being heated the fragment should be examined to see whether, if fused, the glass is clear, full of bubbles (then often called *blebby,* or *vesicular*), or even black; whether it has changed color, even if not fused; and whether it is magnetic (due to iron).

Flame Coloration. In making a test for fusibility, it may be observed that the fusing mineral imparts a color to the flame. There are several elements that have characteristic flame colors, and thus flame coloration becomes a simple and important means of qualitative chemical analysis. Tests for flame coloration are more easily made with a bunsen burner (Fig. 5-13) than with a blowpipe. A small mineral fragment should

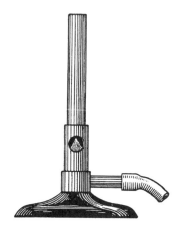

Fig. 5-13. Bunsen burner.

be held in forceps and only its tip inserted into the lower part of the flame. A more decisive test may be obtained when the powdered mineral is introduced into the flame on a platinum wire. Table 5-3 is a list of flame colors produced by some of the common elements.

Sodium compounds are ever present in the laboratory, and the strong yellow color of sodium may mask the flame color of another element. A blue glass filter held in front of the flame will completely absorb the yellow flame and permit the character-

TABLE 5-3   Flame Colorations

Color	Shade	Element
Red	Carmine-red	Lithium
	Purple-red	Strontium
	Yellowish red	Calcium
Yellow	Intense yellow	Sodium
Green	Yellowish green	Barium
	Siskine-green	Boron
	Emerald-green	Copper
	Bluish green	Phosphates
Blue	Greenish blue	Antimony
	Whitish blue	Arsenic
	Azure-blue	Copper chloride
Violet		Potassium

istic flame colors of other elements to be observed. Some minerals contain elements that would normally color the flame but fail to do so until they are broken down by an acid or a flux.

Bead Tests and the Platinum Wire. When dissolved in fluxes, some elements give a characteristic color to the fused mass. The flux most commonly used is borax, $Na_2B_4O_7 \cdot 10H_2O$. Sodium carbonate, Na_2CO_3, is used in the test for manganese. These tests are most satisfactorily performed by first fusing the flux (the bead) on a small loop at the end of the platinum wire (Fig. 5-14). After the flux has fused into a small lens, but before it cools, a few grains of the powdered mineral are picked up in

Fig. 5-14. Loop in platinum wire.

the bead and are dissolved by further heating.* As indicated in Table 5-4, the color of the bead may depend on whether it is heated in the oxidizing or reducing flame.

*All metallic minerals should be roasted before being introduced into platinum wire, to drive off any arsenic that may be present. Small amounts of arsenic make the platinum brittle and cause the end of the wire to drop off.

TABLE 5-4   Borax Bead Colors

Color				
Oxidizing Flame		Reducing Flame		
Hot	Cold	Hot	Cold	Element
Pale yellow	Colorless to white	Grayish	Brownish violet	Titanium
	Colorless to white	Yellow	Pale yellow	Tungsten
	Colorless to white	Brown	Brown	Molybdenum
Yellow	Yellowish green	Green	Green	Chromium
	Yellow-green to colorless	Dirty green	Green	Vanadium
Yellow to orange-red	Yellow	Pale green	Pale green to colorless	Uranium
	Yellow	Bottle green	Pale bottle green	Iron
Pale green	Blue-green	Colorless to green	Opaque red	Copper
Blue	Blue	Blue	Blue	Cobalt
Violet	Red-brown	Opaque gray	Opaque gray	Nickel
	Reddish violet	Colorless	Colorless	Manganese

Reduction on Charcoal. Some elements can be reduced to the metal when heated on charcoal before the blowpipe. The charcoal blocks have dimensions of about $10 \times 2\frac{1}{2} \times 1\frac{1}{4}$ centimeters, and a small pit the size of a small pea is scouped out about 2 centimeters from one end of the block. Although some minerals can be reduced to metal merely by heating on charcoal, others require the use of a flux. The flux, a mixture of sodium carbonate and charcoal in equal amounts, is called the *reducing mixture.* In a reduction experiment, the powdered mineral should be placed in the depression in the charcoal. If the flux is used, it should be mixed with the mineral in the proportions 2 flux to 1 mineral.

Metallic Beads on Charcoal

Antimony	Brittle, gray to white
Bismuth	Brittle to malleable, reddish white
Copper	Malleable, reddish, tarnishes black
Gold	Malleable, yellow
Lead	Malleable, gray, tarnishes white
Silver	Malleable, white, tarnishes black
Tin	Malleable, white
Cobalt, nickel, iron	Magnetic bead

Sublimates on Charcoal. Some elements yield characteristic oxide sublimates when minerals containing them are heated in the oxidizing flame on charcoal. The powdered mineral called the assay is placed in the depression at one end of the block and a blowpipe flame is directed toward the other end. The sublimate will deposit on the surface as in Fig. 5-15. In the test for molybdenum, a plaster tablet is preferable to charcoal, for its dark sublimate is better displayed on the white background.

Sublimates on Charcoal

Arsenic	Volatile white deposit away from assay, garlic odor
Antimony	White deposit close to assay
Selenium	Volatile white deposit, odor of horseradish
Tellurium	Dense, white volatile coating
Zinc	Nonvolatile white sublimate with soda flux
Tin	Metal globules with soda flux
Molydenum	Volatile white deposit
Lead	Metal bead, yellow close and white away from assay
Bismuth	Yellow close and white away from assay

Open-Tube Tests. Glass tubing about 12 centimeters long and 5 millimeters in inside diameter is used in making open-tube tests. In making the test, a small amount of the powdered mineral is placed in the tube about 2.5 centimeters from an end. The tube is inclined at as sharp an angle as possible, with the mineral near the lower end. The upper part of the tube is then heated in a bunsen burner flame to cause air to

Fig. 5-15. *Antimony sublimate on charcoal.*

Fig. 5-16.
Open-tube test.

move up the inclined tube (Fig. 5-16). The tube is then moved so that the flame is just above or beneath the mineral. In a current of air, the mineral will oxidize if such a reaction is possible. The oxides formed will either escape as gases at the end of the tube or condense as sublimates on its walls. Several elements give characteristic open-tube tests, but for them all except sulfur there is an equally good or better test. The open-tube test for an unoxidized sulfur compound is as follows: On heating, the colorless gas, SO_2, issues from the upper end of the tube. It can then be detected by its pungent and irritating odor. But a more positive test is obtained by placing a moistened blue litmus paper in the upper end of the tube. Owing to the acid reaction of sulfurous acid, the paper turns red.

Closed-Tube Tests. The closed tube test is used to determine what takes place when a mineral is heated in the absence of oxygen. The tube, made of soft glass, should have a length of about 8 centimeters and an inside diameter of 4 millimeters. Two tubes are made from a piece of tubing about 15 centimeters long by fusing it in the center and pulling apart. In making the test, the mineral is placed in the closed end of the tube in small fragments or as a powder and is heated.

Closed-Tube Deposits

Sulfur	Red deposit when hot, yellow when cold
Arsenic	Silver-gray mirror, black ring above
Antimony	Reddish-brown coating near bottom of tube
Mercury	Black coating, sometimes a mirror of mercury
Water	Droplets of water

Chemical Tests

In addition to the assay tests made with the blowpipe, there are many other chemical tests that can be made, such as solubility, spot, and stain tests. Of these, the tests for solubility have been used traditionally because they involve only the simple reagents water, hydrochloric acid, and nitric acid and test tubes.

Both the spot and stain tests can be very useful, especially in prospecting, but require the use of a wide variety of chemical reagents. For example, a spot test for nickel consists of dissolving a mineral grain on a watch glass in nitric acid, then making the solution alkaline with ammonium hydroxide, and finally, adding an alcohol solution of dimethygloxine to obtain a scarlet precipitate diagnostic of nickel. A simple test for cassiterite consists of placing some grains of the mineral in 50% hydrochloric acid to which granular zinc is added. The dissolving zinc releases hydrogen, which reacts with the surface of the cassiterite, reducing the tin oxide to metallic tin and coating the surface of the grains with metallic tin. A wide variety of such tests exist for special purposes and can be obtained from the mineralogical literature.

Solubility Tests

Whether a mineral is soluble in one of the acids named is often very important. To test the solubility, hydrochloric acid is generally used, except with metallic sulfides and some other metals containing predominantly one of the heavy metals (lead, copper, silver, etc.); for these, nitric acid is usually better. In general, the mineral should be pulverized as finely as possible in the agate mortar and introduced into a large test tube, some acid poured on, and the whole carefully heated over the Bunsen flame.

It must be remembered here that it is injurious to breathe the acid fumes in the air and that they will act corrosively on brass surfaces nearby; hence such tests can be tried only with caution unless the conveniences of the laboratory are at hand.

Various results may be noted during this trial:

1. The mineral may dissolve quietly with or without coloring the solution; this is true, for example, of hematite, also of many of the sulfates and phosphates.
2. There may be a bubbling off of effervescent gas. This gas is usually carbon dioxide (CO_2), but it may be hydrogen sulfide (H_2S).
3. There may be a separation of some insoluble substance, such as sulfur or silica.

Solubility Guidelines

Minerals soluble in water
   Halite, sylvite, borax, kernite
Soluble minerals that effervesce carbon dioxide in HCl
    Easily: calcite, aragonite, strontianite, witherite, smithsonite
    With Difficulty: dolomite, siderite
Soluble minerals that effervesce hydrogen sulfide in HCl
    Stibnite, galena, pyrrhotite, sphalerite
Soluble minerals that yield chlorine in HCl
    Pyrolusite, manganite, romanechite

Tests for Anion Groups

As minerals are classified according to their anion or anionic group, chemical tests for the various anions are often helpful in placing minerals in the proper classification.

Silicates. Although there is no convenient chemical test for silica, the best test for silicates is actually their hardness as most minerals with a nonmetallic luster and harder than 5½ are either silicates or oxides. Most of the softer silicates have a prominent platy or micaceous cleavage. Some silicates yield silica gel when treated with HCl.

Oxides. As with the silicates, there is no direct test, and again most oxides, whether metallic or nonmetallic in luster, have a hardness greater than 5½.

Hydroxides. A closed tube test will readily detect any mineral that has hydoxyl ions or waters of hydration.

Chlorides. A drop of silver nitrate yields a white precipitate when added to a solution of a chloride mineral. For this test the mineral must be dissolved in water or in nitric acid.

Fluoride. When heated in a closed tube with soda, a fluoride mineral will etch the sides of the tube.

Sulfide. When heated in an open tube, a sulfide mineral turns blue litmus paper red. The anion groups closely associated with sulfides, such as antimony, arsenic, and tellurium, can be detected by use of the closed tube.

Carbonates. The carbonates effervesce with hydrochloric acid. The related borates are detectable using flame coloration.

Sulfates. When dissolved in either water or hydrochloric acid, the sulfates give a white precipate when a drop of barium chloride solution is added. Closely related anions such as tungstate and molybdate require a borax bead test.

Phosphates. When dissolved in nitric acid, the phosphates yield upon addition of ammonium molybdate solution a yellow precipitate of ammonium phosphomolybdate. The related vanadate and arsenate ions are detected using a borax bead and as a sublimate on charcoal, respectively.

6

Mineral Genesis

Mineral genesis involves many complicated processes and it has taken mineralogists many decades of work to unravel much of the detail involved in the various ways in which minerals form. A generalization of these processes can be divided into three major types: crystallization from a fluid, chemical reactions, and internal reorganizations of a mineral, which will be called *solid-state transformations*. These are summarized below together with specific ways that these take place in nature.

1. Crystallization from a fluid
   a. Silicate melts
   b. Hydrothermal solutions
   c. Evaporation
   d. Recrystallization
2. Chemical reactions
   a. Oxidation–reduction
   b. Hydration–dehydration
   c. Decarbonatization
   d. Recombination
3. Solid-state transformations
   a. Polymorphism
   b. Order–disorder
   c. Exsolution

None of these processes take place under static conditions. Rather, some kind of driving force has been imposed that has brought about a change in conditions. Thus a change of temperature as from intrusion of a magma, a change of pressure as from deep burial, or a change in the chemical environment as from exposure to the atmosphere is required to initiate any of these mineral-forming processes. Minerals such as feldspar and quartz in a granite crystallize from a silicate melt only when the melt cools from a high temperature to a lower temperature. Aragonite forms when rocks rich in calcite are buried at great depth and the high pressure converts the calcite to aragonite. The minerals pyrite and marcasite are stable as long as they are buried beneath the surface of the earth. But when they are exposed, either by erosion or by mining, to the oxygen and moisture in the atmosphere, they are rapidly converted to goethite. As each of these mineral-generating processes is discussed, you will find that there are specific environments under which these processes take place that yield certain suites of minerals.

CRYSTALLIZATION FROM A FLUID

Minerals crystallize from fluids over a temperature range which varies from that of a molten basaltic magma at over 1100°C, to temperatures that exist at the earth's surface. For example, feldspar crystallizes at high temperatures from a cooling magma, chalcopyrite precipitates in veins at intermediate temperatures from cooling hydrothermal solutions, and halite crystallizes as beds of salt by evaporation at surface temperatures from brines in saline lakes.

Silicate Melts

A magma is a hot fluid mass of rock material which on cooling and solidification forms an igneous rock. In general, we can think of a magma as a silicate melt which, in addition to the 10 constituent oxides (SiO_2, TiO_2, Al_2O_3, FeO, CaO, MgO, MnO, Na_2O, K_2O, P_2O_5) contains water and other elements in minor amounts. As the magma cools, there is, in general, a definite order of mineral crystallization. If the magma cools slowly, the minerals first to crystallize remove from the melt MgO, FeO, and CaO and enrich the melt in Na_2O, K_2O, SiO_2, and water. The resulting series of minerals that form is known as *Bowen's reaction series*. Study of rocks in both the field and laboratory show that removal of MgO and enrichment in water results in the discontinuous series on the left, while removal of CaO and enrichment in Na_2O gives the continuous reaction (plagioclase) series on the right.

Bowen's Reaction Series

High temperature	Olivine	Bytownite
	Pyroxene	Labradorite
	Hornblende	Andesine
	Biotite Oligoclase	
	Alkali-feldspars	
	Muscovite	
Low temperature	Quartz	

The reaction series is based on the most common type of magma, one that has the composition of a basalt. As such a magma cools, olivine and bytownite are first to crystallize. If they remain in place, the melt will crystallize on additional cooling to form pyroxene and labradorite, and the resulting rock will be a basalt or gabbro. (The mineralogy of the major types of rocks is given in the section "Common Rock Types" later in the chapter.) If, however, some of the early crystallized olivine is removed by crystal settling, the composition of the melt becomes enriched in water and Na_2O, and hornblende crystallizes instead of pyroxene, and andesine instead of labradorite. Those igneous rocks rich in FeO and MgO are termed *mafic* and are characterized by the presence of dark-colored minerals such as pyroxene, while those rich in SiO_2, Na_2O, and K_2O are termed *felsic* and are characterized by light-colored minerals such as feldspar and quartz. It is thus possible to have a single parent magma give rise to an entire sequence of rocks. The series begins with rocks of mafic composition such as basalt/gabbro and progresses toward rocks of more felsic composition such as andesite/diorites, to rhyolites/granites, and ending with pegmatites, where a transition is made from silicate melt to hydrothermal aqueous solutions.

Although this mafic–felsic sequence of igneous rocks is the most common, a number of important deviations are known. If silicate melts are deficient in silica, crystallization leads to nepheline-rich rocks. If they are are low in silica and high in potassium, crystallization leads to biotite- or leucite-rich rocks. In addition, there are a variety of rare rocks that are thought to have separated from silicate melts through liquid immiscibility. These include igneous carbonate melts known as carbonatites, and igneous sulfide melts which can be found as layers in certain gabbro intrusions such as Sudbury, Canada. All of these unusual rocks can be of great economic and mineralogical interest because of certain uncommon minerals present in them.

Pegmatites are of special interest to the mineralogist not only because of the occurrence of many rare minerals but because of the very large crystal sizes attained. Pegmatites form from silicate melts which are very rich in water and may be enriched in lithium, boron, and a host of other rare elements. They are thought to have formed either by the collection of incipient melt during high-grade metamorphism or by the concentration of rare constituents during the final stages in the crystallization of a silicate magma. Pegmatites are characterized by their very large grain size (i.e., over 2 centimeters), which is the result of very low viscosity from the included water. Most small pegmatites are simple ones consisting of feldspar, muscovite, and quartz, but large complex pegmatites may have zones of giant crystals and a core of quartz. In large pegmatites, as crystallization proceeds a stage is reached when the melt becomes frothy, exsolving water. It is at this stage that it is believed that gigantic crystals are formed. Most of these giant crystals are of the rock-forming minerals: feldspar, quartz, and mica; but some are of rare minerals such as beryl, spodumene, and tourmaline. These giant crystals are commonly filled with bubbles and considerable occluded matter. During the final stage of crystallization, the expelled water-rich solutions lead to the formation of a quartz core near the center of the pegmatite. Under the right circumstances, the formation of fine gem-quality tourmaline, topaz, and beryl in pockets is associated with the quartz core.

Hydrothermal Solutions

Just as with igneous rocks, hydrothermal solutions show a temperature-dependent depositional sequence. The sequence begins with rock-forming silicates where the igneous sequence ends. It is then dominated by various sulfide minerals and ends at low temperatures with calcite, aragonite, and cinnabar. Quartz may crystallize anywhere during the sequence but is dominated by gray quartz at high temperature, milky quartz in the middle part of the sequence, and chalcedony at the end. The ideal hydrothermal depositional sequence is:

High temperature
>   Albite and muscovite
>   Gray quartz
>   Wolframite, cassiterite, molybdenite
>   Main sulfides
>>   Arsenopyrite
>>   Chalcopyrite
>>   Pyrite
>>   Sphalerite
>>   Galena
>   Milky quartz
>   Calcite
>>   Stibnite, cinnabar
>   Aragonite-chalcedony
Low temperature

Hydrothermal solutions arise in five ways: as fluid separation from a silicate melt, by descending rainwater interacting with an intrusive magma, by seawater contacting hot igneous rocks beneath the seafloor, by the expulsion of water during the processes of folding and faulting of sedimentary rocks, and by dehydration of sedimentary rocks from the high temperatures experienced during deep burial.

During the crystallization of silicate melts, an aqueous (hydrothermal) fluid is released and will escape from the magma into the surrounding rocks. If the immediately surrounding rocks are slates or schists, these fluids, carrying dissolved silica escaping along fractures in these rocks, will deposit quartz in the form of veins. In addition to quartz, these escaping fluids may deposit tourmaline, cassiterite, and wolframite. Farther away from the intrusion, various sulfides, such as arsenopyrite, chalcopyrite, and galena, may be deposited. If the immediately surrounding rocks are limestones or marbles, the escaping fluids from the magma will react chemically with these carbonate rocks and assist in the decarbonatization process (see p. 126), which results in the formation of *skarn*. The escaping hydrothermal fluids may be rich in boron, fluorine, and metals such as iron, molybdenum, copper, lead, and zinc.

These skarns are noted for a number of exotic minerals and for the many small but very rich ore bodies which commonly form where these escaping fluids have passed through the skarn and into bodies of limestone.

Rainwater may seep downward and come into contact with a buried intrusive magma. The heated rising waters, as illustrated on Iceland and New Zealand and at Yellowstone, Wyoming, may return to the surface as hot springs and geysers. In some locations these rising hot waters have been tapped for electric power generation. The circulating waters will hydrate any rocks with which they come into contact. Because of the high temperatures, the rocks through which these waters pass will be leached of their silica and metals, such as manganese, iron, lead, and zinc. Upon cooling these will be redeposited as quartz, rhodochrosite, ankerite, galena, and sphalerite.

The intrusion of igneous silicate melts is not confined to the continents. Extensive extrusion of basalts and intrusion of gabbros occurs along deep-sea ridges in the midocean, while extrusion of felsic andesites and intrusion of diorites as volcanic mountain chains occurs parallel to oceanic trenches. Seawater will penetrate downward and will come into contact with any intruded magma. The magma will heat the contacting seawater to very high temperatures at high pressure, and these waters emerge as hot springs on the ocean floor. Because of the high temperatures, pressures, and salinity, these solutions are powerful solvents. In these systems, the basalts and andesites are altered to chlorite and epidote, while more mafic rocks rich in olivine and enstatite are altered to serpentine and talc. Any iron present reacts with the sulfate dissolved in the seawater. During this reaction oxygen anions are stripped from the complex sulfate anion, and in this process the sulfur is reduced from S^{6+} to S^{2-}; in compensation ferrous iron, Fe^{2+} is oxidized to ferric Fe^{3+}. The solvent action of the chloride-rich seawater leaches any lead, zinc, copper, iron, manganese, and barium from the rocks they penetrate, and these upwellings are deposited on the seafloor as *exhalites,* a sediment consisting of laminae of pyrite, chalcopyrite, pyrrhotite, sphalerite, galena, and barite.

When sedimentary rocks are deformed, such as during processes of mountain building, into folds or faulted blocks, the compressive forces create pressures that will expel the waters from these rocks by a squeezing action. The water expelled during folding and faulting may be termed *formation water,* as it is the water collected and held in the pores of the rocks between the grains of sand and clay when the sediment was originally deposited. These expelled waters will migrate along permeable beds of sandstones, along fractured limestones, and along brecciated zones near active faults. Also, when sedimentary rocks are buried at progressively greater depths, they experience progressively higher temperatures from the internal heat of the earth. This water driven out of clays by the earth's heat may be termed *metamorphic water.* Although they are hydrothermal, the solutions in both cases tend to be much lower in temperature than those associated with magmas. The most common dissolved material is silica, but often it is deposited as chalcedony instead of as quartz. These expelled waters may carry magnesium, fluorine, zinc, and lead. It is now thought that the important galena–sphalerite deposits of the Missisippi Valley were formed by such solutions.

Evaporation

Evaporation occurs in the desert regions of the world, where rainfall is low relative to the temperature. Most commonly, these regions are centered along the 30th parallel of latitude both north and south of the equator. Here numbers of enclosed basins may be found in which there is insufficient rainfall to fill the basin. Instead of filling the basin and overflowing such that it can carry the dissolved salts into the sea, the evaporating water concentrates them instead. Usually, the salts in these basins are dominated by sodium carbonate. Well-known examples include Searles Lake, California and Lake Magadi, Kenya. If the enclosed basin has ancient salt beds, the waters may be rich in halite and sylvite, as in the Dead Sea, or if the basin has large numbers of volcanics or hot springs, the basin waters may be enriched in borates, as has occurred in Turkey, California, and Argentina, where deposits of borax, colemanite, and ulexite have been mined.

The most important mineral formation by evaporation have been those in which large basins have been flooded periodically by the sea. Evaportaion of seawater generally results first in the deposition of carbonates of calcium and magnesium, followed by the deposition of sulfates of calcium, then the chlorides, first of sodium and then potassium. At the extreme end of the sequence the very soluble salts of magnesium and other unusual salts may crystallize. The specific sequence of mineral deposition as evaporation proceeds is:

Calcite and dolomite

Gypsum

Halite

Sylvite

Polyhalite

As a result of the evaporation, thick beds of halite and gypsum, often dehydrated to anhydrite, are found in many places, such as western New York, Texas, Arabia, and China. Only in certain places did evaporation proceed to the point where sylvite was deposited, as at Stassfurt, Germany or at Carlsbad, New Mexico. The most important world reserves of sylvite are in Saskatchewan, Canada.

Finally evaporation may take place on surfaces not exposed to rainfall. Thus evaporite minerals may be found under the overhangs of cliffs, in caves, and in mine tunnels, yielding minerals such as calcite, gypsum, epsomite, melanterite, or chalcanthite.

Recrystallization

Wherever rocks have fractures or pores, water can penetrate and saturate them. Originally, this formation water was simply incorporated when the sediment was deposited in a lake or in the sea. Later it may have been modified by water that percolated downward from rain falling on the surface or upward from the squeezing of sediments below. If the rocks are buried so that the temperature and pressure are increased or if the original material is chemically unstable, as is typical in biological sediments or the glass in volcanic ash, the formation waters will recrystallize the unstable biological compounds and any glass to form new minerals. This process is known as *diagenesis*. One of the reasons that sedimentary rocks are hard is because unstable biological carbonates and opal recrystallize to stable calcite and quartz, which cement together the grains of sand. Aragonitic shell material will dissolve and recrystallize as calcite. Glass and biological silica will dissolve and repreciptate as opal, chalcedony, or a host of zeolites. Upon decomposition, the cell walls of dead organisms will release phosphate into solution, which will react with any calcite or aragonite present to form apatite. The bacterial decomposition of organic matter in the presence of sulfate solution reduces sulfate to sulfide and will produce pyrite.

CHEMICAL REACTIONS

Large numbers of minerals are the products of chemical reactions. Most of these minerals are fine grained and inconspicuous and must be studied by microscopic examination. Nevertheless, under the right circumstances, these processes may result in pure masses, precipitation of coarse grains, or the growth of individual crystals. Most of the minerals formed during weathering, during the burial of sediments, and during metamorphic processes involve various chemical reactions. Each of the major types of processes is described briefly with common mineralogical examples.

Oxidation and Hydration

Weathering is a very important but special case in which minerals form by the separate and combined action of oxidation and hydration when rocks and especially sulfide ores are exposed to atmospheric conditions. All minerals formed in this way will be termed *secondary*.

Rock Weathering. Some simple examples of hydration are illustrated by the weathering of volcanic glass or common rock-forming minerals such as plagioclase and orthoclase to clay:

$$\text{glass} + \text{water} = \text{montmorillonite} + \text{solution}$$

$$\text{feldspar} + \text{water} = \text{kaolinite} + \text{solution}$$

The water is supplied by rain, which leaches out calcium, sodium, potassium, and silica and carries these as dissolved ions into streams and rivers, leaving a residue of clay. If this process continues for long periods, even the silica of the clay may be leached away:

$$\text{kaolinite} + \text{water} = \text{gibbsite} + \text{solution}$$

leaving a lateritic soil such as *bauxite* rich in alumina.

While the feldspars are being converted to clays, iron-bearing minerals such as augite, hornblende, and biotite are not only being converted to clay but the iron is being oxidized to form either goethite or hematite. Thus

augite + water + oxygen = kaolinite + goethite + solution

biotite + water + oxygen = vermiculite + hematite + solution

Both the water and oxygen are supplied by rain, which acts as a solvent leaching out the calcium, magnesium, and silica while the dissolved oxygen converts ferrous iron to ferric iron, which precipitates, depending on the precise chemical conditions, either as hematite or goethite. If the process is long enough such that most of the silica is leached away, the iron, along with aluminum, is retained to form an iron-rich laterite. Some laterities are sufficiently rich in iron that they have been used as iron ores.

As most iron minerals in rocks carry small amounts of manganese, this also is converted by the dissolved oxygen, often with the aid of bacteria, from Mn^{2+} to Mn^{4+}. Much of the black desert varnish and the black coatings observed on pebbles in streams consist of thin coatings of pyrolusite and various hydrous manganese oxides such as manganite or romanechite.

Weathering of Sulfides. A large number of interesting minerals, including many of the carbonates and sulfates, such as malachite, smithsonite, and anglesite, are formed during the weathering of sulfide ore minerals, such as chalcopyrite, galena, and sphalerite. In many instances well-formed crystals develop in cavities as a result of descending surface waters. As the most abundant naturally occurring sulfide is pyrite, the weathering of pyrite leaves distinctive residual deposits of goethite/limonite known as *gossans*. Weathering of pyrite also contributes acid and sulfate to streams:

pyrite + water + oxygen = solution (ferrous sulfate + sulfuric acid)

Additional oxygen then causes conversion of the ferrous sulfate to ferric sulfate with the formation of additional sulfuric acid. If these iron-rich acid solutions are diluted either by rainwater or by stream water, goethite/limonite is precipitated and coats all exposed surfaces in the beds of streams. A spectacular example of this occurs at Rio Tinto, Spain. In some areas, especially where coal rich in pyrite has been mined, acid waters and iron precipitation in streams have become major environmental problems. Similarly, when burned to produce electricity, coal rich in pyrite can produce rain enriched in sulfuric acid. The weathering of pyrite and other sulfides in the surface capping over ore bodies can mobilize copper, lead, and zinc, leading to the the formation of minerals such as malachite, cerussite, smithsonite, and hemimorphite near the surface, while cuprite and native copper may form just above the water table. Reduction and neutralization below the water table of these downward-percolating acidic waters results in the *supergene* precipitation of chalcocite.

Below the water table, iron tends to be reduced and go into solution. When descending oxygenated and slightly acid soil waters come into contact with the reduced and basic groundwater, any reduced iron in solution may be precipitated as goethite or hematite. If this process goes on over long periods of time, a zone of iron-stone or iron concretions may be formed. In places, these iron deposits and similarly formed bog-iron deposits have been mined.

Hydration

While hydration is an important component of weathering, it is an integral part of hydrothermal processes. When water is released from a magma or when surficial or oceanic water comes into contact with magma, these thermal waters may hydrate any rocks with which they may come into contact.

Close to the magma, the temperatures in the hydrothermal solution are so high that anhydrous minerals are deposited from solution. This region is characterized by orthoclase, biotite, and anhydrite, and these are often interlaced with a network of quartz veinlets known as a *stockwork*. This region is called the *potassic zone*. It is of considerable economic importance because of the molybdenite and chalcopyrite ores that may be found here.

Just beyond the potassic zone, temperatures are sufficiently low that hydrous minerals become pervasive. Here is encountered a region dominated by sericite–quartz–pyrite known as the *phyllic* (or sericitic) *zone*. If the system is low in silica, pyrophyllite may be the dominant mineral. Ores of pyrite, chalcopyrite, bornite, magnetite, and rutile may be found. Sometimes a mantle of clay minerals called the *argillic zone* is found surrounding the phyllic zone. The composition of the clay grades outward from kaolinite to montmorillonite. Farthest from the magma is the *propylitic zone*. It is characterized by the presence of chlorite with associated epidote, calcite, and pyrite. Ores of specularite, sphalerite, galena, and rhodochrosite may be found here.

In some cases, the hydrothermal solutions, as in the Mother Lode of California, are rich in carbon dioxide. In these instances, the alteration includes not only hydration but also the process of *carbonatization,* in which earlier minerals such as feldspar and biotite are converted to calcite, dolomite, or ankerite. Other special cases may arise. If the solutions are rich in sodium, there may be *albitization,* if rich in boron, *tourmalinization,* if rich in fluorine, *greisenization,* in which the original feldspar is converted to coarse-grained muscovite, and because all hydrothermal solutions readily dissolve quartz, *silicification.*

Instead of being on the continents, thermally active areas may be on the ocean floor, where there is active extrusion of basalt and other submarine volcanic activity. When heated, seawater is a highly corrosive solvent. Any rocks encountered by heated seawater will undergo rapid hydration and oxidation from the percolation of seawater through cracks and fractures in these rocks. Seawater is particularly rich in magnesium, and rocks encountered will be altered to chlorite and dolomite with accompanying epidote. The dissolved sulfate in seawater converts the ferrous iron in the basalts to ferric iron and the sulfate is converted to dissolved H_2S. The corrosive chloride solutions will strongly leach iron, manganese, barium, copper, lead, and zinc from the basaltic or andesitic rocks. If penetration is deep, the underlying olivine-rich rocks are converted, depending on the temperature, to serpentinite or steatite.

When these enriched thermal brines reemerge on the ocean floor, they undergo rapid cooling, which causes the rapid precipitation of pyrite, pyrrhotite, chalcopyrite, sphalerite, and galena. If conditions are right, these *exhalations* may be preserved as sedimentary layers (*exhalites*) of various sulfide minerals. In the past such accumulations have formed great ore deposits such as copper on Cyprus and lead–zinc at the Sullivan mine in British Columbia, Canada.

Reduction

Most rocks below the earth's surface are already in a fairly reduced state relative to those at the surface. Thus the environments where reduction processes are active are limited. When sediments are deposited, their oxidation state at the time of deposition is preserved and changes only slowly with time. Thus redbeds rich in hematite have been preserved even in ancient rocks. However, if at the time of deposition large quantities of organic matter are also deposited, rapid reduction takes place, with the precipitation of pyrrhotite and pyrite. This reduction is usually the result of sulfate-reducing bacteria removing oxygen from the sulfate of the included seawater and using it to feed on the included organic matter. A result of this process is the formation of H_2S, a gas with the smell of rotten eggs, frequently noticed in coastal marshes where these reduction processes are active.

The other important place where active reduction takes place is in soil profiles below the water table. This has proven to be of great economic importance in relation to the weathering of sulfide ore bodies. In the zone of oxidation above the water table, weathering converts pyrite to goethite/limonite and mobilizes copper and other metallic elements. When these metal-rich solutions percolate downward from the water table, the copper in the solution is actively reprecipitated to form a zone called the *supergene zone,* enriched in the sulfides chalcocite or covellite.

Dehydration

The weathering products produced by oxidation and hydration are ultimately carried into the sea where they accumulate along with skeltons, shells, tests and organic remains produced by biological activity in the ocean. As the sediments accumulate, they are buried progressively deeper and this process may be enhanced by natural movements of the earth's crust. As a result, sediments are subjected to progressively higher pressures and higher temperatures. As a result sandstones, shales, limestones, and altered basalts undergo progressive transformations into quartzites, slates, marbles, and amphibolites. These changes are called *metamorphism,* and the minerals formed depend not only on the original composition of the rocks being metamorphosed, but also on the sequence in which the pressure and temperature are applied. An outline of these changes is shown in Fig. 6-1.

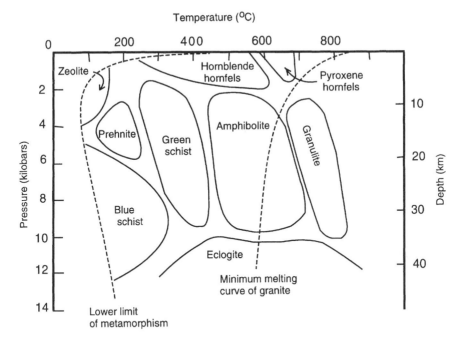

Fig. 6-1. *Metamorphic facies.*

In some cases, as the prehnite facies or zeolite facies, the meaning is relatively self-evident, in that rocks which carry these particular minerals belong to that facies. In the other facies, it is assumed that one is familar with the minerals which characterize that facies. The *blueschist* facies is characterized by glaucophane. The *greenschist facies* is characterized by the presence of green minerals such as chlorite, actinolite, and epidote. Other nongreen minerals, such as albite and quartz, are usually present. The *amphibolite facies* is characterized by the presence of hornblende. In some cases it may be the only mineral present, but it may also occur with plagioclase, biotite, garnet, staurolite, or kyanite. At the highest temperatures one has the *granulite facies*. It is characterized by the anhydrous minerals plagioclase, quartz, and pyroxenes. The *eclogite facies* represents very high pressure conditions and is characterized by pyrope garnet, diopside-jadeite, and kyanite. At low pressures and high temperatures, the zone of fine-grained rock baked by an intrusive igneous rock is known as a *hornfels*. If the original rock is a shale, the hornfels will contain microsopic cordierite, andalusite, biotite, feldspar, and quartz. If the original rock is calcareous shale, microscopic calcic-plagioclase, diopside, and actinolite will be found.

As one moves progressively from zeolite to blueschist to greenschist to amphibolite and finally to granulite, the process of metamorphism is one dominated almost entirely by dehydration. Examples of sequences of minerals showing progressive loss of water are illustrated as follows:

Kaolinite to pyrophyllite to andalusite (or kyanite)

Serpentine to talc to anthophyllite to enstatite or forsterite

Prehnite to zoisite to plagioclase

A full understanding of rock metamorphism requires a detailed knowledge of the complex reactions that take place among the various rock-forming minerals as dehydration progresses with increasing temperature and pressure.

As dehydration proceeds, the excess water with dissolved silica escapes from the rocks. Thus schists are accompanied nearly always by concordant quartz veins as a result of these processes. Usually, these veins are barren of other minerals and, without vugs, rarely yield crystals. At high temperatures, these quartz veins are replaced by small pegmatites formed as a result of incipient melting. As with the quartz veins, they rarely carry interesting minerals or yield crystals.

Decarbonatization

Relative to hydrous minerals, the carbonate minerals are stable to much higher temperatures. Thus reactions analogous to those found above in the dehydration sequence will begin only at the higher metamorphic grades. Although decarbonatization can and does occur at high metamorphic grades, decarbonatization more commonly occurs where igneous intrusions have come into contact with limestones and marbles to produce *contact metamorphism*. At these high temperature, reactions occur such as:

$$calcite + quartz = wollastonite$$

$$dolomite + quartz = diopside$$

$$magnesite + quartz = forsterite$$

$$calcite + anorthite + quartz = grossularite$$

These and other new minerals formed by these reactions are found in rocks known as *skarns*. Many rare and interesting minerals occur where this has taken place, such as at Crestmore, California or Laurium, Greece.

Skarns are of special interest to the mineralogists not only because of the wide variety of rare minerals that may be collected but because they are often the sites of valuable ores of copper, lead, and silver. The most important skarns form where limestones have been intruded by large bodies of felsic magmas. Along the margin between the carbonate rocks and the magma there will be not only large thermal gradients but also large chemical gradients. These gradients produce major interchanges in chemical constituents between the magma and the carbonate rocks. In addition to the formation of minerals by decarbonatization reactions giving the skarn assemblage of wollastonite, garnet, scapolite, and epidote, fluids rich in boron and fluorine will yield a host of rare minerals, such as vesuvianite, datolite, tourmaline, and chondrodite, while as mentioned earlier, escaping hyrothermal fluids can deposit high-temperature minerals such as magnetite, chalcopyrite, molybdenite, pyrrhotite, and scheelite, while farther away the low-temperature minerals galena and sphalerite may be deposited.

Recombination

If temperatures are hot enough or pressures high enough, or if minerals are held at particular conditions for thousands of years, the original minerals will react with each other to form entirely new suites of minerals. Thus at high temperatures, grossular + quartz will react to form anorthite + wollastonite or anorthite + calcite can react to form scapolite (meionite). Similarly, but at high pressures, albite reacts to form a mixture of jadeite + quartz or *coesite,* a high-pressure polymorph of silica.

SOLID-STATE TRANSFORMATIONS

By *solid-state transformations* we simply mean that as a result of the external influences from a change in temperature, pressure, or both, the atoms within a mineral reorganize themselves into minerals with a configuration compatible with the new conditions. The specific kinds of rearrangements include polymorphism, order–disorder, and exsolution.

Polymorphism

When a chemical substance crystallizes in two or more types of structures, the phenomenon is known as *polymorphism.* The various arrangements range from relatively minor ones, in which the directions between atoms are simply distorted, to complete reconstructions. In the case of graphite and diamond, very high temperatures and pressures are needed to bring about reconstruction of the covalent bonds. At the other extreme, high-temperature (beta) quartz, which is the stable form above 573°C, is never observed at room temperature. As the temperature falls below the transition temperature, high-temperature quartz inverts instantaneously to low-temperature (alpha) quartz because only a very slight adjustment in the atoms is required. Most other examples fall between these extremes.

Order–Disorder

This situation arises in cases where solid solution between two elements occurs at a particular structural site within a crystal. The degree of ordering within a crystal is usually determined by the temperature at which the mineral originally formed, but it is also determined by its temperature history. By far the best example of different minerals resulting from order–disorder are microcline–orthoclase–sanidine. All three of these minerals have the overall composition $KAlSi_3O_8$, the only chemical difference among them being the precise nature of the ordering of the aluminum. The framework structure of these feldspars may be thought of in terms of two parallel sets of Si_4O_8 rings in which one of the four Si atoms in each ring is replaced by an Al atom, to give an $AlSi_3O_8$ ring. If the one Al atom can occur randomly anywhere within the ring (disorder), the mineral is known as sanidine. It typically occurs in volcanic rocks, where the high temperatures can completely randomize the Al. This difference can be expressed by writing its chemical formula as $K(Al_{0.25},Si_{0.75})_4O_8$. In igneous rocks formed at lower temperatures, such as granites, the usual form is orthoclase. In this case, aluminum is randomized only among two of the four possible silicon positions in each ring (partial disorder). This difference can be expressed by writing the chemical formula for orthoclase as $K(Al_{0.5},Si_{0.5})_2Si_2O_8$. At lower temperature, in igneous rocks such as pegmatites, the aluminum is fully ordered and a chemical formula expressing this order is $K(AlSi_3)O_8$. As the aluminums are located diagonally between each pair of parallel rings, the monoclinic mirror plane is violated and microcline is triclinic. Order–disorder is also known in albite and several other minerals, but only in the potassium feldspars are the differences so striking in their occurrence and external morphology.

Exsolution

In some minerals, such as olivine and plagioclase, solid solution exists in all proportions between two pure isostructural end members. In the case of olivine, these end members are forsterite and fayalite, while in plagioclase they are albite and anorthite. In the alkali feldspars orthoclase and albite, the extent of solid solution (one element substituting for another) is strongly dependent on temperature. Thus at high temperature, large amounts of Na are present in the potassium feldspar structure. On cooling, Na can no longer be tolerated in the structure and segregates by diffusion to form a mineral domain of its own, a Na-rich feldspar (exsolution). The easiest case for the beginner to observe is in the case of the alkali feldspars. In many specimens of microcline from pegmatites, one can observe lamellae of albite which have exsolved from the potassium-rich host. The lamellae may be up to several millimeters thick and several centimeters in length. This sort of intergrowth is called *perthite*. Most other occurrences of exsolution are microscopic and must be observed either with a petrographic or an electron microscope.

In the case of Ti in magnetite, the solid solution at low temperature may exsolve into an intimate fine-grained mixture of ilmenite and magnetite. Because of these fine-grained intergrowths, some magnetites are too high in titanium to be used as iron ore, and some ilmenites are too high in iron to be used as a titanium ore. By quantifying the details of these processes, it has been possible to determine the temperature of formation at which the original mineral crystallized, as well as its cooling history.

FORMATION OF CRYSTALS

Although all minerals are solids with a crystalline structure, it is only under certain favorable conditions that minerals grow so they clearly exhibit the plane natural surfaces of a crystal. While geologists study in great detail the textural relations among the crystalline grains of a rock to understand their genetic history, the morphology of crystals themselves is of little importance. However, to the mineralogist, the collecting and study of well-formed crystals showing a myrid of faces can be one of its great pleasures. As so many of the minerals in the earth are massive or aggregates of tiny crystalline particles composing rocks, what are the special conditions necessary for a mineral to grow so as to show the regular geometric shape of a crystal? In general, two circumstances favor the formation of crystals: (1) growth in open spaces and (2) growth by replacement.

Open-Space Growth

In open-space growth the mineral is free to grow such that its growth is not inhibited or interferred with by the presence of earlier formed minerals or by the simultaneous growth of other minerals.

The growth of well-formed crystals in silicate melts is fairly rare. However, the very first mineral to crystallize may find itself free to form faces. Thus olivine crystals may form in basalts, and doubly terminated crystals of orthoclase or β-quartz may form in certain rhyolites. In this case, because of the close specific gravity between crystals and liquid and the high viscosity of the melt, the crystals find themselves free to grow in all directions. Some evidence suggests that diamonds may grow deep within the earth as crystals floating at the interface between dense sulfide melt and a lighter silicate melt. Occasionally, vapor bubbles trapped in an igneous melt will give the free space needed for crystals to grow. Aside from these rare instances, the high viscosity and simultaneous growth of other minerals generally prevents the formation of crystals in most igneous rocks. The one important exception is in pegmatites. In such silicate melts, the presence of water and other volatiles greatly reduces the viscosity of the melt, and the grain size of minerals in such rocks is greatly enlarged. It is now thought that giant crystals in pegmatites such as beryl or spodumene form when boiling erupts as fluid is expelled as the silicate melt cools. Although quite large, and making nice museum displays, such crystals are filled with occuluded matter, which generally renders them useless as gemstones. Crystals formed this way in igneous rocks are very difficult to collect because of the very hard tough matrix which has grown around them.

Although certain crystals have a clear association with silicate melts, most of the spectacular examples to be seen in museums are the result of crystal growth in open spaces from an aqueous fluid. These open spaces have a multiplicity of origin, and a wide variety of minerals have formed this way, with many of gem quality. Examples include pockets (cavities) in pegmatites that have yielded fine tourmalines; cavities in skarns with beautiful epidotes (Fig. 7-178) or garnets; gas cavities in basalt flows with geodes filled with agate and quartz crystals or fine crystals of zeolites such as natrolite or heulandite; leached cavities in the zone of weathering over sulfide ores with crystals of wulfenite or hemimorphite; fracture cavities along faults with crystals of quartz, amethyst, or smoky quartz; caverns in limestones with huge crystals of calcite (Fig. 7-79), celestite, dolomite, or gypsum; collapse breccias with magnificent crystals of galena or fluorite; and crystals of halite or borax on the bottom of evaporating pools.

The one defining feature of these crystals is their single termination. That is, the crystal faces occur on only one end of the crystal. This is because the crystallization is initiated on the walls of the open space and the crystals have grown outward by deposition of dissolved material from the aqueous fluid occupying or flowing through the open spaces around the crystals such that material is deposited on all but one side.

Replacement Growth

Crystal growth by replacement means that the mineral material originally present flowed, dissolved, and diffused away as the growing crystal took its place or was incorporated into the growing crystal (Fig. 3-147 to 3-149). The defining feature of such growth is the presence of double terminations. That is, well-developed crystal faces are at both ends of the crystal. As gypsum crystals (Fig. 7-117) grow in mud, the mud surrounding the crystal is displaced. In a restricted sense the growth of olivine in basalt or orthoclase in rhyolite is very similar in its nature. The fine doubly terminated quartz crystals found at Herkimer, New York, were formed in a quartzite in which the crystals grew as the enclosing sand grains slowly dissolved away. A great variety of doubly terminated crystals are to be found in metamorphic rocks where diffusion and chemical recombination act together to form a wide variety of crystals. Crystals formed under these conditions have much occuluded matter and can rarely be used as gemstones. Good examples include andalusite crystals in slate, garnet crystals in chlorite schists (Fig. 7-183) and staurolite, kyanite (Fig. 7-193), or corundum in mica schists.

COMMON ROCK TYPES

Geologists divide rocks into three broad categories: igneous, sedimentary, and metamorphic. Igneous rocks are those that have formed by the crystallization of minerals from molten rock. Sedimentary rocks are those formed by the deposition of sand, mud, and the remains of organisms. Metamorphic rocks are those formed by the application of high pressure and high temperature to preexisting rocks that have been buried deep within the earth. As we know from earlier discussions, rocks consist of aggregates of minerals and other natural materials and are usually described in terms of their mineral compositions. We give here only a brief summary of the common rock terms that the student is likely to encounter in reading about minerals or in the description of mineral occurrences in Chapter 7.

Igneous Rocks

Most igneous rocks, both fine and coarse grained, are commonly characterized by having a texture in which the mineral grains are randomly oriented. The igneous rocks are described here in order from mafic to felsic. The fine-grained varieties are the result of rapid cooling by the extrusion of the molten rock onto the surface, while the coarse-grained varieties form by much slower cooling upon intrusion.

Basalt/gabbro are composed of augite and labradorite with varying amounts of olivine. If fine grained, the rock is called basalt; if medium grained, *diabase;* and if coarse grained, a gabbro. The term *traprock* has been applied to basalts and to any fine-grained dark rock.

Andesite/diorite are composed of oligoclase, some potassium feldspar, and small amounts of quartz. Minor amounts of augite, biotite, or hornblende are usually present. If it is fine grained with glass present, it is called andesite; if coarse grained, it is called diorite.

Trachyte/syenite are composed of sanidine/orthoclase and oligoclase with small amounts of pyroxene, biotite, or hornblende. Quartz or glass are rarely present. If fine grained, it is called a trachyte, while if coarse grained, a syenite.

Rhyolite/granite are composed of sanidine/orthoclase and quartz along with some oligoclase and minor biotite or muscovite. Glass is common in the fine-grained rhyolite; if coarse grained, it is a granite.

Sedimentary Rocks

Most sedimentary rocks are laminated with the thickness of the laminae or beds often related to the grain size of the sediment being actively deposited. They are classified not only on composition but also on grain size.

Conglomerate is a rock composed of rounded rock fragments that range in size from granules (>2 millimeters), to pebbles, to cobbles, to boulders in a matrix of sand. Conglomerates are usually formed where steep, fast-flowing rivers of a mountain range enter a valley.

Sandstone is a rock composed usually of quartz grains between $\frac{1}{16}$ and 2 millimeters in size. The sand is usually deposited at the mouths of rivers and along the seashore. Other minerals, such as feldspar, and resistant minerals, such as zircon, garnet, and rutile, may be present in small amounts.

Shale is a rock that consists of material less than $\frac{1}{16}$ millimeters. These may include fine-grained quartz, kaolinite, and other phyllosilicates, along with fine-grained tests and organic remains of organisms. It is often finely laminated and may split parallel to the original sedimentary layering.

Limestone usually consists of the recrystallized shells of both large and microscopic marine organisms, but may be chemically precipitated.

Metamorphic Rocks

Metamorphic rocks all require the presence of recrystallization, usually with dehydration. During the recrystallization process, metamorphic rocks usually develop a fabric in which the grains are parallel to one another. For the beginner, this cleavage or banding can sometimes be confused with the laminae or beds of a sedimentary

In the description of many species no mention is made of certain properties which are relatively unimportant in these particular cases. Thus if the cleavage is not mentioned, it is because it either is not observed or is too imperfect to be important. Since nearly all minerals are brittle, it is unnecessary to repeat this word in each mineral description, but if the mineral is not brittle but malleable or sectile, this property is stated and should be noted carefully. If the streak is not given, it is understood to be *white* or nearly white, like that of most nonmetallic minerals, even when the mineral itself has a deep color. All minerals having a metallic luster are opaque.

The student will find it easier to remember the properties of different minerals if, after studying the descriptions in the book and comparing them with specimens to which the student has access, a tabular list is made of the properties for each species, somewhat similar to the suggested form shown in Table 7-1.

TABLE 7-1   Suggested Form for Tabulating the Properties of Minerals

	Diamond	Graphite	Galena	Sphalerite
Crystal system	Isometic	Hexagonal	Isometric	Isometric
Common form	Octahedron	Tabular	Cube	Tetrahedon
Aggregates	—	Foliated	Granular-cleavable	Granular-cleavable
Cleavage	Octahedral	Platy (flexible)	Cubic	Dodecahedral
Hardness	10!	1–2!	2½–3	3½–4
Specific gravity	3.5	2.2	7.5!	4
Luster	Adamantine	Metallic	Metallic	Resinous
Color	Colorless, yellow	Black	Lead-gray	Yellow, brown, black, etc.
Streak	White	Black	Dark gray	White to brown
Composition	Carbon	Carbon	PbS	ZnS
Tests	Infusible	Infusible	Easily fusible	Infusible

It is easy to arrange a notebook for this purpose by ruling a series of parallel vertical columns, and to avoid writing the list of properties on each page, they may be written on the edge of the first left-hand page and the corresponding strip from a sufficient number of the subsequent sheets may be neatly cut off. When a property is particularly important it should be underscored or followed by an exclamation point. It is not worthwhile to repeat in tabular form the entire description in the text; a little experience will soon show how much may advantageously be written down.

Filling out a similar column of properties determined from the specimen itself and comparing it with the list in the student's notebook made out from the text is a useful exercise. If the species was unknown at first, the list of properties will often suffice to determine it.

It is not necessary to *learn* by sheer effort of memory all the properties at once; this would be difficult and tiresome. The important properties can be learned (such as chemical composition), but knowledge of most of the physical properties will be acquired gradually by repeated handling of the specimens themselves.

CLASSIFICATION

Several methods have been used to classify minerals, and each has its value depending on the properties one wishes to emphasize. In the nineteenth century it was common to arrange minerals according to the prominent elements of which the minerals are compounds. In this way all the iron minerals, such as hematite, goethite, magnetite, and siderite, were grouped together. Similarly, the zinc minerals—sphalerite, smithsonite, hemimorphite, and willemite—were together. This may seem to the beginner the best and most logical way to classify minerals, and indeed it might be if we were considering minerals only from an economic viewpoint as the source of the metals they contain. For those who wish to consider them in this manner, a list of the minerals arranged according to the most important element is given in Appendix I.

To the trained mineralogist the presence of a certain metal is not nearly as important as the type of compound and the properties of the compound. For example, the native elements copper, silver, and gold, besides being metals, are isometric, with similar habits and physical properties. The group of carbonates known as the *calcite group* is an even better example:

Calcite, $CaCO_3$	Rhodochrosite, $MnCO_3$
Magnesite, $MgCO_3$	Smithsonite, $ZnCO_3$
Siderite, $FeCO_3$	

All these minerals have the same type of crystal structure, and the angles between the corresponding crystal faces are nearly the same, varying from 105 to 107° over the edges of the rhombohedron. Hence they are called *isomorphs,* a term that originally meant that they had similar external form and analogous chemical composition, but today usually means that they have similar internal arrangement of their constituent atoms, *isostructural.* If the carbonates were classified according to the metal present in each, they would be scattered through the book, and these similarities would not be brought out. Calcite would be found with apatite, calcium phosphate; with fluorite, calcium fluoride; with anhydrite, calcium sulfate; and with others with which calcite has nothing in common except the presence of calcium. All the other carbonates would also be scattered. Furthermore, it is difficult to place some minerals; for example, dolomite could be classified equally well with either the calcium or the magnesium minerals.

For these reasons the minerals described later in this book are grouped according to a chemical classification in which all the sulfides, all the oxides, all the sulfates, and so on, are grouped together. This is the classification used by all modern advanced textbooks on mineralogy. Furthermore, it seems wise for the beginner to start with a classification that will be required when more advanced work is undertaken.

The chemical classes into which the minerals are grouped are as follows:

1. *Native elements.* A few of the elements occur in nature uncombined and are hence called native elements, such as native gold and native sulfur.

2. *Sulfides.* The sulfides are compounds of a metal with sulfur, such as galena, PbS; sphalerite, ZnS; pyrite, FeS_2. Similar to the sulfides and included with them are the rare tellurides and arsenides, such as calaverite, $AuTe_2$, and nickeline, NiAs. Also included with the sulfides are those minerals called sulfosalts. They are compounds composed of a metal in combination with sulfur and arsenic, antimony, or bismuth: for example, enargite, Cu_3AsS_4; and proustite, Ag_3AsS_3.

3. *Oxides.* The oxides are composed of a metal in combination with oxygen, such as hematite, Fe_2O_3; cuprite, Cu_2O; cassiterite, SnO_2. In addition to the metal and oxygen, the hydrous oxides contain water or the hydroxyl (OH), as, for example, goethite, FeO(OH).

4. *Halides.* The halides include compounds with chlorine, fluorine, rare bromides, and iodides. Halite, NaCl, and fluorite, CaF_2, are examples.

5. *Carbonates.* The carbonates are minerals with anion group $(CO_3)^{2-}$. Calcite, $CaCO_3$; smithsonite, $ZnCO_3$; and cerussite, $PbCO_3$, are examples. Borates are included here because they have $(BO_3)^{3-}$ as a fundamental anion group in all their compounds.

6. *Phosphates.* The phosphates are minerals with an anion group $(PO_4)^{3-}$. Amblygonite, $LiAlFPO_4$, and pyromorphite, $PbCl(PO_4)$, are examples. The arsenates and vanadates are also included here because of the similarity of the anion groups $(AsO_4)^{3-}$ and $(VO_4)^{3-}$ with the phosphate anion group.

7. *Sulfates.* The sulfates are minerals with an anion group $(SO_4)^{2-}$. Examples are barite, $BaSO_4$, and anglesite, $PbSO_4$. The tungstates and molybdates are also included here because of the similarity of the anion groups $(WO_4)^{2-}$ and $(MoO_4)^{2-}$ with the sulfate anion group.

8. *Silicates.* The silicates form a very large and complex chemical class of minerals which are subdivided according to the manner in which silica tetrahedral, $(SiO_4)^{4-}$, have been linked together, *polymerizered* in the form of frameworks, sheets, chains, rings, or left as islets of unlinked tetrahedra. Examples are orthoclase, $KAlSi_3O_8$, and rhodonite, $MnSiO_3$.

NATIVE ELEMENTS

Of the 88 natural elements, only about 20 are found in the native state, that is, uncombined with other elements. These are divided into three groups: (1) metals, (2) semimetals, and (3) nonmetals.

The Native Elements

Metals	Semimetals
Gold, Au	Arsenic, As
Silver, Ag	Bismuth, Bi
Copper, Cu	Nonmetals
Platinum, Pt	Sulfur, S
Iron	Diamond, C
Taenite, (Fe,Ni)	Graphite, C
Kamacite, Fe	

The Metals

The atoms in metals arrange themselves in three fundamental patterns: face-centered cubic, body-centered cubic, and hexagonal. Among native metals, the hexagonal pattern is never found, and the body-centered pattern occurs only in kamacite, Fe. All of the other native metals: gold, silver, copper, platinum, and taenite (Fe,Ni) have the face-centered-cubic arrangement. Despite the isostructural nature of these metals, in general they occur separately in nature.

Gold, Au

Since the beginning of historic time, gold has been the most highly prized of all the metals. It is only within relatively recent years that other metals, such as platinum and radium, have had a value equal to or greater than that of gold. It was used by early peoples because it occurred in the native state and thus did not require an elaborate metallugical process to extract it from the ore. Furthermore, it could be worked easily and fashioned into durable ornaments of pleasing color. Gold was used very early as a medium of exchange and has been used as such by all civilized people. It is interesting to note that although there are other gold minerals, most of the gold of the world has been obtained from the native metal.

Habit. Gold is isometric and at a few localities has been found in well-formed octahedral (Fig. 7-1) or dodecahedral crystals. However, crystalline gold is usually in plates or in wirelike forms (Fig. 7-2). The plates are flattened parallel to an octahedron face and may show triangular markings, whereas the wires are usually elongated parallel to a 3-fold symmetry axis. Crystals of any sort are the exception, and most gold is found in irregular masses that if large are called *nuggets* (see Plate I-3, 4).

Fig. 7-1.
Malformed gold octahedron.

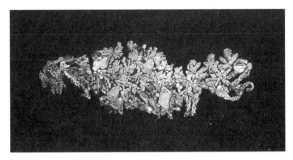

Fig. 7-2. *Gold crystals, Trinity County, California.*

Physical Properties. H = 2½–3; G = 19.3. The physical properties of gold are so characteristic that they serve readily to identify it. It can be easily scratched by a knife, leaving a shining groove. It is one of the heaviest substances and has a far greater specific gravity than any of the "fool's golds" or any of the other commonly occuring heavy minerals so that a skilled gold-panner can positively identify individual specks of gold by their immobility while being washed by a circular current of water. However, this property is of little help in identifying gold in the average specimen, for the tiny flakes that are present are greatly overshadowed by the large bulk of quartz or other minerals with which it is associated. Native gold may contain some silver, which lowers the specific gravity and makes the deep yellow of pure gold somewhat lighter. It is highly malleable and ductile and can be hammered out into sheets so thin that they will transmit a faint greenish light. As gold is the only yellow mineral that is malleable, it can be distinguished from all others. Chalcopyrite has a yellow color, but, if struck a blow with a hammer, it is crushed to a powder and not flattened like gold. Furthermore, if chalcopyrite is scratched by a knife, it yields a green powder, not the smooth groove of gold.

Composition. Most native gold contains some silver and may contain small amounts of copper and iron. Silver may be present in any amount; gold from California carries between 10 and 20%. When silver is present in amounts of 20% or greater, the mineral is called *electrum.* Gold has long been known as a *noble metal,* since it is not attacked by ordinary acids. It is dissolved in *aqua regia,* a mixture of one part nitric and three parts hydrochloric acids.

Occurrence. Gold occurs in all kinds of rocks of all ages on all continents. There are thus many types of gold deposits, but most commonly it is in veins associated with quartz. This gold-bearing quartz is often milky and may show little particles of gold scattered through it, but more often, no gold is visible until it is crushed to a powder and washed to remove the lighter material. This is a method of recovering gold that has been in use since the time of the early Egyptians as recorded by Diodorus Siculus, a first-century B.C. Greek historian.

In some places nature has been performing this crushing and washing process for many centuries, for when gold-bearing rocks are disintegrated by weathering, the broken material is washed into the neighboring streams. The small particles of the lighter minerals are carried away, while the heavier gold works its way toward the bottom of the streambed, forming a placer* deposit. Much of the world's gold, particularly in the early days of gold mining, has come from the sands and gravels of streambeds, when the miner washed the gravels in a pan. Some placer and paleo-placer deposits as in the Sierra Nevada mountains of California have been worked on a large scale by throwing a powerful stream of water against the gravel bank, a procedure that washes the gravels through a sluice, where the gold is caught behind riffles while the lighter minerals are carried away. Because of the adverse environmental impact, this method of mining was abandoned late in the nineteenth century. Placer deposits located in valley bottoms such as along the Yukon River in Alaska have been worked by giant dredges through which pass thousands of cubic yards of gravel a day and from which the gold is washed mechanically. In some places there are ancient gold-bearing gravels that are now conglomerates called *fossil placers.* This is the case for the world's greatest gold-producing district, the Witwatersrand or "The Rand" in the Transvaal and the Orange Free State of South Africa. Some mines now extend to a depth of over 4000 meters. In this case the gold ore is crushed to a fine powder and the gold dissolved in a weak solution of cyanide.

---

*Placer rhymes with *passer.*

Silver, Ag

Silver is one of the precious metals, equally useful for ornaments of many kinds, for utensils, and for money. The silver coins produced in the the United States between 1837 and 1965 contained 9 parts of silver and 1 part of copper. One of the largest uses of silver is in light-sensitive compounds used in photography. Native silver is not uncommon, although the world's supply of the metal comes chiefly from other minerals.

Habit. Silver is like gold in its habit. It is found in some places, though rarely, in distinct isometric crystals, and more frequently in arborescent or branching groups, in plates and scales or wirelike forms (Figs. 7-3 and 7-4). Most native silver, however, shows no crystal forms and is in irregular masses conforming to the shape of the cracks in the rocks in which it is found.

Physical Properties. $H = 2\frac{1}{2}–3$, $G = 10.5$. Silver is highly malleable and ductile. Its specific gravity is higher when silver is alloyed with gold, as it often is. Of the metals silver is the best-known conductor of both heat and electricity. When it is fresh, the color is tin-white with its surface having the appearance of a well-polished silver spoon. However, on exposure to the air, a tarnish forms, which may be bronze to dull black.

Fig. 7-3. Native silver. Kongsberg, Norway.

Fig. 7-4. *Native silver. Channarcillo, Chile.*

Composition. Silver may contain some copper, gold, and mercury, and more rarely, traces of other metals. It is readily dissolved by nitric acid, forming silver nitrate. If hydrochloric acid is added to this solution, a white, curdy precipitate of silver chloride results. This reaction is a very delicate test for silver.

Occurrence. Native silver occurs in many places, but in only a few has it been found abundantly. The mines at Kongsberg, Norway, which were worked along calcite veins for several hundred years, have yielded outstanding specimens of crystallized wire silver. In the early part of the twentieth century at Cobalt, Ontario, great slabs of native silver up to 5 feet in length were recovered. These were found in calcite–dolomite veins along with rare cobalt and nickel arsenides. More recently it has been mined at Great Bear Lake, Northwest Territories, where it is associated with uraninite. In the United States, native silver has been found with native copper in the Lake Superior copper mines.

Copper, Cu

Copper is one of the most useful metals and has been employed since early times in many ways, both as a metal and in alloys. Today its greatest use is in electrical equipment because it is an excellent electrical conductor. Nearly all electrical wiring in our homes and factories is copper; the windings on motors and other familiar electric appliances are also copper. Copper is used in many alloys, of which brass (copper and zinc) and bronze (copper and tin) are the best known. Much of the magnesium metal that is now coming into wide use is alloyed with small amounts of copper to give it more desirable characteristics. There are many other uses for copper, so many in fact that, next to iron, copper is used more extensively than any other metal.

Habit. Native copper is isometric, and octahedral and dodecahedral crystals have been found, but crystallized copper is usually in distorted arborescent groups (Fig. 7-5). It is usually found in irregular masses which show no crystal forms.

Fig. 7-5. *Native copper, Lake Superior, Michigan.*

Physical Properties. $H = 2\frac{1}{2}-3$, $G = 8.9$. Like gold and silver, copper is highly ductile and malleable. The reddish hue so characteristic of copper is called copper-red. However, this color is seen only on fresh surfaces, and as a bright new penny soon becomes dull, so does native copper on exposure to the air.

Composition. Native copper may contain small amounts of silver, bismuth, mercury, arsenic, and antimony. The copper from Lake Superior contains extremely small percentages of these impurities, and thus *Lake copper* for many years was considered the standard for high purity. Copper is easily dissolved by nitric acid, giving a blue solution, and an excess of ammonia added to this solution turns it a deep azure blue.

Occurrence. The most celebrated locality for native copper is on the Keweenaw Peninsula in Michigan on the shores of Lake Superior, where mining began in the middle of the nineteenth century. Today mining activity has almost ceased there, but in the past, production was large. Beautiful crystallized specimens were found there in ancient vesiculated lava flows and beds of conglomerate associated with calcite, datolite, and a number of zeolites. In some specimens the copper is enclosed in crystals of calcite which imparts the bright red color to the crystals. Great masses of native copper have been found there, one of which weighed 420 tons.

Native copper is widely distributed in small amounts in the oxidized zones overlying deposits of copper sulfides, where it is associated with cuprite, limonite, malachite, and azurite. In this association it has been found in Arizona, New Mexico, and northern Mexico.

Platinum, Pt

Platinum is considered with gold a noble metal and, like gold, is not attacked by any of the single acids. Unlike gold, however, it is a relatively new metal, for it was not discovered until 1735. In that year it was found in placer workings in Colombia, South America, and was given the name *platina*, from the word *plata*, Spanish for silver, since it was regarded as an impure ore of that metal. One of the first uses for platinum was in counterfeit gold coins. A center of platinum was surrounded with gold, and because of its high specific gravity, the fraud was difficult to detect. From such a lowly beginning, platinum has taken its place among the most valued metals. Today it is has a value equal to or greater than that of gold.

Between the years 1828 and 1845, platinum coins were in circulation in Russia, but they were recalled and the experiment has not been repeated. Today, platinum has many uses, based mostly on the fact that it melts at very high temperatures and is not attacked by ordinary chemical reagents. It is used in the chemical laboratory for crucibles, dishes, spoons, and other types of laboratory equipment. A few of its many other uses are in electric apparatus, jewelry, dentistry, and in making photographic prints. For many of its applications it is alloyed with other metals of the platinum group. Thus platinum jewelry commonly contains about 10% of iridium.

Habit. Native platinum is isometric but is rarely found in crystals; its usual occurrence is in small grains, scales, and occasionally as nuggets. Some nuggets weighing as much as 10 kilograms have been found in placer workings in the Ural Mountains.

Physical Properties. H = 4–4½, G = 21.5. Its hardness is high for a metal. Pure platinum, such as that used in the laboratory, has a specific gravity of 21.5, but for native platinum, alloyed with other metals, it is lower, 14 to 19. It is malleable and ductile but less so than gold, silver, and copper. The color is steel gray. Some platinum rich in iron is magnetic.

Composition. Native platinum may contain considerable amounts of iron and also a number of rare metals of the platinum group, such as iridium, osmium, ruthenium, rhodium, and palladium. Like gold, platinum does not readily combine with other elements, the only important compound known in nature being an arsenide, $PtAs_2$, called *sperrylite.*

Occurrence. From the time that platinum was discovered on the eastern slope of the Ural Mountains in 1822 until 1934, most of the world's platinum came as the native metal from placers in the district centered around the town of Nizhe Tagil, Russia. The placer deposits are now exhausted, but platinum is recovered on a smaller scale by mining it in place, where it is associated with chromite in the olivine rock, peridotite. In 1934, Canada became the leading producer of platinum because of the large amounts recovered as a by-product from the great ore body at Sudbury, Ontario. Here it is associated with the massive nickel–copper ores, consisting of pyrrhotite, pentlandite, and chalcopyrite. However, beginning in 1954, South Africa became the world's major producer. Most of the South African production comes from the Merensky reef, a 15- to 30-centimeter-thick layer in the gigantic Bushveld igneous complex in the Transvaal. The reef extends for many kilometers, with a platinum content of about 15 grams per ton of ore. The platinum is associated with other platinum metals and chromite in a pegmatitic pyroxinite composed of bronzite and labradorite.

Iron, Fe

Terrestrial native iron, that is, iron occurring in rocks in the metallic state, is known only as a great rarity and hence is of no practical importance. However, meteorites, which occasionally fall to earth, often consist entirely of metallic iron; others, which have a stony appearance may contain particles of metallic iron distributed through them. The iron of meteorites always contains some nickel. This is important to remember, for many a piece of old iron is found each year that the finder suspects of being a meteorite. A test showing the absence of nickel is proof that the specimen is not a meteorite. If an iron meteorite is polished and etched, a triangular pattern known as Widmanstätten structure (Fig. 7-6) is revealed. This consists of large alternating barlike grains of intergrown kamacite rimmed with taenite, which suggests exsolution at low temperature.

Fig. 7-6. *Iron meteorite, polished and etched, Altonah, Utah. The dark patch is troilite, FeS.*

Kamacite (H = 4, G = 7.3–7.9), also known as α-*iron,* Fe, is usually massive with cubic or rare octahedral crystals but may show a distinct cubic cleavage. It is strongly magnetic. It has a body-centered-cubic arrangement of the atoms and in this arrangement commonly contains 2 to 7% nickel with minor amounts of cobalt and copper. It dissolves readily in hydrochloric acid.

Taenite (H = 5, G = 7.8–8.2), also known as γ-*iron,* (Fe,Ni), is usually massive with no cleavage. It has a face-centered arrangement of the atoms and because nickel has a high preference for the same atomic arrangement, it may contain from 13 to

65% nickel. It is strongly magnetic, although low-nickel varieties can be nonmagnetic. It dissolves readily in hydrochloric acid.

Occurrence. Native iron occurs primarily in the form of meteorites and has been noted also in a few places in terrestrial rocks, but only two occurrences are especially noteworthy. At Disko, Greenland, iron fragments ranging in size from small grains to masses weighing tons have been found in basalts. The other locality is at Josephine County, Oregon, where nuggets are found in stream placers.

The reason that native iron is not found more abundantly is that it oxidizes so easily. Just as an iron implement left outdoors will, after a short while, rust and fall to pieces, so will native iron, for the "rusting" is merely the combining of the iron with oxygen and water. Iron is our most important industrial metal, and the ores from which it comes, chiefly the oxides or iron, are discussed later (see the section "Hematite.").

The Semimetals

Of the elements that occur in the native state, some, like gold, silver, and copper, are true metals; others, like diamond and graphite, the two forms of carbon, and sulfur are nonmetals. Intermediate between these two groups are tellurium, arsenic, antimony, and bismuth, known as the semimetals. Minerals of tellurium and antimony are extremely rare and will not be discussed here. As minerals the semimetals are relatively unimportant, and the sources of these elements are for the most part other minerals.

Arsenic, As; Bismuth, Bi

The semimetals arsenic and bismuth are isostructural, having a rhombohedral atomic arrangement, and form a single mineral group. This similar atomic pattern means that they have similar physical properties.

Arsenic (H = 3½, G = 5.7) when native, is usually found in reniform and stalactite masses. When crystals are found, they are rhombohedral with a perfect basal platy cleavage. Arsenic has a metallic luster and tin-white color, but it soon tarnishes on the surface to a dull gray. Most of the arsenic used is obtained as a by-product in the smelting of ores for other metals.

Bismuth (H = 2–2½, G = 9.8) of the native semimetals, is the most important since it is a major source of that element. The sulfide *bismuthinite* is a rare mineral, although some bismuth is obtained from it. Bismuth fuses at a low temperature, and alloys of it with tin, lead, and cadmium fuse at lower temperatures, even below the boiling point of water. For this reason they are used for electric fuses and safety plugs in water sprinkling systems. Lead shot is currently being replaced by bismuth. Much of the bismuth produced is used in medicine and cosmetics.

Native bismuth is rarely found in distinct crystals but is usually granular or present in small veinlets. Artificial crystals are rhombohedral, pseudocubic. Good basal platy cleavage is seen even in granular aggregates. Bismuth is silver-white with a reddish tinge and has a bright metallic luster. It is sectile but at the same time brittle. Bismuth is a relatively rare mineral and the metal is obtained mostly as a by-product in the smelting of gold and silver ores.

The Nonmetals

There are no unifying characteristic among the nonmetals; each must be treated individually.

Sulfur, S

Native sulfur is the chief source of that element, and large deposits of it are found in various parts of the world. Not only does sulfur occur in the native state, but it is also present in nature as sulfides and sulfates. Although these two groups of minerals contain vast quantities of sulfur, only pyrite, iron sulfide, has been utilized as a source; and the world's supply comes mostly from native sulfur.

Sulfur has many uses, but most of it is consumed in the chemical industry in the manufacture of sulfuric acid. It is also used in fertilizers, insecticides, explosives, paper, rubber, matches, and gunpowder.

Habit. Sulfur has three polymorphic forms. Two are monoclinic and are very rare in nature; the third type, most common as a mineral, is orthorhombic. It is found in crystals usually showing rhombic dipyramids (Figs. 7-7 to 7-9). It occurs also in masses and as fine granular aggregates.

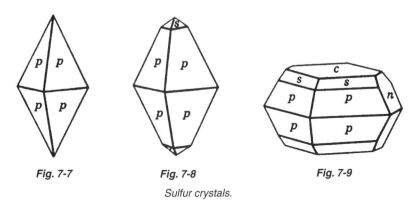

Fig. 7-7   *Fig. 7-8*   *Fig. 7-9*

Sulfur crystals.

Physical Properties. $H = 1\frac{1}{2}$–2, $G = 2.05$. Sulfur is soft and though brittle when struck by a hammer, it is easily cut by a knife. Its specific gravity is distinctly lighter than normal. It has a resinous luster and a bright yellow color and streak. The color is so characteristic that it is known as sulfur-yellow. Crystals are often clear and transparent, and some may have a greenish cast. Sulfur is such a poor conductor of heat that if a crystal is heated gently by being held in the hand, the surface will expand and spall off. One should therefore not handle good crystals.

Composition. Sulfur, S, is usually pure, but it may contain impurities of clay or asphalt. It is remarkable among minerals because when heated it takes fire and burns with a pale blue flame, giving off sulfur dioxide, which has a very characteristic suffocating odor. Because of this, it was early given the name *brimstone* (burn + stone). This is the best test for sulfur.

Occurrence. Sulfur is found on the rims of the craters of some recent volcanoes by direct deposition from vapor by the reaction of escaping H_2S upon exposure to oxygen in air. In this form it has been found in Japan, Argentina, and Chile in large amounts. It is also found in beds associated with gypsum, as at the famous locality near Girgenti, Sicily (see Plate I-2). Most of the world's sulfur today comes from Louisiana and Texas, where it is associated with the cap zone of salt domes. This massive sulfur is found in limestone cavities overlying beds of anhydrite and was probably produced by bacterial reduction of sulfate. It is not mined but is recovered by pumping superheated water down to melt the sulfur in place, then bringing it to the surface by compressed air.

Diamond, C

Of all the minerals described in this book, diamond may be the most familiar to the average person. Everyone has seen cut diamonds in jewelry, but few have seen diamonds in the rough. It is in the uncut state that they are of the greatest interest to the mineralogist. One of the rarest mineral specimens is a diamond embedded in the rock in which it grew.

Habit. Diamonds are usually found in distinct, isolated crystals, most of them ranging from 1 to 10 mm, but some are known to be as large as English walnuts. The crystals are usually octahedrons (see Plate I-1) and rounded dodecahedrons (vicinal tetrahexahedron), but cubes and other isometric forms (Figs. 7-10 to 7-12) are not uncommon. Manufactured diamonds are now so common that they might accidentally be confused with very small natural diamonds; however, the cubo-octahedral habit and the lack of any natural etching of most synthetic diamonds is diagnostic. The natural crystals frequently have rounded edges and curved faces and may show etch pits as described on page 14. On octahedral faces these consist of triangular pits called *trigons;* they are scalloped shaped on dodecahedral faces and are square on the

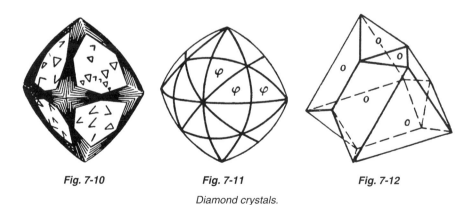

Fig. 7-10   Fig. 7-11   Fig. 7-12

Diamond crystals.

cube faces. There are also crystals with irregular habit, occasionally as round as peas. Flattened crystals known as *macles,* twinned on the octahedron, such as shown in Fig. 7-12, are common. In Brazil a black massive variety known as *carbonado* or *carbon* has been found. It is tougher and less brittle than the crystals.

Physical Properties. H = 10, G = 3.5. The hardness of a diamond is greater than of any other known substance either natural or artificial. Although diamond stands next to corundum in the scale of hardness, it is actually many times harder; the hardness difference between corundum and diamond is greater than that between talc (1 in the scale) and corundum (9). The specific gravity is high for a nonmetallic mineral, and diamonds can be separated from quartz and most rock fragments by gravity separation. When the surface of diamond is clean, it has the remarkable property of adhering to stiff petroleum jellies in preference to water; this property is utilized in its economic recovery. The crystals have perfect octahedral cleavage, and despite its high hardness, diamond is brittle because it breaks easily along the cleavage planes. Diamond has an extremely high luster known as *adamantine.* The term is derived from this species and means the brilliancy of the diamond. The brilliancy is increased when a stone is cut with many facets. Many natural crystals have a peculiar greasy appearance. The most highly prized gemstones are colorless and water-clear, but they are not as common as those that are pale yellow in color. Pale shades of red, orange, green, blue, and brown also are observed. Deeper shades of these colors are highly prized and known as *fancies.* Most manufactured diamonds have a distinctive yellow-green color.

Composition. Diamond is pure carbon and thus has the same composition as a piece of charcoal. It is infusible, as is charcoal, and is not acted upon by acids, but unlike charcoal, it does not burn readily. When heated very hot in the presence of oxygen, it is slowly consumed, forming, like burning charcoal, carbon dioxide.

Occurrence. All the diamonds that were known from early times until the beginning of the eighteenth century came from India, where they were found in stream gravels. Once placer diamonds were discovered in Brazil in 1725, India has been of little importance as a producer. Following the finding in 1866 of diamonds in the stream gravels of the Vaal River in South Africa, it was soon discovered that diamonds were embedded nearby in a special kind of rock called *kimberlite,* consisting of brecciated fragments entrained in a fine-grained matrix of serpentine which occurred in bodies having the shape of a pipe. It is now known that these breccias, including the diamonds, along with pyrope and magnesia-ilmenite, originated at depths of over 100 kilometers below the earth's surface. In addition, in the Argyle pipe, discovered in 1976, in Western Australia, diamonds are found embedded in a sandy tuff consisting of olivine *lamproite.* This rock has affinities with kimberlite but is clearly igneous in origin and is rich in potassium and water. The lamproites are notable in that they carry leucite, phlogopite, magnesia-chromite, and a large variety of other rare minerals.

Some of these great pipes have been mined to thousands of feet in depth and contain diamonds uniformly but sparsely distributed through them. It is estimated that of the material that makes up the pipes, 1 part in 14 million is diamond! During the first half of the twentieth century, South Africa was the only country with important diamond pipes. However, they have since been found in Siberia, Tanzania, Botswana, Australia, and northern Canada. In addition to the diamonds mined, great quantities of diamonds continue to be produced from the coastal placers of Namibia and the stream placers of western Africa, including Angola, Ghana, Sierra Leone, and Congo (formerly Zaire), which has traditionally been the chief producer of industrial diamonds.

Diamonds have been reported from many parts of the United States, but mostly as isolated stones here and there. Until recently, when diamonds were discovered in Colorado and Wyoming, the only place in the United States that can be called a diamond deposit is at Prairie Creek near Murfreesboro, Arkansas, where in 1906 a diamond pipe similar to those of Africa was discovered. Although intermittent mining there has produced some 40,000 stones, the concentration is relatively low. Interestingly, this deposit is now classified with the lamproite type of deposits.

Uses. Of all the diamonds that are mined, only about 20% are suitable for cutting into gemstones; the others are used for industrial purposes. Industrial uses for the off-colored or flawed stones, known as *bort,* are many. Drills set with diamonds are used by miners to obtain samples of rock in advance of workings to reveal the extent and grade of the ore. Larger diamond drills are used to drill thousands of feet through solid rock. Metal disks impregnated with diamonds are used for sawing rocks and other hard material. Diamond dies are used for drawing most fine wire. For this purpose a hole is drilled through the diamond and the wire is drawn through it, thus reducing the wire to the diameter of the hole. As natural diamonds smaller than 1 mm are generally uneconomical to recover and must be produced by crushing of bort, manufactured diamonds, which consist of small but perfect crystals, are replacing these crushed cleavage fragments in many industrial applications, such as for cutting glass and industrial grinding. Diamond powder is used for grinding diamonds and other hard gems. Today, the annual production of synthetic diamonds is about 300,000 carats,* which is three times the weight of natural diamonds recovered.

Everyone knows the use of the diamond for jewelry, for which its brilliancy, hardness, and comparative rarity make it the most important of all gemstones. The clear, colorless or "blue-white" stones are in general the most valuable; a faint yellow color, often present, detracts from their value. However, deep shades of yellow, red, green, or blue, called fancies, can have great value.

Graphite, C

The name *graphite* comes from the Greek meaning *to write* because of its use in "lead" pencils. It has also been called *plumbago* and *black lead.* Both these names were given because it was confused with galena, the common black lead sulfide. Only in its black color does it resemble galena, however; the other properties of both minerals are quite distinct.

*A carat is 200 milligrams.

It is interesting to compare the two natural forms of carbon—diamond and graphite—for it is difficult to picture two substances of similar composition with such vastly different properties. Diamond is the hardest of minerals, light in color, and high in specific gravity; graphite is one of the softest of minerals, black and opaque, with a low specific gravity. It is the difference in the arrangement of the atoms in the two minerals that gives rise to the different properties.

Habit. Graphite is hexagonal, and some crystals have a hexagonal outline with a prominent basal plane. Usually, it appears massive, but may be separated easily into thin leaves or plates; hence it is said to be *foliated*. It may also be finely granular and compact.

Physical Properties. $H = 1–2, G = 2.23$. Graphite is sectile and so soft that it makes a mark on paper and feels greasy to the hand. Its specific gravity is very low for a mineral with a metallic luster. The color is iron-black to steel-gray. A good platy cleavage parallel to the base permits the easy separation of the flakes.

Composition. Graphite is pure carbon, C, but some deposits may be impure, mixed with clay, iron oxide, or other minerals.

Occurrence. Graphite is commonly found in small scales scattered through rocks, such as marbles, crystalline limestones, schists, and gneisses, where it was formed as the result of metamorphism of original organic matter. Such are the occurrences of massive flake graphite in the gneisses of Madagascar or in the quartz schists in the Adirondack region, particularly at Ticonderoga in the United States. At Sonora, Mexico, an igneous rock has intruded a coal bed and converted adjacent coal to graphite. It is mined most extensively on the island of Sri Lanka, where high-grade graphite occurs in veins with quartz and calcite.

To the beginning of the twentieth century all the graphite used in industry was the natural mineral and was mined. Since that time graphite has been manufactured by heating coke to an intense heat in an electric furnace. At present more synthetic than natural graphite is used.

Uses. Graphite mixed with clay is the so-called *lead* in lead pencils. It is also mixed with clay for making crucibles for handling molten metal, because it is infusible, is not affected by the heat of an ordinary furnace, and will not react with the molten metal. It is an excellent lubricant and as such is frequently mixed with oil. It is also used in electroplating, foundry facings, stove polish, and in many other ways.

SULFIDES

The sulfides form an important group of minerals, for among them are many of the ores of the common metals and many of the less common nonmetals. In them, sulfur or other nonmetal, such as selenium, tellurium, arsenic, antimony or bismuth, is combined with one or more metals. Upon heating one of the common sulfides, the sulfur unites with oxygen and is driven off as sulfur dioxide gas. It can be detected by its pungent and irritating odor. Distinctive reactions in the open tube or before a blowpipe can also be used to identify the other nonmetallic elements occurring with the sulfur.

The structure of the sulfide minerals is quite complex and it is possible here to give only the simplest classification based on the ratios of sulfur to metal. In this scheme, when sulfur is combined with a group V element such as arsenic, antimony, or bismuth to form compounds called *sulfosalts,* these minerals are placed immediately after the common sulfide to which they are structurally related.

Sulfides

A_2X	A_2X_3
Acanthite, Ag_2S	Oripiment, As_2S_3
Chalcocite, Cu_2S	Stibnite, Sb_2S_3
A_3X_2	AX_2
Bornite, Cu_5FeS_4	Pyrite, FeS_2
AX	Cobaltite, $CoAsS$
Galena, PbS	Skutterudite, $CoAs_3$
Sphalerite, ZnS	Marcasite, FeS_2
Chalcopyrite, $CuFeS_2$	Arsenopyrite, $FeAsS$
Enargite, Cu_3AsS_4	Molybdenite, MoS_2
Tetrahedrite, $Cu_{12}Sb_4S_{13}$	Calavarite, $AuTe_2$
Nickeline, $NiAs$	Sylvanite, $(Au,Ag)Te_2$
Pyrrhotite, $Fe_{1-x}S$	A_3X_4
Millerite, NiS	Pyrargyrite, Ag_3SbS_3
Cinnabar, HgS	Proustite, Ag_3AsS_3
Realgar, AsS	

Acanthite, Ag_2S

Acanthite historically has been called *argentite* (silver glance), so named from the Latin *argentum,* meaning silver. Above 173°C, Ag_2S is isometric (argentite), but on cooling, it inverts to a monoclinic form (acanthite). Acanthite is not a common mineral but in places it has been a valuable ore mineral, because when pure, it contains 87% silver.

Habit. Acanthite is monoclinic. However, all crystals are twinned polymorphs of the high-temperature isometric form and show the cube, octahedron, and dodecahedron. Most commonly, it is massive or in coatings on other minerals.

Physical Properties. $H = 2$, $G = 7.3$. It is eminently sectile and can be cut with a knife almost as easily as lead. It can be flattened to some extent under the hammer, whereas almost all other sulfides, being brittle, break into many fragments. The luster is metallic, and the color and streak grayish black.

Composition. The formula is Ag_2S. When heated by a blowpipe flame on charcoal, the sulfur is easily roasted off and a little silver ball is left behind. The ball can be tested chemically by dissolving it in nitric acid and adding a drop of hydrochloric acid. A white, curdy precipitate of silver chloride will be formed.

Occurrence. Acanthite is usually found associated with the ruby silver minerals, pyrite, chalcopyrite, or galena in veins with chalcedony, calcite, alunite, or barite. Such occurrences include Pachuca and Parral, Mexico and in the United States at Virginia City and Tonopah, Nevada. It can also occur as a supergene mineral from the alteration of primary silver. Much of the argentiferous galena consists of microscopic acanthite intergrown with the galena.

Chalcocite, Cu_2S

Chalcocite, or *copper glance,* is one of the most valuable ores of copper, for when pure it contains about 80% of the metal. If one reads the history of copper mining, first one mineral and then another was the chief copper ore. For example, during the middle part of the nineteenth century, native copper from the Lake Superior district was the most important in the United States. Then supergene chalcocite formed by the alteration of primary copper minerals became most important. Today, most of the copper is produced from primary ores present as tiny grains, slightly enriched by chalcocite, scattered through large bodies of rock called porphyry copper ores. The mining of these ores is carried out on a gigantic scale, for they may carry as little as 0.4% copper.

Habit. Chalcocite is monoclinic, pseudo-orthorhombic, but crystals are very rare and usually small. It is most commonly fine-grained and massive.

Physical Properties. $H = 2\frac{1}{2}–3$, $G = 5.6$. It is brittle when struck with a hammer but can be cut to some degree with a knife. Miners sometimes identify it by this property of being imperfectly sectile. It is black or bluish black and when fresh has a brilliant metallic luster. The streak is grayish black. When exposed to the air, it becomes dull and tarnished on the surface.

Composition. The formula is Cu_2S. When heated on charcoal with a blowpipe flame the sulfur is easily roasted off, leaving behind a little ball of metallic copper.

Another, less common copper sulfide contains only 66.4% of copper with the formula CuS. This is *covellite.* Like chalcocite, it forms as a secondary mineral, but it is easily distinguished from chalcocite because of its indigo-blue color.

Occurrence. Most chalcocite is secondary, that is, it was formed by supergene enrichment from the alteration of some earlier copper mineral by solutions working downward from the surface. In such a way large masses of solid massive chalcocite have formed, such as those at Clifton and Morence, Arizona. Elsewhere, as at Bingham, Utah, primary pyrite, chalcopyrite, and bornite disseminated in an altered quartz monzonite have been coated with supergene chalcocite and covellite to form workable deposits called *porphyry copper* ore. Such deposits are known worldwide and rarely contain more than 1 or 2% of copper and may be as low as 0.4%. Because they can be worked on such a large scale, they furnish most of the copper produced in the world today. Late-stage primary chalcocite occurred with enargite, bornite, and chalcopyrite in quartz-pyrite veins cutting altered granodiorite at the great copper deposit at Butte, Montana. Crystals of chalcocite are rare, but fine specimens have come in the past from cavities in quartz veins at Bristol, Connecticut and Cornwall, England.

Bornite, Cu$_5$FeS$_4$

Bornite was named after the Austrian mineralogist von Born, but it has a variety of other names, such as *purple copper ore, variegated copper ore,* and *peacock ore.* All these suggest a property by which it is easily recognized: the bright iridescent tarnish of the surface.

Habit. Bornite is tetragonal but is isometric above 228°C. At a few places it has been found in rough cubic crystals, but usually it is massive as embedded particles or irregular grains.

Physical Properties. H = 3, G = 5.07. The variegated purple to blue color on the exposed surface of bornite is its most characteristic feature. However, exposed surfaces of other copper minerals may be somewhat similar, so that, in identifying a specimen, one should always break off a small fragment and observe the fresh fracture, which is brownish bronze in color. The luster is metallic, and the streak is grayish black.

Composition. Bornite contains both copper and iron and has the formula Cu$_5$FeS$_4$. When heated with a blowpipe flame on charcoal, the sulfur is driven off and the fragment becomes magnetic, showing the presence of iron. If the fragment is touched with a drop of hydrochloric acid and is again heated, the azure-blue flame of copper chloride shows the presence of copper.

Occurrence. Bornite is a widespread and important ore of copper. At most of the copper mines of the western United States, bornite is found associated with chalcopyrite and chalcocite. It has been an important ore mineral at Superior, Arizona; Butte, Montana; Magma mine, Arizona; and Engles mine, California. It is an important mineral in the sedimentary ores of the Kupferschiefer, Germany, and it is often found as disseminated grains and globules with augite and plagioclase in gabbros and diabases.

Galena, PbS

Galena is the most important ore of lead and one of the commonest sulfide minerals. Other ores of lead, such as the sulfate, anglesite, and the carbonate, cerussite, will be considered later. *Native lead* is a very rare mineral, although found occasionally in small amounts.

Lead is one of the most important metals, used for many purposes familiar to all, such as drainpipes, storage batteries, cable coverings, foil, bullets, and shot. It is also used in solder alloyed with tin and in type metal alloyed with antimony. Because of its opacity and ability to form stable organic compounds, the basic carbonate, *white lead,* was formerly used as a paint pigment. This use has been discontinued because lead-base paints are a major source of lead poisoning.

Habit. Galena crystallizes in the isometic system, and well-formed cubic crystals are common. It is found also in octahedrons or in crystals showing a combination of cube and octahedron (Figs. 7-13 to 7-17).

Physical Properties. H = 2½, G = 7.5. Perfect cubic cleavage (Fig. 4-1) is the most outstanding feature by which galena can be identified. A fragment struck with a hammer will be seen to break up into a multitude of perfectly formed rectangular blocks.

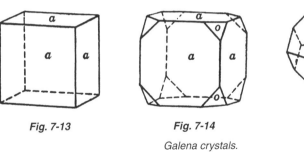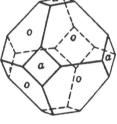

Fig. 7-13   **Fig. 7-14**   **Fig. 7-15**

Galena crystals.

Fig. 7-16. *Galena crystals, Breckenridge, Colorado.*

Fig. 7-17. *Galena and sphalerite on dolomite, Joplin, Missouri.*

Even when a specimen of galena is fine-grained and appears almost massive, the eye can catch the light reflected from the myriad of tiny cleavage faces. The specific gravity of 7.5 is nearly as high as that of metallic iron. Lead is a metal of such high density (specific gravity 11.4) that all its compounds are heavy. The luster of galena is metallic and usually very brilliant; the color and streak are bluish lead-gray, but the exposed surface may be somewhat dull from tarnish.

Composition. Galena is lead sulfide, PbS, but it may contain small amounts of impurities such as bismuth and antimony. One of the most important impurities is silver, probably in the form of microscopic acanthite. When it is present in quantity sufficient to justify its being worked for the precious metal, galena is regarded as a silver ore and is called *argentiferous galena.* When galena is heated on charcoal before a blowpipe, it fuses easily and on continued heating will yield a globule of metallic lead.

Occurrence. Galena is virtually the only ore mineral that is mined for lead worldwide. Galena is commonly associated with sphalerite, pyrite, hemimorphite, smithsonite, and chalcopyrite in veins with quartz, calcite, barite, and fluorite. Galena occurs in fine laminated layers alternating with laminae of sphalerite, pyrite, or pyrrhotite in exhalitive deposits such as the Broken Hill and Mount Isa deposits in Australia or the Sullivan mine in British Columbia, Canada. Galena occurs as replacement bodies in contact zones with limestone, as around Darwin, California; Bingham, Utah; Park City, Utah; and Laurion in Greece. It may be found in veins with siderite, as in the Coeur d'Alene district of Idaho, where galena is mined for its rich silver content. The major source of lead in the United States is in southeastern Missouri, where galena is disseminated along fractures in the limestone. This is but one of the many occurrences of lead and zinc that have come to be known as the Mississippi Valley deposits. Included in these deposits is one of the most famous lead–zinc mining districts of the world, the Tri-State district of Missouri, Kansas, and Oklahoma. Here galena and sphalerite are associated with dolomite, chalcopyrite, marcasite, and calcite in breccia pipes, veins, and caverns in limestone. Although mining has essentially ceased in the area, it is represented in most mineral collections because of the many beautiful specimens of mineral crystals that have grown into the open spaces of the deposits.

Sphalerite, ZnS

Sphalerite is named from a Greek work meaning *treacherous.* Certain varieties of it are hard to identify, and the young mineralogist, after misidentifying it several times, will think it well named, because it often occurs with and is mistaken for the more easily recognized lead ore, galena. The miner's names *black jack* and *false galena* refer to the same fact. Another name is *zinc blende,* chosen, for the same reason, from the German word meaning *blind* or *deceiving.* Light-colored fine-grained varieties are often confused with siderite.

Sphalerite is the most important ore of zinc, a useful metal in industry and the arts. Iron in sheets and in wire is frequently coated with zinc to protect it from rusting and is said to be *galvanized.* Zinc is used in storage and dry-cell batteries, and alloyed with copper it forms brass. Large amounts are used in the pigments zinc oxide and lithopone (a mixture of zinc sulfate and barium sulfide).

Habit. Sphalerite is isometric and when it is well crystallized, is found in tetrahedrons. The crystals, however, are usually twinned and in distorted aggregates, and it requires a trained and skillful eye to understand them (Figs. 7-18 and 7-19). It is usually found in coarse- to fine-granular cleavable masses.

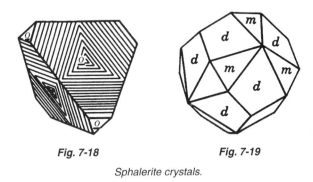

Fig. 7-18   Fig. 7-19

Sphalerite crystals.

Physical Properties. H = 3½–4, G = 4. Sphalerite has perfect blocky cleavage, and from coarse masses it is possible to cleave an almost perfect dodecahedron. Even if the sphalerite is fine granular, the cleavage surfaces are usually prominent, although some types are so fine-grained and closely compacted that they show no cleavage.

In the rare perfectly pure specimen, sphalerite is clear and nearly colorless and has a resinous luster. Commonly, it contains some iron, and the color deepens with increasing amounts of iron, from yellow to yellowish brown (the most common) to black. Some crystals, known as *ruby zinc,* are red.

The resinous luster in most specimens of sphalerite is so distinct that mineralogists come to depend on it to enable them to identify the mineral. The streak is white, pale yellow, or brownish, becoming deeper the darker the color of the mass, but always lighter in color than the massive mineral.

Composition. Although pure sphalerite is zinc sulfide, ZnS, iron, as stated above, may be in solid solution, and black sphalerite, (Zn,Fe)S, may contain as much as 50% iron. Manganese and the rare element cadmium are usually present in small amounts. In some places the rare mineral *greenockite,* cadmium sulfide, CdS, is found in earthy crusts on sphalerite.

The tests for sphalerite are poor. Before a blowpipe it does not fuse, but if powdered, mixed with sodium carbonate, and heated on charcoal, it gives a zinc coating that is yellow when hot and white when cold. When warmed in a test tube with hydrochloric acid, it effervesces, giving off bubbles of gas that one might mistake for carbon dioxide, except that the disagreeable odor shows it to be hydrogen sulfide, H_2S.

Occurrence. Sphalerite is one of the commonest metallic minerals and is frequently associated with galena, chalcopyrite, pyrite, and other sulfides in veins with quartz, dolomite, calcite, barite, and fluorite. As mentioned above, it occurs in fine laminated layers alternating with laminae of galena, pyrite, or pyrrhotite in exhalitive deposits such as the Broken Hill and Mount Isa deposits in Australia or the Sullivan mine in British Columbia, Canada. Sphalerite occurs as replacement bodies in contact zones with limestone, as around Darwin, California; Bingham, Utah; Balmat, New York; and Park City, Utah. It may be found in veins with siderite, with which it can be confused, as in the Coeur d'Alene district of Idaho. One of the great zinc mining districts of the world, known as the Tri-State district, which centers on Joplin, Missouri, contains what are known as the Mississippi Valley deposits. Here sphalerite associated with galena, marcasite, and dolomite has been deposited along fractures, breccia pipes, and in caverns in limestone. If sphalerite ore is exposed to weathering, it goes into solution and may be reprecipitated as smithsonite (zinc carbonate) or hemimorphite (zinc silicate).

Chalcopyrite, $CuFeS_2$

Chalcopyrite, or *copper pyrites,* as it is sometimes called, is an important ore of copper. The same elements are present as in bornite but in different proportions. The color of chalcopyrite is a beautiful deep brass-yellow, so golden that chalcopyrite is frequently mistaken for gold, especially when it is scattered in small particles through a mass of quartz. Although it can easily be distinguished from gold, as we shall see, the name "fool's gold," which it shares with pyrite, is not inappropriate. It is interesting to compare the percentages of copper in the four copper minerals considered thus far; native copper, 100%; chalcocite, 79.8%; bornite, 63.3%; chalcopyrite, 34.5%.

Habit. Chalcopyrite belongs to a low-symmetry class of the tetragonal system, and crystals are usually sphenoids. The crystals appear isometric, and it is difficult with the unaided eye to distinguish them from tetrahedrons (Figs. 7-20 and 7-21). Chalcopyrite is usually massive and thus may be found in large specimens or in tiny specks in the enclosing rock.

Physical Properties. $H = 3\frac{1}{2}$–4, $G = 4$. The luster is brilliant metallic, and the color, as we have seen, deep brass-yellow; the streak is greenish black. A tarnish often devel-

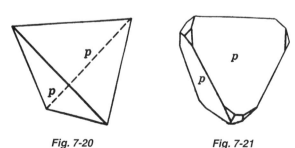

Fig. 7-20   Fig. 7-21

Chalcopyrite crystals.

ops on the surface, deepening the color or giving it a variegated appearance similar to that of bornite. To determine the color, one should always examine a fresh fracture.

Chalcopyrite can be readily distinguished from pyrite, for unlike pyrite, it can be scratched with a knife easily and its color is deeper. It is distinguished from gold by being brittle. When scratched with a knife, it yields a greenish powder, whereas the scratch in gold is a shiny groove.

Composition. The formula is $CuFeS_2$. Chalcopyrite is a structural derivative of sphalerite in which copper and iron alternate for the zinc, and as a consequence, two cubic cells are required, which changes the habit from isometric to tetragonal. When heated on charcoal, a fragment fuses to a black mass that is strongly magnetic. Dissolved in nitric acid, it gives a blue solution that turns azure-blue when an excess of ammonia is added.

Occurrence. Chalcopyrite is a very common mineral, the most widespread of the copper sulfides, and one can expect to find small amounts of it almost anywhere. It occurs as a magmatic sulfide with pyrrhotite in a matrix of hypersthene and plagioclase in norite at Sudbury, Ontario, Canada; as alternating laminae with pyrite in sedimentary exhalitives as on Cyprus, Rio Tinto in Spain, and Ducktown, Tennessee; and as limestone replacements with epidote and garnet at Morenci, Arizona. It is common in most of the porphyry copper and in most hydrothermal vein deposits. In these deposits it is commonly associated with pyrite, bornite, sphalerite, or molybdenite, usually in a quartz matrix but occasionally with other minerals, such as tourmaline or chlorite. When chalcopyrite is weathered, it forms in the zone of oxidation an array of secondary copper minerals, including azurite, malachite, cuprite, native copper, chrysocolla, and turquoise. When pyrite is also present, all of the copper minerals may be dissolved, leaving a limonite gossan or capping. At the water table, the copper may be reprecipitated chiefly as chalcocite.

Enargite, Cu_3AsS_4

Habit. Enargite is orthorhombic, and crystals are usually elongate prisms with striations parallel to *c*. It is most commonly found in bladed or granular aggregates.

Physical Properties. H = 3, G = 4.43–4.45. A perfect prismatic cleavage is one of the outstanding features of enargite, for it is the only common mineral with two prismatic cleavage directions having metallic luster and black color and streak. Careful observation of cleavage faces will reveal traces of pinacoidal cleavages as well.

Composition. Enargite is copper sulfarsenide, Cu_3AsS_4. Enargite is a complex derivative of sphalerite in which because of the presence of arsenic, some of the sulfur atoms have been omitted.

It is easily fusible on charcoal and gives the volatile white sublimate of arsenious oxide and characteristic garlic odor. If roasted on charcoal and moistened with a drop of hydrochloric acid and again heated, it gives an azure-blue copper chloride flame.

Occurrence. Although enargite is a comparatively rare mineral, it can be an important component of copper ores such as it was at Butte, Montana, and at Bingham Canyon, Utah, where it occurred with bornite and chalcopyrite in quartz–pyrite veins.

Tetrahedrite, $Cu_{12}Sb_4S_{13}$

Habit. Tetrahedrite is isometric and receives its name because the crystals are commonly tetrahedral in habit, although often highly modified (Figs. 7-22 and 7-23). As with many metallic minerals, good crystals are rare, and mineralogists often must be content with massive pieces.

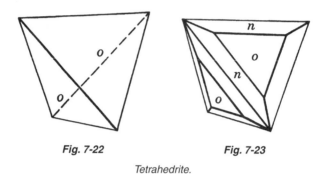

Fig. 7-22   **Fig. 7-23**

Tetrahedrite.

Physical Properties. H = 3–4½, G = 4.6–5.1. Massive tetrahedrite can usually be recognized by the brilliant metallic luster and dark grayish-black color and streak. It is frequently called *gray copper* by miners. Although it can be confused with magnetite, tetrahedrite is easily scratched by a knife and is nonmagnetic, whereas magnetite is too hard to be scratched and is magnetic. The specific gravity varies from 4.6 to 5.1, depending on the proportion of the various elements present.

Composition. Ideally, the only metal present in tetrahedrite is copper, but it is really a complex sulfantimonide of copper, iron, zinc, and silver, $(Cu,Fe,Zn,Ag)_{12}(Sb,As)_4S_{13}$. As with enargite, tetrahedrite is a complex derivative of sphalerite in which because of the presence of antimony and arsenic, some of the sulfur atoms are missing. Copper is always the predominant metal. If arsenic takes the place of all the antimony, the mineral is called *tennantite.* The variety that is highly rich in silver is known as *freibergite.*

Tetrahedrite is easily fusible on charcoal and usually gives tests for both antimony and arsenic; if the remaining globule is touched with a drop of hydrochloric acid and heated again, it gives the azure-blue flame that indicates copper.

Occurrence. Tetrahedrite is widespread and is the most common of the sulfosalts. It is usually associated with chalcopyrite, pyrite, sphalerite, galena, and acanthite and occurs in veins with calcite, dolomite, siderite, barite, fluorite, or quartz. For example, it occurs with enargite and chalcopyrite in quartz–pyrite veins at Butte, Montana. When it is high in silver, it becomes a valuable ore of that metal.

Nickeline, NiAs

Nickeline, formerly called niccolite, is often called *copper nickel* from the German *kupfernickel,* the first name given to the mineral. It is called copper nickel only because of its conspicuous copper color; it contains no copper but is a minor ore of nickel. It is the mineral in which the element nickel was discovered.

Habit. Nickeline occurs rarely in hexagonal crystals and is usually massive or reniform with columnar habit.

Physical Properties. $H = 5–5\frac{1}{2}$, $G = 7.8$. As noted, the color is pale copper-red; the streak is brownish black.

Composition. Nickeline is nickel arsenide, NiAs. It is rarely pure and usually contains a little iron, cobalt, sulfur, and antimony. When it is heated on charcoal, dense white fumes of arsenious oxide form and the characteristic garliclike odor of arsenic is given off.

Occurrence. Nickeline accompanied by pyrrhotite and chalcopyrite as disseminated grains and globules is usually found associated with augite and plagioclase in norites. It is also found in veins with cobaltite and skutterudite, as at Cobalt, Ontario, where it occurs with native silver, acanthite, and arsenopyrite in veins of calcite and dolomite.

Pyrrhotite, $Fe_{1-x}S$

Pyrrhotite takes its name from a Greek word meaning *reddish* because of its peculiar reddish-bronze color; its color is a very important property as an aid in its identification. It is called also *magnetic pyrites,* which refers to its being magnetic, a more striking property. Other sulfides containing iron are magnetic after heating, but pyrrhotite alone is attracted by a magnet without heating. In places where it is mined for its associated copper and nickel, the sulfur is converted into sulfuric acid and the excess iron into high-grade ore.

Habit. Pyrrhotite crystallizes in the hexagonal system, but good crystals are rare (Fig. 7-24). When found, they are tabular showing a hexagonal outline. It is usually found in irregular masses.

Physical Properties. $H = 4$, $G = 4.6$. The luster is metallic, and the color, as noted before, a peculiar reddish bronze quite different from pyrite and marcasite. The streak is black.

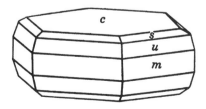

Fig. 7.24. *Pyrrhotite.*

Composition. Pyrrhotite is a sulfide of iron expressed by the unusual formula $Fe_{1-x}S$, where x lies between 0 and 0.2. This means that it is not quite equivalent to the simple sulfide troilite (FeS), where Fe and S are in the ratio 1:1, but there is a deficiency in iron that differs in different specimens. It has been placed immediately following nickeline because it is a complex derivative.

Occurrence. Pyrrhotite has a melting temperature equal to that of basalt, so it occurs as a comagmatic minor constituent with augite and plagioclase in mafic igneous rocks such as gabbros and diabases. At Sudbury, Ontario, pyrrhotite with associated pentlandite, $(Fe,Ni)_9S_8$, and chalcopyrite occurs in very large masses at the base of the mafic intrusion, where it is mined for the nickel content in pentlandite. It is also an important constitutent of sedimentary exhalitive deposits, where it can be found as alternating laminae and masses with other sulfides, such as galena at the Sullivan deposit, British Columbia, with pyrite and chalcopyrite on Cyprus, at Rio Tinto in Spain, and at Ducktown, Tennessee. As the high-temperature decomposition product of pyrite, it can be found in sulfur-rich metamorphic rocks and pegmatites. Because it it is readily precipitated from hydrothermal solutions, it is often found in veins with quartz or calcite. Finally, the variety *troilite,* FeS, is found as nodules with kamacite in iron meteorites (Fig. 7-6).

Millerite, NiS

Millerite is remarkable among minerals because of its occurrence in very fine hairlike or capillary crystals, and for this reason it is called *capillary pyrites.*

Habit. Millerite is rhombohedral, and the crystals are usually greatly elongated parallel to the c crystal axis, forming, in places, masses that resemble a wad of hair. In other localities, delicate, radiating crystals or crusts with a fibrous habit are found in cavities.

Physical Properties. $H = 3-3\frac{1}{2}$, $G = 5.5$. The luster is metallic, and the color, pale brass-yellow with a greenish tinge when the crystals are in fine hairlike masses. The streak is greenish black.

Composition. Millerite is nickel sulfide, NiS. When it is heated before a blowpipe, the sulfur is driven off and the remaining fragment, after heating in the reducing flame, is attracted by a magnet. It should be remembered that nickel, like iron, is magnetic, although to a lesser intensity.

Occurrence. As stated before, millerite occurs in hairlike crystals in cavities with calcite, dolomite, or fluorite in limestone or dolomite or with chrysotile in cavities in serpentine. In places, such as near St. Louis, Missouri, it occurs in geodes in limestone where the fibers are matted together like a wad of hair. It is found as tufts of extremely delicate radiating crystals in cavities in siderite at Glamorgan, Wales and with ankerite in cavities of hematite at Antwerp, New York. At the Gap mine, Lancaster County, Pennsylvania, where it occurs as coatings on pyrrhotite, it has a compact fibrous habit. In some places, millerite fibers are found as inclusions in other minerals, such as calcite as at St. Louis, Missouri. A major exception to the fibrous habit is at the Marbridge mine, Lamotte Township, Quebec, where coarse cleavable masses are mined as a nickel ore. The iron–nickel sulfide *pentlandite* is more important as an ore.

Cinnabar, HgS

Cinnabar is the source of the world's supply of mercury, or *quicksilver,* and is thus a very important mineral. If pure, it is easily recognized by its extremely high specific gravity and its red color.

Native mercury has been found in small amounts associated with cinnabar, but such occurrences are rare. Mercury is the only metal that is liquid at ordinary temperatures, for it becomes a solid only when cooled to −39°C. This property gives it many uses. We are familiar with it in clinical and household thermometers, and it is used also in barometers, pressure gauges, electrical switches, and various other scientific apparatus. A "silver" filling in a tooth consists of mercury mixed with silver to form an amalgam. One of the methods of recovering gold and silver is to use mercury, which forms an amalgam with these metals. Other important uses are in the manufacture of fulminate of mercury for detonating high explosives, and in the preparation of drugs.

Habit. Cinnabar is rhombohedral but is usually found in fine-granular masses or disseminated through the rock in which it occurs. Crystals that are occasionally found are usually rhombohedral or prismatic.

Physical Properties. $H = 2\frac{1}{2}$, $G = 8.1$. Cinnabar has perfect prismatic cleavage, which gives granular aggregates a sparkle as the light is reflected from the many brilliant faces. The specific gravity, 8.1, is higher than that of metallic iron of 7.8. The great weight cannot escape the observer and is a striking characteristic. In some specimens, however, the cinnabar is not pure but is scattered through a clayey gangue and gives its color to the rock. In such cases the density of the whole is much lower. The color is vermilion-red; the streak, scarlet; the luster, adamantine.

Composition. Cinnabar is mercury sulfide, HgS. Because of the high atomic weight of Hg, 200.6, cinnabar contains 86.2% mercury. When heated on charcoal, a piece of pure cinnabar is volatilized entirely. In an open tube, if it is heated very slowly, so that the sulfur has time to oxidize, a ring of metallic mercury is formed on the cold part of the tube. If the cinnabar is heated too rapidly, a ring of black mercury sulfide will form that has the same composition as that of the original mineral (see p. 110).

Occurrence. Cinnabar is usually found in near-surface veins and hot spring deposits associated with shallow recent igneous intrusions. By far the single most important locality is Almaden, Spain, which has been operating continuously since Roman times and continues to be the world's single most important producer of mercury. Here cinnabar is found both as disseminations and massive bands in veins of quartz with accompanying sericite. At occurrences in Arkansas, cinnabar associated with metacinnabar, $Hg_{1-x}S$, stibnite, and pyrite are found in fault breccias with quartz, opal, siderite, and dickite. Similar occurrences are known at Idria, Italy and at New Almaden and New Idria in California.

Realgar, AsS; Orpiment, As₂S₃

Realgar and orpiment can be considered together, since both are compounds of arsenic and sulfur and are almost invariably associated with each other. They are rarely used as ores, as most of the industrial arsenic is obtained from smelter smoke produced from the recovery of copper and lead ores containing enargite or arsenopyrite.

Realgar ($H = 1\frac{1}{2}-2$, $G = 3.5$) is found in transparent monoclinic crystals and massive aggregates which have a beautiful aurora-red color. It is soft and sectile. The luster is resinous, and the streak is red to orange.

Realgar is arsenic monosulfide, AsS. It is easily fusible and, when heated on charcoal, gives a volatile white sublimate of arsenious oxide with a characteristic garlic odor.

Orpiment ($H = 1\frac{1}{2}-2$, $G = 3.5$), named from the Latin *auripigmentum,* meaning *golden paint,* is a beautiful golden yellow. It is monoclinic, but distinct crystals are rare, and it is generally found in foliated or columnar masses. It has a perfect platy cleavage parallel to the side pinacoid, so that it splits easily into thin flexible leaves. The luster is pearly on this cleavage face.

The composition of orpiment is As_2S_3 (arsenic trisulfide), and it is often formed as a weathering product of realgar or other arsenic minerals. When it is heated on charcoal, its behavior is like that of realgar; it volatizes completely and gives the characteristic garlic odor of arsenic and white fumes of the oxide, As_2O_3.

Occurrence. Realgar and orpiment are usually found together, frequently in near-surface veins and hot spring deposits such as Steamboat Springs, Nevada, where they are related to recent igneous intrusions. They may be deposited as volcanic sublimates and from geyser water, as at Yellowstone National Park. They are often associated with stibnite, as at Manhattan, Nevada, where they are in veins of calcite, barite, or chalcedony. Fine crystals have been found in cavities in calcite at Mercur, Utah. Sometimes they are found lining fractures or cavities in sedimentary limestones, dolomites, or shales.

Stibnite, Sb₂S₃

Stibnite, sometimes called *antimony glance,* is the commonest and most important ore of antimony, a metal that is very useful in the arts. One part of antimony is alloyed with 3 parts of lead to form type metal. Antimony has the unusual property of expanding on cooling and causes the alloy to fill out the mold and give sharp clean letters. Babbitt metal, used in bearings, is an alloy of antimony with tin and copper.

Habit. Stibnite is orthorhombic and is frequently found in vertically striated prismatic crystals elongate parallel to *c*. Terminations are often with spear-shaped (Figs. 7-25 and 7-26). Some crystals may be curved or bent. Stibnite is also found in radiating groups as well as granular aggregates.

Physical Properties. $H = 2$, $G = 4.6$. There is perfect platy cleavage parallel to the side pinacoid, and the cleavage surfaces are smooth and appear highly polished, sometimes showing striations parallel to the front pinacoid. Even in the granular aggregates, the cleavage is usually visible. The hardness of 2 permits it to be easily scratched and to leave a mark on paper. It should not be confused with graphite,

Fig. 7-25. Stibnite.

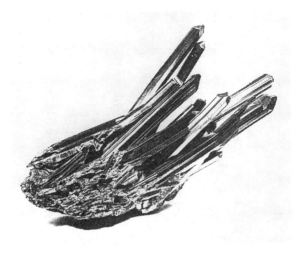

Fig. 7-26. Stibnite crystals, Japan.

which is much softer and greasy to the touch and marks paper without the slightest tendency to tear it. The luster is metallic and, on a fresh cleavage surface, is very brilliant. The color and streak are lead-gray to black.

Composition. Stibnite is antimony trisulfide, Sb_2S_3. It fuses very easily in a match flame (1 in the scale of fusibility). When heated on charcoal, it gives off fumes of the oxide, Sb_2O_3, which form a thick coating at a little distance; after a few moments, the fragment is entirely volatilized. If the reducing flame is thrown for a moment on the coating, it is burned off with a greenish-blue flame (see p. 108).

Occurrence. Stibnite occurs in shallow veins and in hot spring deposits such as Steamboat Springs, Nevada. It is commonly associated with cinnabar, orpiment, galena, and sphalerite in quartz veins. Most of the world's supply comes from Hunan, China. Fine crystals were found at the Manhattan Mine in Nevada, but the most perfect and beautiful crystals have come from cavities in quartz veins cutting black phyllite at the Ichinokawa Mine on Shikoku Island, Japan.

Pyrite, FeS_2

Pyrite, or *iron pyrites,* is one of the minerals known as "fool's gold." It is the most common, as well as the most striking, sulfide mineral. It forms under an extremely wide range of conditions and is thus associated with many minerals, most commonly with chalcopyrite, sphalerite, and galena.

Massive pyrite is mined in some places for the gold and copper associated with it but only under unusual circumstances for its constituent elements alone. When thus mined, it is usually for the sulfur rather than the iron, for sulfur makes up 53.4% of pyrite. In a few countries where there is none of the richer oxide ores, it has been mined on a small scale as an ore of iron.

Habit. Pyrite is often found in cubic crystals the faces of which usually show fine lines or striations parallel to one pair of edges only (Fig. 7-27) and at right angles to those on the adjoining faces. These striations are the result of an oscillatory combination of the cube faces with those of the pyritohedron. The pyritohedron (Fig. 7-29), named from this species, is also common; and it, like the cube, may show fine striations (Fig. 7-28). The two forms may be present more or less equally well developed

Fig. 7-27. Pyrite.

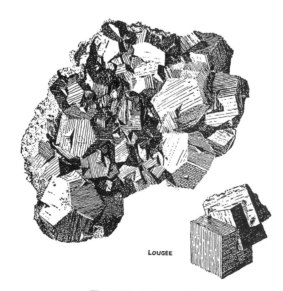

LOUGEE

Fig. 7-28. Pyrite crystals.

(Fig. 7-30). More rarely, one may find octahedrons of pyrite, but this form is more common in combination with the cube or the pyritohedron (Figs. 7-31 and 7-32). Pyrite is also found in massive or granular aggregates, which in some places form huge lenses that may be mined for their sulfur content.

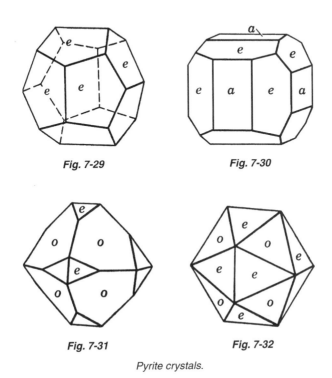

Fig. 7-29   Fig. 7-30

Fig. 7-31   Fig. 7-32

Pyrite crystals.

Physical Properties. H = 6, G = 5.02. The hardness of pyrite is a little over 6, so that it scratches glass and is not scratched by the knife blade. Its hardness, therefore, high for a sulfide, distinguishes it from chalcopyrite (easily scratched by the knife) with which it may be confused. The luster is brilliant, metallic; and the color, light brass-yellow, growing a little darker when tarnished. The streak is greenish black.

Composition. Pyrite is iron disulfide, FeS_2. As mentioned previously, chalcopyrite or small amounts of gold may be found as intergrowths. When these occur, pyrite is mined for the copper or small amount of gold that can be recovered when it is processed. Before a blowpipe it fuses on charcoal to a black magnetic bead and gives off abundant sulfur, which burns, producing the suffocating fumes of sulfur dioxide.

Occurrence. Pyrite is such a common mineral that its presence is almost universal in veins carrying metallic sulfides, in some of which it is the most abundant mineral. It is found also in crystals in schists and in concretions in coal. The most important deposits of pyrite are large granular masses, such as are mined at Rio Tinto and elsewhere in Spain. Similar large bedded deposits of pyrite accompanied by pyrrhotite and some chalcopyrite occur at Ducktown, Tennessee. Massive pyrite along with chalcopyrite and bornite occur in skarn at Clifton, Arizona. Fine crystals come from talc-chlorite schists at Chester, Vermont and occur with hematite in skarn on the island of Elba.

Cobalite, CoAsS; Skutterudite, CoAs₃

Cobalt is a rare element produced chiefly from cobaltite and skutterudite. The mineral *erythrite,* or cobalt bloom, is a bright rose-red arsenate of cobalt that forms as an oxidation product on these two minerals. The early use of cobalt was chiefly as *smalt,* a cobalt glass, with an ultramarine blue color that was ground up and used as a pigment in coloring glass and pottery. Today the principal use of cobalt is as an alloy in making permanent magnets, high-speed tool steel, and stainless steel.

Cobaltite (H = 5½, G = 6.33) is isometric and commonly is found in cubes or pyritohedrons, thus resembling pyrite; but it is distinguished from that mineral by its silver-white color and a cleavage that forms perfect rectangular blocks. The composition of CoAsS makes it a derivative of pyrite in that arsenic substitiutes for one sulfur atom; it usually contains some iron and nickel. When heated on charcoal it gives white arsenious oxide with the characteristic garlic odor.

Skutterudite (H = 5½–6, G = 6.5), like cobaltite, is isometric and silver-white in color. However, unlike cobaltite, it lacks cleavage and is rarely in crystals, but it is usually massive. Skutterudite is CoAs₃, but it frequently contains some nickel and sulfur. Arsenic deficient varities are called smaltites.

Occurrence. Cobaltite and skutterudite occur together usually associated with nickeline, native silver, bismuth, and arsenopyrite in veins with calcite as in Saxony, Germany and at Cobalt, Ontario. Fine crystals of skutterudite have been found with nickeline and chalcopyrite in calcite veins in serpentinite at Bou Azzer, Morocco. Today, much of the world's cobalt comes from exhalitive sedimentary copper deposits in eastern Congo and Zambia, where it occurs as the mineral *carrolite,* Co₂CuS₄, with chalcopyrite and bornite.

Marcasite, FeS₂

Marcasite, or *white iron pyrites,* as it is sometimes called, has the same chemical composition as the more common mineral pyrite. It differs from it, however, in its crystal form and physical properties. The two are consequently distinct minerals, and the compound FeS₂ is said to be *dimorphous* (see p. 101).

Fig. 7-33. Marcasite.

Habit. Marcasite is orthorhombic and is commonly in tabular crystals flattened parallel to the side pinacoid. It is often twinned, and to the grouping that results the fanciful names of *spear pyrites* (Fig. 7-33) and *cockscomb pyrites* (Fig. 7-34) have been given. The crystal habit of marcasite is so characteristic that it is usually easy to distinguish the crystals from the cubes and pyritohedrons of pyrite. It is also found in nodules, in spherical masses, and as stalactites.

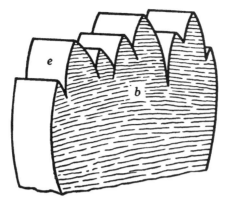

Fig. 7-34. Marcasite.

Physical Properties. $H = 6–6\frac{1}{2}$, $G = 4.9$. The hardness is about the same as that of pyrite, but the specific gravity is slightly lower, 4.9 instead of 5.02. Unless the specimen is very pure, this difference may not be detected easily, but a distinction can usually be made on the basis of color. The name *white iron pyrites* has been given to marcasite, because on fresh surfaces it is a paler yellow than pyrite, a difference that becomes quite apparent after comparing a few specimens. It tarnishes easily, however, and one should be certain that a fresh surface is being examined.

Composition. Marcasite is iron disulfide, FeS_2, the same as pyrite, but it tends to disintegrate more easily, and specimens usually fall to pieces after a short exposure to the air, so that special precautions are necessary for both storage and display. Specimens that appear to be stable are usually pyrite pseudomorphs after marcasite.

Occurrence. Marcasite is much less common than pyrite and forms under a much narrower range of conditions. The exact conditions required for its formation are poorly known, but it is believed to be a near-surface, low-temperature mineral which forms under more acid conditions than pyrite. Marcasite is commonly found with galena and sphalerite in veins with chalcedony, calcite, or dolomite, as in the Mississippi Valley deposits. It occurs in the supergene enrichment zone of copper deposits with chalcocite and covellite, and as concretions in clays, shales, and coal beds and replacements of fossils in limestones.

Arsenopyrite, FeAsS

Like pyrite, arsenopyrite contains iron and sulfur, but as it contains arsenic as well, it was formerly called *arsenical pyrites. Mispickel* is another name that has been applied to it. It is the most common arsenic-bearing mineral, and thus is the chief ore of that element, although much arsenic is obtained as a by-product in the smelting of copper ores. Arsenic is chiefly used to make the white arsenious oxide, which is used as a poison, a preservative, and a pigment.

Habit. Arsenopyrite is monoclinic, but the small crystals in which it is sometimes found have a definite orthorhombic symmetry (Figs. 7-35 to 7-37) produced by twinning. Another type of twinning produces groups that strongly resemble those of marcasite. It occurs most commonly massive.

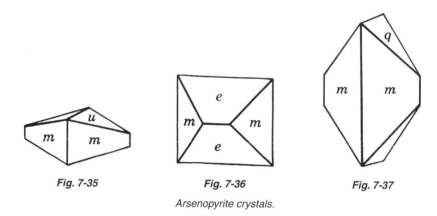

| Fig. 7-35 | Fig. 7-36 | Fig. 7-37 |

Arsenopyrite crystals.

Physical Properties. H = 5½–6, G = 6.1. The color is silver-white when fresh, becoming a little dull and tarnished after exposure. It is much the commonest of the silver-white minerals, and its color readily distinguishes it from the bronze of pyrrhotite or the pale yellow of pyrite and marcasite. The luster is metallic, and the streak is grayish black.

Composition. Iron arsenide sulfide, $FeAsS$, is a derivative of marcasite in which arsenic replaces one of the sulfur atoms. When cobalt takes the place of part of the iron, the name *danaite* has been given to the modification. Before a blowpipe it fuses on charcoal to a magnetic globule and gives a coating of white arsenious oxide and the characteristic garlic odor. If struck with steel, it gives sparks with a garlic odor.

Occurrence. Arsenopyrite is a widespread mineral formed under varying conditions of deposition. It is most commonly encountered in gold-quartz veins such as those at Lead, South Dakota; Dahlonega, Georgia; Mother Lode of California; Deloro, Ontario; and Ashanti district of Ghana. It has been found in quartz veins with cassiterite at Cornwall, England and in calcite veins with native silver and cobalt minerals at Cobalt, Ontario.

Molybdenite, MoS_2

Molybdenite is an important mineral, for it is the only commercial source of the relatively rare element molybdenum. Today, this element has an important place as an alloy in high-speed tool steels. It is used with tungsten to increase the toughness of steel and to permit steel to maintain its cutting edge at a high temperature.

Habit. Molybdenite is hexagonal and is rarely found in distinct hexagonal crystals. It is usually in foliated masses or scales similar to graphite.

Physical Properties. H = 1–1½, G = 4.7. In most of its physical properties, molybdenite resembles graphite, with which it is easily confused. It has a perfect basal platy cleavage. It is very soft with a soapy feel and leaves a trace on paper. However, the specific gravity of 4.7 is higher than that of graphite. It is sectile with a metallic luster. The streak is black when observed in the ordinary manner, but on glazed porcelain it is greenish, whereas that of graphite is black. Although its color is black, as in graphite, with experience it can usually be distinguished from graphite by its bluish tone.

Composition. Molybdenite is molybdenum sulfide, MoS_2. Heated on charcoal it gives off strong sulfur fumes and yields a deposit of molybdic oxide, which is pale yellow or white. This test further distinguishes it from graphite, for graphite gives no reaction on charcoal.

Occurrence. Molybdenite is not a common mineral but is found in many localities as a minor constituent in granites and pegmatites intergrown with feldspar and quartz. It is often found with scheelite and wolframite in high-tempertature veins with quartz, topaz, or fluorite; also, with scheelite and chalcopyrite embedded in epidote, diopside and other silicate minerals formed in contact metamorphic limestones. Most of the world's supply comes from Climax, Colorado, where molybdenite occurs in quartz veins cutting granite and is mined on a large scale. Much molybdenite is produced as a by-product in mining porphyry copper deposits.

Calavarite, $AuTe_2$; Sylvanite, $(Au,Ag)Te_2$

Calavarite and sylvanite until late in the nineteenth century were very rare minerals, but with their discovery at Kalgoorlie, Western Australia, in 1886, and at Cripple Creek, Colorado, in 1891, they became extremely important. They have been the ore minerals at these deposits that have yielded well over $1,000,000,000 in gold. Although deposits of these minerals have largely been worked out, small quantities of these minerals continue to be recognized in many of the world's gold deposits. These minerals are usually found with native gold and pyrite in veins with quartz. Examples of this association include Kalgoorlie, Western Australia; Cripple Creek, Colorado; Nagyag, Transylvania; Calaveras (Mother Lode), California; and Kirkland Lake district, Ontario.

Calavarite (H = 2½, G = 9.35) is monoclinic but is rarely in good crystals and is usually massive. The luster is metallic, and the color is brass-yellow to silver-white. Its composition is $AuTe_2$, which yields 44% gold. It is easily fusible on charcoal with a bluish-green flame, and after the tellurium is thus driven off, a metallic globule of gold remains.

Sylvanite (H = 1½–2, G = 8.0–8.2) is monoclinic, but distinct crystals are rare. It has been called *graphic tellurium* because of the skeleton forms, resembling written characters, that the crystals sometime take on a rock surface. Unlike calavarite, it has good platy cleavage parallel to the side pinacoid.

The luster is metallic; the color, silver-white. The composition is $(Au,Ag)Te_2$, which yields 24.5% gold and 13.4% silver. Its reaction on charcoal is similar to that of calavarite, but the remaining globule is considerably lighter in color because of the presence of silver.

Pyrargyrite, Ag_3SbS_3; Proustite, Ag_3AsS_3

Pyrargyrite and proustite are relatively rare but beautiful minerals known together as the *ruby-silver ores*. They are similar in crystal form and physical properties. The compositions are analogous in that the arsenic of proustite takes the place of the antimony of pyrargyrite.

Pyrargyrite (H = 2½, G = 5.85) is rhombohedral and frequently shows well-developed faces of the prism and the rhomobohedron. It also has a blocky rhombohedral cleavage. The luster is adamantine, almost metallic in some specimens, and the color deep red to black, giving it the name *dark ruby silver*. The streak is red.

Proustite (H = 2–2½, G = 5.55) is like pyrargyrite in crystal form, cleavage, and hardness, but with a slightly lower specific gravity. The color, ruby red, brighter than pyrargyrite, gives it the name *light ruby silver*. The streak is red. Both minerals fuse easily on charcoal, pyrargyrite giving the dense white coating of antimony trioxide, and proustite giving the volatile sublimate of arsenious oxide with the garlic odor.

Occurrence. Of these two minerals, pyrargyrite is the more abundant and in some places is an important silver ore as at Potosi, Bolivia and at Guanajuato and Zacatecas in Mexico. Both are found with galena, sphalerite, argentite, tetrahedrite, and pyrite in veins of chalcedony, quartz, or calcite, such as the Comstock lode, Nevada. They were found with native silver in calcite veins at Cobalt, Ontario. Fine crystals of proustite with arsenopyrite have in the past come from calcite–barite veins at Chanarcillo, Chile.

OXIDES

Oxides	
A_2X	A_2X_3
Ice, H_2O	Hematite, Fe_2O_3
Cuprite, Cu_2O	Ilmenite, $FeTiO_3$
AX	Corundum, Al_2O_3
Zincite, ZnO	AX_2
BA_2X_4	Rutile, TiO_2
Spinel, $MgAl_2O_4$	Cassiterite, SnO_2
Magnetite, $FeFe_2O_4$	Pyrolusite, MnO_2
Franklinite, $ZnFe_2O_4$	Uraninite, UO_2
Chromite, $FeCr_2O_4$	
Chrysoberyl, $BeAl_2O_4$	

Hydroxides	
Brucite, $Mg(OH)_2$	Bauxite
Manganite, $MnO(OH)$	Gibbsite, $Al(OH)_3$
Goethite, $FeO(OH)$	Boehmite, $AlO(OH)$
Limonite, $FeO(OH)\cdot nH_2O$	Diaspore, $HAlO_2$
	Romanechite, $BaMnMn_8O_{18}(OH)_4$

In the oxide group are included several minerals of great economic importance, for the chief ores of iron, manganese, aluminum, tin, and chromium are oxides. As with the sulfides, the most convenient way to arrange the oxides is according to their ratio with oxygen, where the oxygen anion is represented by X. In the general formulas for the oxides, the metal cations, represented by A, occupy octahedral sites in which each metal ion is surrounded by six oxygen atoms. The minerals listed with spinel have in addition to the octahedral sites, tetrahedral sites designated B, in which a metal cation is surrounded by four oxygens. Thus to be consistent with the other oxides, the general formula for spinel is written BA_2X_4, with A occupying the octahedral sites and B the tetrahedral sites. The A cations have a 3+ valence, the B cations usually have a 2+ valence.

The presence of water as hydoxyl ion changes the internal atomic arrangements such that these minerals are listed separately.

Ice, H_2O

To many it may seem out of place to include ice here, but it is just as truly a mineral as diamond or quartz, even if it cannot be preserved in a mineral cabinet.

Habit. Ice occurs in crystalline forms with hexagonal symmetry, often of great complexity and beauty as snow crystals (Fig. 3-1). As discussed on page 12, these are formed in the atmosphere directly from water vapor. The ice that makes the pellets of hail, not infrequently occurring with summer thunderstorms, are also occasionally in clusters of crystals, although usually they show a concentric concretionary structure. The ice of pools and ponds is crystalline, although usually the crystals are separately visible only in the first stages of the process of freezing. The process of crystallization begins, as commonly known, when the temperature falls to 0°C (32°F). As many as six polymorphs of ice have been described, but none of these is observed in nature.

Physical Properties. H = 1½, G = 0.92. The hardness of ice near the freezing point is 1½, but it increases slightly at lower temperatures. The specific grvity is 0.92 and hence ice floats in water with a little more than nine-tenths of its bulk submerged. Water expands on freezing and can exert a great force on confining surfaces. One consequence is the breaking of vessels, water pipes, and so on, when the water they contain is frozen. In nature, ice is, for this reason, a powerful agent in breaking up rock masses; the water seeps into the cracks, especially narrow ones, and when it freezes the rock masses are slowly but surely wedged apart. In some locations this is a permanent condition (permafrost).

Composition. Water and ice consist chemically of hydrogen and oxygen, H_2O.

Occurrence. Ice forms as continuous and discontinuous layers by the crystallization of water (freezing) on the surfaces of ponds and lakes as well as rivers and streams whenever atmospheric temperatures fall below 0°C for any length of time. Sustained lower temperatures will result in ice forming on the surfaces of bays and fjords and even the surface of the ocean, as around the Antarctic continent and across the Arctic Ocean. Ice can be deposited directly as a sublimate (frost) on exposed surfaces, as smoke in the form of a powder, and closer to the melting point as crystalline flakes (snow). If droplets of liquid water (rain) pass through cold air, these may turn to ice before hitting the ground (sleet) or, alternatively, if droplets of rain are carried by strong updrafts to great heights where cold air is encountered, concretions of ice (hail) may form even when surface temperatures are quite warm (e.g., 35°C). If sustained outfall of snow occurs, massive quantities of pure ice may accumulate as a permanent accumulation, such as the higher elevations of mountain ranges (glaciers) or on the island of Greenland and the Antarctic continent (icecaps), where it may be thousands of feet thick.

Cuprite, Cu_2O

Cuprite is called *ruby copper* and *red copper ore* because of the fine red color of the crystals. It has been an important ore of copper in the upper portions of deeply oxidized copper sulfide ore bodies such as Bisbee and Morenci, Arizona.

Habit. Cuprite is isometric, and the crystals often show the cube, octahedron, and dodecahedron forms or combinations of these (Figs. 7-38 to 7-40 and 3-31). In one

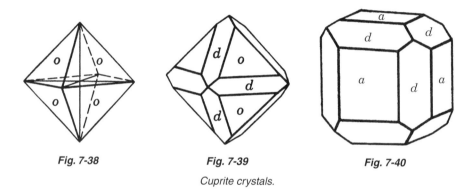

Fig. 7-38   **Fig. 7-39**   **Fig. 7-40**

Cuprite crystals.

type of crystal the cubes are spun out into long threads, forming a matted mass of bright red hairs called *chalcotrichite* or *plush copper.* When they are examined closely with a loupe, the threads are often seen to cross each other at right angles as if trying to build up skeleton cubes, the threads taking the direction of the cube edges. Common cuprite is massive, and the crystals are usually found in its cavities.

Physical Properties. H = 3½–4, G = 6.0. The luster is adamantine but on some dark surfaces may look almost metallic. The color is red of various shades—ruby red in the clear transparent crystals—but the surface is often darkened and may appear nearly black. The streak is always brownish red.

Composition. Cuprite is cuprous oxide, Cu_2O. A fragment heated on charcoal is easily robbed of its oxygen and reduced to metallic copper.

Occurrence. Cuprite is a mineral of secondary origin; that is, it has formed at or near the earth's surface by the oxidation of copper sulfide minerals such as chalcopyrite and bornite. It usually occurs just above the zone of supergene enrichment, often with native copper, azurite, and malachite in a matrix of limonite, silica, and clays as at Bisbee and Morenci, Arizona. Cuprite has been found in fine crystals at Bisbee, Arizona, and at Chessy, France, where cubes and octahedrons of malachite are found that are pseudomorphs after cuprite. Octahedral cuprite crystals coated with malachite are found in the Organja mine, Namibia.

Zincite, ZnO

Zincite is an important mineral at only one locality, Franklin, New Jersey, where it is associated with franklinite and willemite. These three are the zinc-bearing ore minerals at this famous locality, where they were mined for over 100 years.

Habit. Zincite is hexagonal, but crystals are rare. It is usually massive with a platy appearance.

Physical Properties. H = 4–4½, G = 5.5–5.7. There is perfect prismatic cleavage; hence even a massive specimen shows flashing cleavage faces. The color is deep red to orange-yellow, but the streak is always orange-yellow.

Composition. Zincite is zinc oxide, ZnO. The color of zincite is probably due to manganese, which is always present in small amounts, for pure ZnO is white.

Spinel, MgAl$_2$O$_4$

Spinel is a rather rare mineral that in places is found in beautifully colored transparent crystals that are used as gems. For example, the famous gem in the British crown known as "Black Prince's" ruby is actually a red spinel.

Habit. Spinel is isometric and is characteristically found in octahedral crystals (Fig. 7-41). Twinned octahedrons (Fig. 7-42) are common, and thus the name *spinel twin* is given to this type.

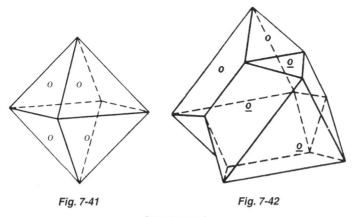

Fig. 7-41   **Fig. 7-42**

Spinel crystals.

Physical Properties. H = 8, G = 3.5–4.1. The hardness of 8 is as great as that of topaz. The color is not diagnostic, for it may be white, red, blue, green, lavender, brown, or black. The red variety has been called *ruby spinel* or *balas ruby* and should not be confused with the true ruby, corundum. It is nonmetallic with a vitreous luster and usually translucent, although the dark varieties appear almost opaque. The streak is white.

Composition. The composition of spinel proper is given above, but a more complete description is given by $(Mg,Fe,Zn,Mn,Zn)Al_2O_4$. Chromium replacing aluminum may also be present in varying amounts. Common varieties include *hercynite,* $FeAl_2O_4$, and *gahnite,* a green spinel in which zinc has taken the place of all the magnesium.

Occurrence. Spinel is a high-temperature mineral usually occurring with forsterite, diopside, phlogopite, and graphite in contact metamorphic limestones. It can also be found in aluminous schists with cordierite and hypersthene and in high-magnesia alumnia-bearing mafic-igneous rocks such as peridotites with olivine or enstatite. Because of its chemical and mechanical stability, it is frequently found as pebbles in streambeds. Thus gem spinel weathered from limestones is recovered along with gem corundum in Sri Lanka, Thailand, and Myanmar.

Magnetite, $FeFe_2O_4$

Magnetite is an important ore of iron and ranks nearly the same as hematite in importance. Its name suggests its most striking characteristic, its magnetism. All kinds are strongly attracted by a magnet, and one variety, *lodestone,* is a powerful magnet itself. It has north and south poles and the power of picking up particles of iron or steel (Fig. 4-12). When suspended it aligns itself with its poles, north and south, like a compass needle. The magnetism of lodestone excited the imagination of the early poets, and they attributed tremendous powers to it, great enough to pull nails out of ships!

Habit. Magnetite is isometric, and crystals usually are octahedral, more rarely dodecahedral (Figs. 7-43 to 7-45). It is most commonly simply massive as a granular aggregate.

Physical Properties. $H = 6$, $G = 5.2$. Although magnetite has no cleavage, it shows on some specimens an octahedral parting that will yield octahedral fragments. The luster is metallic, and on fresh surfaces can be very brilliant. Both the color and streak are iron-black, an important characteristic, for it distinguishes magnetite at once from other black minerals and from hematite, which, although at times iron-black, has a *red* streak. As previously stated, its magnetism is its most important physical property.

Composition. The formula for magnetite proper is either Fe_3O_4, or $FeFe_2O_4$ to show its relation to spinel. This yields 72.4% iron, slightly more than hematite. However, a more complete description would be $(Fe,Mg,Zn,Mn)Fe_2O_4$. Important varieties include *jacobsite,* $MnFe_2O_4$, and titanomagnetite, which occurs in certain high-temperature igneous rocks such as basalt, where Ti^{4+} has substituted for some of the Fe^{3+}.

Occurrence. Like hematite and ilmenite, magnetite is found scattered through a wide variety of igneous, metamorphic, and sedimentary rocks as an accessory mineral. It occurs as massive magmatic segregations with apatite and augite in mafic-

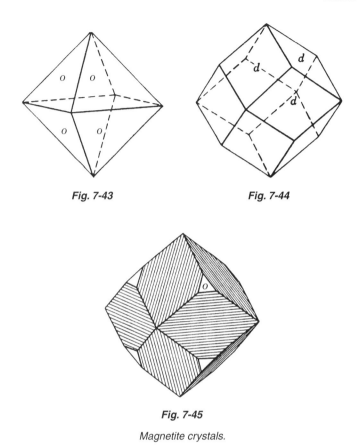

Fig. 7-43　　　　　　　**Fig. 7-44**

Fig. 7-45

Magnetite crystals.

igneous rocks as in the Adirondacks, New York, or with diopside, garnet, and olivine in contact metamorphic limestones at Iron Springs, Utah, and Prince of Wales Island, Alaska. The world's greatest magnetite deposits are at Kiruna and Gillivare in northern Sweden, where it is associated with apatite in sedimentary layers. The Precambrian iron formations of the Lake Superior region carry about 25% iron, mostly in the form of magnetite. These low-grade sedimentary deposits known as *taconite* are a major source of iron in the United States today. The magnetite, separated magnetically from waste material, is peletized to form high-grade shipping ore. In taconite lies the great iron ore reserves of the United States. Lodestone is found as masses and crystals embedded in nephelinite of the carbonatite complex at Magnet Cove, Arkansas.

Franklinite, $(Fe,Zn,Mn)(Fe,Mn)_2O_4$

Franklinite ($H = 6$, $G = 5.2$), so called from its only important locality, Franklin, New Jersey, is a mineral much resembling magnetite in form, color, and general appearance (Fig. 7-46). It is, however, only feebly magnetic, if at all, and has a brown, not a black, streak. Compositionally it is a variety of magnetite, with both zinc and manganese, and hence is valuable as a zinc ore and for making spiegeleisen, an alloy of

Fig. 7-46. *Franklinite crystals in calcite, Franklin, New Jersey.*

iron and manganese employed in the making of steel. Franklinite in collections can usually be identified by its characteristic association with willemite and zincite (see p. 173).

Chromite, FeCr₂O₄

Chromite, or *chromic iron ore,* is the only important source of chromium, an element that in recent years has been in great demand. Its chief use is in making chrome steel, an alloy not only hard and tough but also resistant to chemical attack. It is also used in making stainless steel, resistance wire in electrical equipment, and high-speed cutting tools. The most familiar use is as a material for plating. After being plated with chromium, automobile accessories, plumbing fixtures, and hardware can be polished to a brilliant surface. Because of its refractory nature, chromite can be crushed and then pressed into bricks for lining open-hearth furnaces.

Habit. Chromite is isometric and when in crystals is octahedral resembling magnetite, but crystals are rare, and it is usually massive.

Physical Properties. H = 5½, G = 4.6. The color is iron-black to brownish black; the streak, dark brown. The luster of some specimens is metallic, of others, submetallic with a pitchy appearance, an important characteristic for the sight determination of chromite.

Composition. The formula for chromite proper is given above, but a more complete description is given by $(Fe,Mg)(Cr,Al)_2O_4$. The magnesium-rich variety *magnesiochromite* is much prefered as an ore of chromium, while the iron-rich varieties are used to make the industrial alloy ferrochromium. A small grain of chromite in a borax bead imparts a green color to the bead in both the oxidizing and reducing flames.

Occurrence. Chromite is a high-temperature mineral that crystallizes with olivine, bronzite, and bytownite. It occurs as disseminated grains and as massive segregations with these minerals near the base of large mafic intrusions, such as the Bushveld igneous complex in South Africa, the Great Dike of Zimbabwe, or on a smaller scale in the Stillwater complex in Montana. It also occurs with these minerals again as disseminated grains or as magmatic segregations in the lower ultramafic cumulate portions of ophiolite sequences, as in Oman or in Pakistan. When such rocks are metamorphosed and the olivine and bronzite are converted to serpentine or to talc, the chromite is largely unaffected and the chromite may be found with serpentine, as in the Bare Hills of Maryland and Guleman, Turkey or with talc at Selukwe, Zimbabwe.

Chrysoberyl, $BeAl_2O_4$

Chrysoberyl is a rare mineral occurring in pegmatites with quartz and feldspar as at Newry, Maine and in biotite schists of the Ural mountains. Its only use is as a gem and as such it is recovered from placer deposits in Brazil and Sri Lanka. The variety known as *alexandrite* is highly valued as a gemstone. Originally from the Ural mountains it is of interest because its color changes from emerald-green in daylight to red in artificial light. *Cat's-eye* or *cymophane* mostly from Sri Lanka is a variety that, when cut as a cabochon, shows a narrow beam of light that changes position with movement of the stone.

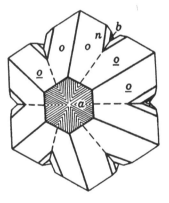

Fig. 7-47.
Chrysoberyl twin.

Habit. Chrysoberyl is orthorhombic and crystals are usually flattened parallel to the front pinacoid. Other crystals are twinned giving them a hexagonal appearance (Fig. 7-47).

Physical Properties. H = 8½, G = 3.65–3.8. Chrysoberyl has prismatic cleavage and thus has two directions of easy breaking. The hardness is 8½ and hence chrysoberyl will scratch topaz. The common type has a greenish-yellow color, slightly resembling beryl, whence it takes its name, for chrysoberyl means *golden beryl.*

Composition. Chrysoberyl is beryllium aluminate, $BeAl_2O_4$, and although it has a chemical formula analogous to the spinels, it is actually isostructural with olivine, in which Be takes the position of Si and Al takes the position of Mg.

Hematite, Fe_2O_3

Hematite is named from the Greek word for blood, because many of its varieties are red and all give a red streak. Although alloys of magnesium and aluminum are for certain purposes replacing iron and steel, we are still living in an age dominated by iron, which makes up 95% of all the metals mined. Several minerals are mined as ores, but hematite heads the list and thus may be considered commercially the most important of all the minerals. In the United States about 100 million tons of iron are produced each year, and much of this comes from hematite.

It seems unnecessary to mention the uses of iron, for they are so numerous and encountered so frequently in our everyday life that they are familiar to all.

Habit. Hematite is rhombohedral, and in certain places beautifully formed crystals showing one or more rhombohedrons are found (Fig. 7-48). Other crystals are flattened parallel to the basal pinacoid (Fig. 7-50) and may appear as extremely thin flakes (Fig. 7-49). In some specimens the plates are grouped in rosette forms (iron

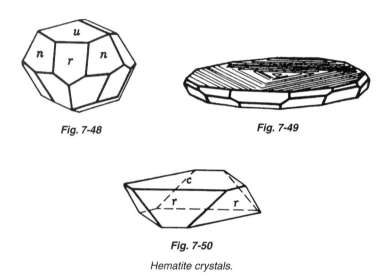

Fig. 7-48   Fig. 7-49

Fig. 7-50

Hematite crystals.

roses) as shown in Fig. 3-153. When the flakes are arranged in a foliated micaceous mass, the aggregate is called *specular hematite* or *specularite.* Radiating aggregates, *kidney ore* (Figs. 3-158 and 3-160), give reniform shapes. Hematite is often earthy and shows no crystal form. The term *martite* or *maghemite* is given to octahedral pseudomorphs of hematite after magnetite.

Physical Properties. H = 6, G = 5.26. The most outstanding characteristic of hematite in most specimens is its red color. However, crystals and the specular variety are black with a brilliant metallic luster, but the streak of these, as of the massive varieties, is red. The earthy kind, dull in luster, is the *red ocher* used for making paint.

The hardness of the crystals is about 6, and they are hence too hard to be scratched by a knife. One should be cautioned in taking the hardness of the specular varieties, for in drawing the knife blade across the specimen it is easy to separate the tiny flakes, and this separation may be mistaken for a scratch. In the red earthy varieties the hardness may be as low as 1. Similarly, the specific gravity for crystals is 5.26, but earthy varieties have much lower values.

Crystallographically hematite is similar to corundum and like it, may show, in coarsely crystallized specimens, a rhombohedral parting. The intersections of these three parting directions are nearly at right angles, and fragments can resemble minerals with cubic cleavage.

Composition. Hematite, which is isostructural with both ilmenite and corundum, has the composition ferric oxide, Fe_2O_3, and if pure contains 70% iron. When heated in the reducing flame it becomes strongly magnetic.

Occurrence. Hematite, along with goethite, is one of nature's most abundant pigments, giving a red color to soils and rocks that have been exposed to the atmosphere. Spectacular examples include the Painted Desert of Arizona and the red limestones and sandstones of the Grand Canyon. Large masses of specular ore are found where sedimentary iron formations have been metamorphosed, as in the Republic mine, Michigan. Oölitic hematite is found in extensive beds of sedimentary origin, as in the Silurian Clinton formation running from New York to Birmingham, Alabama. The chief deposits in the United States, which in the past have supplied a high percentage of the world's iron ore, are grouped around the northwestern and southern shores of Lake Superior in Minnesota, Wisconsin, and Michigan. They were formed by supergene enrichment when Proterozoic sedimentary iron formations consisting of interlayered chert, jasper, hematite, and siderite had most of the silica leached away and the siderite was oxidized to hematite, leaving massive deposits of direct shipping hematite ores.

Hematite may also occur as a rare accessory mineral with quartz and feldspar in granites and in crystalline schists. Beautiful crystals have been found in skarns on the island of Elba, while iron roses have been found in quartz veins in Switzerland (Fig. 3-153).

Ilmenite, FeTiO₃

Ilmenite, also called *titanic iron ore,* is isostructural with hematite, in which titanium takes the place of half of the iron. Tremendous masses of ilmenite that are potential iron ores are known, but because of difficulties in smelting, it is not used extensively for that purpose. It is, however, a source of titanium oxide which has now displaced lead as the dominant paint pigment. It is also an ore of titanium metal which because of its high strength-to-weight ratio, is used in aircraft and space vehicles.

Habit. Ilmenite is rhombohedral with the crystals, usually tabular, showing prominent basal planes. It is most commonly massive.

Physical Properties. H = 5½–6, G = 4.7. The luster is metallic; the color, iron-black; the streak, black to brownish red. It can be distinguished from hematite by the streak and from magnetite by its lack of magnetism. Some specimens may, however, be slightly magnetic without heating.

Composition. Ilmenite is ferrous titanate, $FeTiO_3$. Because it is isostructural with hematite, solid solutions may exist at high temperature with hematite. It becomes strongly magnetic after heating.

Occurrence. Ilmenite is widely distributed as an accessory mineral with augite and plagioclase in mafic-igneous rocks such as gabbros and diorites. Large masses occur as magmatic segregations in association with andesine-rich anorthosites, which typically consist of fine-grained ilmenite–hematite intergrowths as at Allard Lake, Quebec, or as ilmenite–magnetite intergrowths as in the Adrirondacks of New York or at Arendal, Norway.

Plates of ilmenite also are found in aluminia-rich schists with quartz, pyrophyllite, and kyanite as at Graves Mountain, Georgia. Massive crystal plates of ilmenite occur in pegmatites in the Ural Mountains. Ilmenite is also recovered from beach sands as at

Travancore, India; Queensland, Australia; and Trail Ridge, Florida. This "ilmenite" is actually preferred, as much of its iron has been leached away and its actual composition may approach that of rutile, TiO_2.

Corundum, Al_2O_3

Corundum is a mineral that has attracted attention through the centuries, for the clear blue varieites make the *sapphire* and the clear red the highly prized *ruby*. Next to diamond it is the hardest mineral, and for this reason it has long been used as an abrasive. *Emery,* a natural mixture of fine-grained corundum and magnetite, was at first the only type of corundum used as an abrasive, but later coarsely crystalline corundum was mined and pulverized to be made into grinding wheels and similar tools.

The commercial importance of natural corundum is not as great today as it has been in the past, for most of the abrasive materials, such as synthetic corundum (alundum) and silicon carbide (carborundum), are now manufactured. The gem varieties also have been made in the factory and with such success that only the expert can tell the natural ruby and sapphire from the manufactured.

Habit. Corundum is hexagonal and when in distinct crystals it usually shows either the hexagonal prism or tapering pyramids (Figs. 7-51 and 7-52). It is frequently deeply striated as the result of repeated twinning on the rhombohedron.

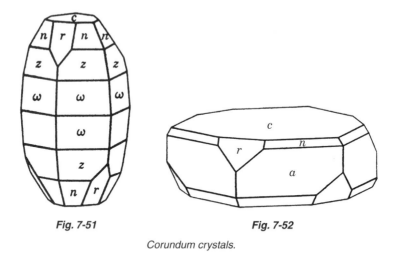

Fig. 7-51   Fig. 7-52

Corundum crystals.

Physical Properties. H = 9, G = 4.0. Corundum has no cleavage, but it may have as many as seven directions in which it breaks easily along parting planes. The most common of these are the three rhombohedral surfaces that intersect at almost right angles and yield fragments that appear cubic. The hardness is 9, and therefore it will scratch every other mineral except diamond. The specific gravity is 4.0, which is high for a nonmetallic mineral and remarkably high for the oxide of a metal of such low density. The oxide of a metal is not often more dense than the metal itself; this density is obviously related to the high hardness. As with most very hard minerals, the luster is brilliant and adamantine in clear crystals; it may be dull in some massive varieties.

The color is gray to brown in many of the common varieties, bright blue in the sapphire, and red in the ruby (see Plate II-1). Today, the name *sapphire,* when used alone, implies a blue gemstone. Corundum gems of other colors are called sapphires and are indicated by their color as yellow sapphire, green sapphire, and so on. When viewed in the direction of the c axis, some rare varieties of natural corundum have a stellate opalescence and when cut into gems are the highly prized *star sapphire* and *star ruby.* As already stated, emery is a black natural mixture of corundum and magnetite, frequently so fine-grained as to appear homogeneous. It can be identified by powdering a small sample and removing the magnetite grains with a magnet, leaving behind the lighter-colored corundum.

Composition. Corundum is Al_2O_3. Small amounts of chromium give the red color of ruby, while small amounts of iron plus titanium give the blue color of sapphire.

Occurrence. Common corundum occurs with feldspar and mica in high-grade schists and gneisses formed by the metamorphism of alumina-rich sedimentary rocks as in Madagascar or in South Carolina. It also occurs where pegmatites have intruded silica-deficient rocks such as dunites in North Carolina and Georgia, gabbros as in South Africa, syenite as in India, or limestone as in Myanmar. It occurs with feldspar and nepheline in igneous rocks which are silica deficient, such as syenites and nepheline syenites as at Craigmont, Ontario. Emery is typically found in contact-metamophic limestones but occurs as bands witn magnetite in aluminous schists at Chester, Massachusetts. Emery originally came from Cape Emeri, Greece, but commerical production is from Naxos, Greece, where it occurs as lenticular masses in limestone.

Corundum is highly resistant to weathering, and much of the world's supply of rubies and sapphires comes from placer deposits in Myanmar, Thailand, and Sri Lanka (see Plate II-4). In the United States, rubies and sapphires have been recovered in Montana, from placer deposits of the Missouri River, and sapphires have been mined from an igneous lamproite dike composed of pyroxene, biotite, and analcime cutting limestones at Yogo Gulch.

Rutile, TiO_2

TiO_2 exists in three polymorphic forms, of which rutile is the most common. The others are *anatase,* which is tetragonal, and *brookite,* which is orthorhombic. Paramorphs of one after the other are common, and it is difficult for the beginner to determine which mineral is being examined.

Rutile, together with ilmenite and titanite, is a source of titanium for paint pigment, for electrodes in arc lights, and for coloring procelain and false teeth. Rutile alone is used for coating welding rods for welding steel. Synthetic rutile is nearly colorless and can be fashioned into very attractive gemstones.

Habit. Rutile is tetragonal and is commonly found in crystals (Fig. 2-3), the forms of which are shown in Fig. 7-53. It is frequently present in elbow twins (Fig. 7-54 and

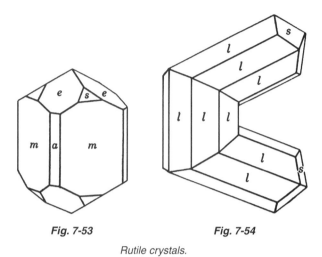

Fig. 7-53   **Fig. 7-54**

Rutile crystals.

Plate II-2). Crystals are sometimes slender, and a network of them may penetrate quartz crystals to form specimens of great beauty.

Physical Properties. H = 6–6½, G = 4.2 The color varies from reddish brown to red; some specimens may be nearly black, but even they let a little reddish light through a thin splinter. The luster is adamantine to submetallic; the streak, pale brown.

Composition. Rutile is titanium dioxide, TiO_2. A small amount of iron may be present. Although rutile is isomorphous with both cassiterite and pyrolusite, virtually no solid solution of tin or manganese occurs under natural conditions.

Occurrence. Rutile is a common, though minor constituent of many rocks, such as granite, granite pegmatite, gneiss, and schist. It is both mechanically stable and chemically inert; hence, on the disintegration of rocks, it may be washed many miles, where it accumulates as a constituent of black sands associated with magnetite, zircon, and monazite. Such accumulations in beach sands in eastern Australia make that country the world's largest producer of rutile. It is mined from apatite veins in Norway, and disseminated rutile has been extracted from syenites of the carbonatite complex at Magnet Cove, Arkansas, and at Amherst and Nelson counties, Virginia where it occurs in dikes of ilmenite and apatite associated with an anorthosite–gneiss complex. Large fine crystals have come from Graves Mountain, Georgia, where it occurs with kyanite, pyrophyllite, and quartz in metamorphic schists.

Cassiterite, SnO_2

Cassiterite or *tinstone* is almost the sole source of tin. The only other ore mineral of tin is the rare sulfide of tin, copper, and iron called *stannite.* Tin is in great demand, and the world's supply appears to be limited. Its chief traditional use was in the manufacture of pewter and *tin plate* to be made into cans for food containers. It is also used with lead in solder, with antimony and copper in babbitt metal, and with copper in bronze.

Habit. Cassiterite is tetragonal and commonly forms in prisms and pyramids (Fig. 7-55). It is found frequently as twins (Fig. 7-56) but is usually massive. Some cassiterite shows a reniform shape with radiating fibrous appearance and is called *wood tin.*

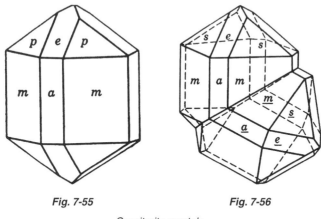

Fig. 7-55   Fig. 7-56

Cassiterite crystals.

Physical Properties. H = 6–7, G = 7. Cassiterite is remarkable for its high hardness and still more for its high specific gravity, which is unusually high for a nonmetallic mineral. The luster is adamantine to submetallic; the color is usually brown to black, more rarely yellow or white. The streak is white.

Composition. Cassiterite is tin dioxide, SnO_2. Pure tin dioxide is white; it is the presence of small amounts of iron that makes the mineral dark. Cassiterite can usually be recognized by its physical properties, but if one is uncertain, the following test can be used. Drop a fragment of the mineral and a piece of metallic zinc into a test tube with a little 50% hydrochloric acid and warm it gently. If the mineral is cassiterite, the specimen will become coated with a dull gray deposit of metallic tin.

Occurrence. Cassiterite is usually closely associated with granitic rocks, where it may occur not only in quartz veins and in pegmatites, but as disseminated grains with quartz and feldspar in the granite itself. These tiny grains can be separated by the same process that nature has used in making stream tin; that is, after the rock has been crushed, the lighter material can be washed away, leaving behind the heavy cassiterite. Indeed, most of the world's tin is obtained by washing sand and gravel that contain cassiterite, known as *stream tin,* washed in from a nearby primary source. The deposits in Malaysia, Indonesia, and Thailand, from which most of our tin comes, are of this type.

Cassiterite is often a minor constitutent of lithium-rich pegmatites, where it occurs with quartz, feldspar, and muscovite and because of its extreme resistance to weathering has been an important guide mineral for the discovery of such pegmatites as those in the Black Hills of South Dakota; King Mountain, North Carolina; and Bikita in Zimbabwe.

Cassiterite also occurs in quartz veins in Malaysia, Bolivia, and in Cornwall, England. These veins, which frequently show distinct zoning, are closely related to adjacent granite, in which the feldspar is strongly altered to muscovite, (*greisen*). At Cornwall the veins closest to the granite carry cassiterite, wolframite, and pyrite. Farther away from the intrusion the cassiterite gives way to chalcopyrite, and still farther away, galena is the principal ore mineral.

Pyrolusite, MnO_2

Pyrolusite is an important mineral, for it is the chief ore of manganese, a vital element in the production of steel. It is used with iron to make spiegeleisen, which when introduced into the steel batch, removes oxygen, thus preventing oxidation of the steel. Ninety percent of the manganese produced is used in the manufacture of steel.

Because of the large amount of oxygen in pyrolusite, it is sometimes used in the laboratory as a source of that gas. The glassmaker also employs it to remove unwanted color from glass, and for this reason it takes its name from two Greek words meaning *fire* and *to wash*. It has many other uses, such as an oxidizer in the manufacture of chlorine and bromine, a drier in paints, and in electric dry cells and batteries.

Habit. Pyrolusite is tetragonal, but it is rarely in crystals; when they are found they go under the name of *polianite*. Pseudomorphs of pyrolusite after manganite are common. It is usually massive or in radiating fibers and often forms dendrites (Fig. 3-164) along cracks in rocks.

Physical Properties. H = 1–2, G = 4.75 (A hardness of 1–2 is soft enough to soil the fingers). The luster is metallic; the color and streak, iron-black. One can distinguish it from the other manganese oxides by the streak, for that of manganite is brown and that of romanechite is brownish black.

Composition. Pyrolusite is manganese dioxide, MnO_2, and perhaps a little water, although less than that in manganite and romanechite. When pyrolusite is powdered and dissolved in a bead of sodium carbonate, the bead becomes bluish green, indicating manganese.

Occurrence. Like the other manganese oxides, pyrolusite is usually a secondary mineral formed by the oxidation of manganese-rich minerals such as rhodochrosite, rhodonite, and spessartine. Examples of this include alteration of the spessartine in the "gondite" schist at Madhya Pradesh, India; both spessertine and rhodochrosite lenses in crystalline rocks at Serra do Navio, Brazil; and with the hematite ore at Negaunee, Michigan. However, it is thought that the Tertiary age oölitic pyrolusite deposits of Nikopol, Ukraine were primary sedimentation.

Uraninite, UO_2

Prior to August 1945, uraninite would have been mentioned as a rather rare mineral mined as an ore of uranium and of the very rare element radium. Since that date, however, uraninite has taken on peculiar significance, for it was the chief raw material used in the construction of the atomic bomb and for the fuel used in nuclear reactors in the production of electrical power. Until the advent of the nuclear age, interest in uraninite lay in the fact that it was the source of radium, which was in considerable demand as a phosphorescent. Its extraction was a slow and costly process, for about 750 tons of ore must be mined to yield about 1 gram of radium. Today, large quantities of uraninite and other uranium minerals are mined for the production of uranium metal enriched in the fissionable isotope U-235 for use in nuclear reactors. Such reactors cannot only be used to produce electrical power but to produce the fissionable isotope U-233 from thorium and plutonium-239 from uranium as well as fusionable tritium.

Habit. Uraninite is isometric, and crystals, though rare, show faces of the octahedron and the dodecahedron. It is usually massive.

Physical Properties. H = 5½, G = 9–9.7. The specific gravity is unusually high but what one would expect, for uranium is one of the heaviest metals (G = 18.86). The luster is pitchlike, and because of this the mineral is commonly called *pitchblende.* The color is black, and the streak is brownish black.

Composition. Uraninite is essentially UO_2 and UO_3, and although it has a formula analogous with rutile, it is actually isostructural with fluorite, in which U takes the position of calcium and O takes the position of F. Uraninite always contains small amounts of thorium and other rare elements, and small quantities of radium, lead, and helium are present as the result of radioactive disintegration. Although uranium itself is only weakly radioactive, some of its disintegration products are strongly so and are readily detected with a scintillation counter.

Occurrence. With the great importance that uraninite has assumed since 1945, many new localities have been discovered and exploited. Even in the nineteenth century, small amounts of uraninite had been found at many localities associated both with pegmatites and in veins. Thus in the United States, isolated crystals were known from the pegmatites of New England and North Carolina, while in England it was found at Cornwall in quartz veins with cassiterite, chalcopyrite, and arsenopyrite. The three most important sources prior to 1945 were from the following deposits: a breccia pipe in dolomite at Shinkolobwe (eastern Congo), in quartz–dolomite veins with acanthite at Joachimsthal, Czechoslovakia, and in quartz–calcite veins with pyrite, chalcopyrite, and native silver at Great Bear Lake, Canada.

Most of the current production of uranium comes from sedimentary deposits. These include paleoplacer deposits of the Archean, where uraninite and gold occur as detrital grains in quartz conglomerates as in the Witwatersrand, South Africa and at Blind River, Canada. Uraninite has also been precipitated from groundwaters along with pyrite and elevated amounts of selenium, vanadium, and copper as a cement in porous sandstones at oxidation–reduction fronts. Such deposits are known from Wyoming, the Colorado Plateau, the Texas Gulf coast, and the Athabasca formation in Canada. The presence of uranium is usually detected first by its brightly colored alteration products. The commonest of these in pegmatites are yellow *autunite* and *uranophane,* and orange *gummite,* a mixture of several minerals. Yellow *carnotite,* a uranium vanadate, is found as a cement in porous sandstones of the Colorado Plateau and can be especially rich where it has replaced buried logs in ancient stream channels.

Hydroxides

Brucite, Mg(OH)₂

Until recent years brucite has been of interest to the mineralogist only and has had no particular commercial significance. It is, however, precipiated in large quantities from seawater by the addition of calcined dolomite. Both precipitated and mined brucite is heated to produce MgO for use as refractory bricks for furnace linings.

Habit. Brucite is rhombohedral, but rhombohedrons are not prominent crystal forms, and crystals are usually tabular parallel to the basal pinacoid. It is commonly in foliated or massive aggregates.

Physical Properties. $H = 2\frac{1}{2}$, $G = 2.4$. There is perfect platy cleavage which somewhat resembles that of mica, but the folia are not elastic and, when bent, will not return to their initial position as in mica. The luster on the base or cleavage face is pearly. Elsewhere or on a massive specimen it is waxy. The color is white, pale gray, or light green. Brucite is sectile.

Composition. Brucite is magnesium hydroxide, $Mg(OH)_2$. Although infusible, brucite becomes chalky white when heated in a blowpipe flame. It gives abundant water in the closed tube.

Occurrence. Brucite is commonly a secondary mineral formed by the decomposition of magnesium-bearing rocks, usually serpentine, and is precipitated along joints and fractures with magnesite and hydromagnesite, such as at Low Chrome mine, Rock Springs, Pennsylvania and Krubat, Austria. It also occurs as an alteration product of periclase, MgO, which has been formed in contact metamorphosed dolomite or magnesite as at Crestmore, California; the Gabbs mine, Nevada; or Predazzo, Austria.

Manganite, MnO(OH)

Like pyrolusite, manganite is a manganese oxide usually formed by secondary processes. The two minerals are thus found together and with romanechite. Manganite can be considered a minor ore of manganese.

Habit. Manganite is monoclinic (pseudo-orthorhombic), and crystals of it are common in brilliant prisms (Fig. 7-57) and in fibrous, radiated masses. One should not be hasty about identifying manganite by crystal form alone, for it commonly alters to pyrolusite.

Physical Properties. $H = 4$, $G = 4.3$. There is a perfect platy cleavage parallel to the side pinacoid. The color is steel-gray to iron-black; the streak, dark brown. It is by means of the streak that one is able to distinguish manganite from pyrolusite, which gives a black streak.

Composition. Manganite is a basic manganese oxide, MnO(OH). If manganite is powdered and dissolved in a sodium carbonate bead, the fusion is bluish green, indicating manganese.

Occurrence. As noted above, manganite is associated with the other manganese oxides, pyrolusite and romanechite, which may either be primary sedimentary as at

Fig. 7-57. *Manganite crystals, Ilfeld, Germany.*

Nikopol, Ukraine or as in central India may be of secondary origin; that is, they have formed by the weathering and oxidation of earlier manganese minerals, such as rhodochrosite or rhodonite. It may also be found in veins with barite or calcite, as at Ilfeld, Germany or Cornwall, England. Manganite frequently is found altered to pyrolusite, and hence one must test the hardness and the streak rather than rely on crystal form alone for identification.

Goethite, FeO(OH); Limonite, FeO(OH)·nH₂O

Goethite and limonite can be dealt with together, for most of their properties are similar. For a long time goethite was considered the minor of the two minerals. Today, however, it is known that goethite is the common mineral and limonite is relatively rare. The distinction between the two minerals is made almost entirely on the basis of crystal structure. If the substance is crystalline, even if an x-ray study is necessary to determine it, it is goethite. Only if it is amorphous, showing no crystal structure at all, is it called limonite. The name *limonite,* however, is frequently used as a field term for any rock composed of brown iron oxide that does not have the distinct habit of goethite.

At certain localities goethite is an important iron ore. It is the principal constituent of the iron ores of Alsace-Lorraine. In the Mayari and Moa districts of Cuba, minable deposits of goethite have been formed by the alteration of iron-rich rocks.

Habit. Goethite is orthorhombic. Crystals are rare but when found are usually prismatic with vertical striations. It is usually massive, reniform, or stalactic (Fig. 3-162) and commonly has a radiating habit (Fig. 7-58 and Plate II-3). Limonite is amorphous with no crystal structure.

Physical Properties. H = 5–5½, G = 4.37. There is perfect platy cleavage parallel to the side pinacoid. In the fibrous aggregates the cleavage can be seen parallel to the length of the fibers. Goethite is frequently impure with a specific gravity as low as 3.3.

Fig. 7-58. Radiating goethite, Negaunee, Michigan.

The luster is adamantine to dull. The color is yellowish brown to dark brown; the streak, yellowish brown. Both goethite and limonite are characterized by a yellowish-brown streak.

Composition. Goethite is FeO(OH), which gives 62.9% iron, a smaller percentage than in hematite and magnetite. Limonite is of similar composition but has additional water in indefinite amounts; it may also contain manganese oxides and clay. Both minerals become strongly magnetic after heating.

Occurrence. Goethite, along with hematite, is one of nature's most abundant pigments, giving a yellow-brown color to soils and rocks that have been exposed to the atmosphere. Spectacular examples include the Painted Desert and Grand Canyon of Arizona. Goethite forms under a wide variety of conditions and is usually present where iron minerals such as pyrite and siderite have been oxidized by the processes of weathering. Thus it forms the iron-rich surface rock known as gossan, over ore bodies containing pyrite, while the iron ores of Cuba are largely goethite formed by the alteration of serpentinite. In some of the iron mines of the Lake Superior district, goethite is a dominant mineral. As with hematite, the goethite was formed by supergene enrichment when Proterozoic sedimentary iron formations consisting of interlayered chert, jasper, hematite, and siderite had most of the silica leached away and the siderite was oxidized to goethite, leaving massive deposits of direct shipping ore as at Superior mine, Marquette, Michigan. Similar deposits of goethite are found in the Steep Rock district of Canada. Oölitic limonite is found in extensive beds of sedimentary origin as in the Jurassic sedimentary formations of the Lorraine–Luxemborg region.

The word *limonite* comes from the Greek meaning *meadow* because it is often found in marshy places; in fact one type is called *bog iron ore*. A careful study of such material would probably show that it is mostly goethite. Deposits of bog iron ore were mined during the early days in the United States in Massachusetts, New York, Pennsylvania, and Virginia but were low grade because of the clay and other impurities present. When pyrite is exposed to atmospheric conditions, it tends to alter to limonite and thus one of the commonest types of pseudomorphs is limonite after pyrite.

Bauxite: Gibbsite, Al(OH)₃; Boehmite, γAlO(OH); Diaspore, αAlO(OH)

These three minerals are commonly found together in varying proportions with small amounts of goethite in a rock known as bauxite, whose name is taken from the important district at Baux, France. At one time mineralogists thought that bauxite was a definite mineral species; it is the ore of aluminum. The uses of aluminum are so many and so common that it is hardly worth noting them. For purposes for which it is necessary to have strength in a lightweight material aluminum has found many uses. It is used in the construction of aircraft, in automobiles, and in railroad cars. All are familiar with the cooking utensils, other household appliances, and furniture made from aluminum. It is a good electrical conductor and each year is replacing copper increasingly for transmission lines. Other uses are in aluminum foil and in paints.

Gibbsite (H = 2½–3½, G = 2.4) is monoclinic and crystals are tabular parallel to the basal pinacoid. It has perfect platy cleavage that has a pearly luster. It is softer than diaspore but when fine grained is nearly indistiguishable from boehmite.

Boehmite (H = 2½–3½, G = 3.0) is orthorhombic and crystals as in gibbsite are tabular parallel to the basal pinacoid. Similarly, it has perfect platy cleavage.

Diaspore (H = 6½–7, G = 3.4) is orthorhombic and is usually in thin crystals flatten parallel to the side pinacoid. As with gibbsite, it has perfect platy cleavage with a pearly luster. The color of all of these minerals is white but may be gray, yellowish, or greenish. Diaspore differs from gibbsite and boehmite by its hardness.

Composition. All three minerals are essentially aluminum hydroxide with varying amounts of water. Bauxite grades into iron laterite as more goethite is added. The considerable water present can be driven off by heating in a closed tube. If bauxite is heated on charcoal and then touched with a drop of cobalt nitrate and heated again, it assumes a blue color, the test for aluminum in infusible substances.

Habit. When these minerals occur together as bauxite, they usually form in pisolitic aggregates and in round concretionary grains (Fig. 7-59); they also occur in claylike masses. Each of these minerals is so finely divided that its identity can only be recognized by the use of x-ray diffraction or with the electron microscope.

Physical Properties. In the form of bauxite, the hardness varies from 1 to 3; the specific gravity, from 2 to 2.5. The luster is usually earthy; the color, white, gray, yellow, or red.

Occurrence. Gibbsite, boemhite, and diaspore generally occur as intimate intergrowths in bauxite a rock of secondary origin formed by the weathering of silica-deficient rocks such as syenite under tropical or subtropical conditions. Recently formed bauxites are composed of gibbsite but become richer in boehmite with age. In this process the silica of the feldspars is carried away and the hydrous aluminum

Fig. 7-59. *Pisolitic bauxite, Bauxite, Arkansas.*

oxides are left behind. If iron is present as in biotite, it also will remain behind, usually as limonite, which if present in large amounts, renders the material unusable as an ore of aluminum. In the United States, bauxite occurs in Arkansas from the deep weathering of nepheline syenite. Although some bauxite has been mined in the United States, most of the bauxite is imported from Guyana and Suriname. Primary deposits were formed by the deep weathering of granite and gneisses of Precambrian shield rocks, while secondary deposits were formed by deep weathering of sedimentary kaolinite deposits.

At times gibbsite may be found as a late-stage alteration of the felspars in syenite or nepheline syenites in the wall rocks adjacent to hydrothermal veins. Diaspore is a rare mineral formed usually as a decomposition product of corundum and associated with that mineral, as at Chester, Massachusetts.

Romanechite, $BaMn^{2+}Mn_8^{4+}O_{18}(OH)_4$

Romanechite is a secondary mineral usually associated with pyrolusite and manganite in a mixture that was formerly called *psilomelane.*

Habit. Romanechite is monoclinic, but crystals are extremely rare and very small. It usually is found massive or in botryoidal or stalactitic aggregates.

Physical Properties. H = 5–6, G = 3.7–4.7. It is much harder than pyrolusite or manganite. The color is black, but the streak is brownish black, thus distinguishing it from the other manganese oxides.

Composition. Romanechite, formerly psilomelane, is essentially a hydrous manganese oxide $BaMn^{2+}Mn_8^{4+}O_{18}(OH)_4$, but it contains many other elements, particularly barium. Varying amounts of magnesium, calcium, nickel, cobalt, and copper may be present. When a small fragment is fused with sodium carbonate, it gives a bluish-green bead, the test for manganese.

Occurrence. Romanechite associated with pyrolusite and manganite usually forms as a result of the weathering and oxidation of rhodochrosite, rhodonite, spessartine, and other managanese minerals. It is found with limonite as in the Dharwar schists at Madras, India. However, it is thought that the Tertiary age oölitic romanechite deposits of Tchiaturi, Georgia were primary sedimentation. It also can be precipitated with clay and organic matter in bogs and swamps.

HALIDES

The halides are compounds whose anion is one of the halogens: fluorine, chlorine, bromine, or iodine. Many of them are very rare, and only three chlorides and two fluorides are described here. As with the sulfides and oxides, the most convenient way to arrange the halides is according to their cation/anion ratios.

Halides

AX	A_mX_p
Halite, NaCl	Cryolite, Na_3AlF_6
Sylvite, KCl	AX_2
Chlorargyrite, AgCl	Fluorite, CaF_2

Halite, NaCl

Halite is familiar to us all as common salt used in cooking and on the table. Since it is essential to human life, it has been sought, traded, and fought for throughout history. Besides its culinary use, salt is used as a preservative, in fertilizers, as a weed killer, and in salting icy highways. It is also a major raw material of the chemical industry as a source of sodium and chlorine. A major use of sodium is in the manufacture of sodium carbonate (soda ash) a compound used in making soap and glass as well as in dyeing, bleaching, and refining oil. The major use of chlorine is in making hydrochloric acid, but in addition, chlorine compounds have many other uses, including bleaching materials.

Habit. Halite is isometric and crystallizes in fine clear cubic crystals. It is most abundantly found in granular, cleavable masses known as *rock salt.* Sometimes the crystals have the skeleton or hopper-shape illustrated in Fig. 7-60.

Physical Properties. H = 2½, G = 2.16. There is perfect cubic cleavage, which yields clear rectangular blocks and which is easily seen in the aggregates as well as in the crystals. It has a vitreous luster and is colorless when perfectly pure. Impurities may give it various shades of red and yellow; occasional patches of a fine deep blue are seen in the clear crystals.

Fig. 7-60. *Halite.*

Composition. Halite is sodium chloride, NaCl. The deep yellow color that it gives to a blowpipe flame is the characteristic test for sodium. Halite is one of the few important minerals that are readily soluble in water and hence give a decided taste.

Occurrence. Salt is commonly found in beds as a sedimentary rock, where it has formed by the evaporation of seawater; thus it is found with other minerals, such as gypsum and sylvite, that form in the same way. The ocean is hence the important source as well as the great storehouse of the salt of commerce. Sodium chloride is present, and in even greater concentration, in some inland seas such as the Great Salt Lake in Utah, the Dead and Caspian Seas, and many others. In some places salt is actually mined from sedimentary layers, as in Austria and Louisiana; in other locations wells are drilled to the salt bed, from which brine is pumped up and evaporated to yield the salt. Salt is produced from such wells in New York, Michigan, and Ohio. Along the Gulf coast of Texas and Louisiana and far out into the Gulf itself, many vertical pipelike bodies of salt have apparently punched their way upward from an underlying bed of salt. These *salt domes* are searched for because of the petroleum frequently associated with them.

Sylvite, KCl

Sylvite is a salt similar to halite in its crystal form, occurrence, origin, and mineral association, but it can be distinguished from halite by its more bitter taste. It is a very important mineral, for it is the chief source of potassium compounds, which are used extensively as fertilizers. It is also used as a substitute for halite in low-sodium table salt.

Habit. Sylvite is isometric, and crystals frequently show the cube and octahedron in combination. It is most commonly found in granular, cleavable masses.

Physical Properties. H = 2.0, G = 1.99. As in halite there is perfect cubic cleavage, but sylvite is distinctly more sectile (i.e., less brittle when scratched with the point of a knife). It is colorless or white when pure but may be colored various shades of blue, yellow, or red by impurities. It is more soluble in water than halite and can be distinguished by its bitter taste.

Composition. Sylvite is potassium chloride, KCl. It is isostructural with halite, but there is little solid solution because of the large difference in the size of the Na^+ and K^+ ions. It fuses easily before a blowpipe, imparting a violet color to the flame, due to potassium. If sodium is present, the violet color will be seen only if the yellow sodium flame is filtered out by means of a piece of blue glass.

Occurrence. Like halite, sylvite is formed in beds resulting from the evaporation of seawater. Inasmuch as sylvite is more soluble than halite, it is precipitated later and may crystallize above earlier-formed halite. If evaporation does not continue to almost complete dryness, no sylvite may be precipitated. It is, therefore, a much less common mineral than halite. Sylvite has been mined in large amounts from bedded sedimentary layers at Stassfurt, Germany, and near Carlsbad, New Mexico. The largest world reserves are in bedded deposits 1000 meters below the surface in Saskatchewan, Canada.

Chlorargyrite, AgCl

Chlorargyrite is named for its chemical composition. Formerly, it was called cerargyrite, which, translated from its Greek roots, means *horn silver,* a name applied to it because of its hornlike appearance and the ease with which it is cut by a knife. In certain localities it has been an important ore of silver.

Habit. Chlorargyrite is isometric, but crystals are rare. It is usually found in scales, plates, or masses resembling wax.

Physical Properties. H = 2–3, G = 5.5. It is remarkable for being perfectly sectile; that is, it can be cut with a knife like a piece of lead or wax. The luster is adamantine; the white or pale gray color darkens rapidly to violet-brown on exposure to light.

Composition. Chlorargyrite, silver chloride, AgCl, and halite are isostructural, although little solid solution exists. Before a blowpipe on charcoal it fuses very easily, yielding a globule of silver.

Occurrence. Chlorargyrite is a secondary mineral and is found in the upper, near-surface portions of silver veins such as the Comstock Lode, Nevada and at Leadville, Colorado, where it occurs with limonite, jarosite, and cerussite.

Cryolite, Na₃AlF₆

Cryolite takes its name from two Greek words that mean *ice* and *stone,* because blocks of the mineral often have the appearance of slightly clouded blocks of ice. It is easily fusible and, when in the molten state, is an excellent solvent. For this reason it is used to clean metal surfaces and to dissolve the aluminum oxides used in the electrolytic process for the production of aluminum. Most of the cryolite used today is manufactured from fluorite. During the latter part of the nineteenth century when metallic aluminum was a rarity, cryolite was the ore of that metal.

Habit. Cryolite is monoclinic, although crystals have nearly cubic angles (Fig. 7-61). It is most commonly found in massive form.

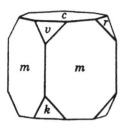

Fig. 7-61. *Cryolite.*

Physical Properties. H = 2½, G = 2.95–3.0. Cryolite has no cleavage but three directions of parting, which, unless examined carefully, may be confused with cubic cleavage. The luster is vitreous to greasy; the color, snow white.

Composition. Cryolite is sodium aluminum fluoride, Na_3AlF_6. It fuses easily in a candle flame, coloring the flame the intense yellow of sodium.

Occurrence. Cryolite is a rare mineral, and the only locality where it occurs in quantity is at Ivigtut in southwestern Greenland. Here it is found in a pegmatite with microcline, quartz, siderite, and fluorite. It has also been found in minor amounts with quartz and feldspar in pegmatite at the foot of Pikes Peak in Colorado.

Fluorite, CaF$_2$

Fluorite or *fluor spar* is one of the most beautiful minerals, occurring in well-formed crystals of many different colors. It is sometimes cut into gems, but because of its rather low hardness, the stones are not durable. The name comes from the Latin word meaning *to flow,* since it melts more easily than other gems which it resembles as cut stones.

Over 50% of the fluorite produced is used by the chemical industry chiefly in making hydrofluoric acid. Another major use of fluorite is as a flux in making steel, but it is also used in making opalescent glass and in enameling cooking utensils. Small amounts of fluorite of high purity are used in making prisms and lenses for optical equipment. Most optical fluorite is grown synthetically today.

Habit. Fluorite is isometric, and cubic crystals and groups of crystals are common (Fig. 3-2). It is found in penetration twins also, such as those shown in Fig. 3-118 or Fig. 7-62, in which the angles of one cube project from the faces of another. The position of the crystals is such that if one of them were revolved 180° about a line joining opposite angles, they would be brought into parallel position. On the edges of the cubes, pairs of narrow faces of the tetrahexahedron may be seen on some crystals (Fig. 7-63). Less commonly, six small faces of the hexoctahedron are seen on the solid angles (Fig. 7-64).

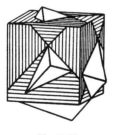

Fig. 7-62

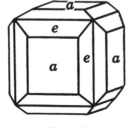

Fig. 7-63

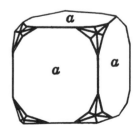

Fig. 7-64

Fluorite crystals.

Octahedral crystals are also found, but when they are examined closely it can be seen that most of them are built up of minute cubes. The cubic habit, however, is so characteristic that when this form is encountered in a nonmetallic mineral, fluorite is at once suggested to the careful mineralogist. Fluorite also occurs in fibrous or columnar aggregates and in one variety with the colors arranged in bands, known as *blue john,* is used as an ornamental stone. There are granular and closely compact varieties also.

Physical Properties. H = 4, G = 3.2. There is perfect blocky octahedral cleavage so that fine crystals must be handled carefully to prevent breaking off the corners of the cubes. Close examination of a single cleavage face will reveal triangular cleavage lines. The great variety in color, embracing many shades of green, purple, yellow, and red, has already been mentioned. It may also be colorless, dark brown, and black. Crystals are usually transparent, and some show a banding due to the deposition of successive layers of differently colored material.

Some fluorite shows a beautiful fluorescence when exposed to ultraviolet light. The phenomenon of fluorescence was first observed in fluorite and takes its name from the mineral. Some fluorite exhibits *thermoluminescence* also, the property of emitting visible light when heated. The variety *chlorophane* is so named because of the beautiful green light that it emits when heated. In some varieties the blow of a hammer is enough to make a mass yield a faint but beautiful light (triboluminescence) for hours afterward.

Composition. Fluorite is calcium fluoride, CaF_2. When powdered and warmed with sulfuric acid, it gives off hydrofluoric acid, which will etch the test tube containing it. It is an interesting experiment to cover a watchglass with a layer of wax, then with a fine point make a design by removing part of the wax. If a fluorite–sulfuric acid mixture is placed on the wax, the design will be etched into the glass. Fluorite usually flies to pieces violently when heated before a blowpipe, but when pulverized, it can be fused easily.

Occurrence. Fluorite is a widespread mineral but is usually found in veins with quartz, calcite, and sulfide minerals such as galena and sphalerite. In England, beautiful specimens have come from veins in Cumberland (Fig. 7-65) and the fibrous variety *blue john* from Derbyshire and the Freiberg mining district of Saxony. Although fluorite is not found as a primary mineral in igneous rocks, it is commonly found in a late-stage alteration of granite known as *greisen* and in derivative quartz veins with cassiterite, wolframite, and chalcopyrite as at Cornwall, England and Panasqueiria, Portugal. In a very similar manner, fluorite is sometimes found associated with the quartz cores of pegmatites. Fluorite also commonly occurs with galena and sphalerite along with marcasite and dolomite in fractures of the Mississippi Valley type deposits. In this setting, the major occurrence of fluorite in the United States is in southern Illinois in veins at Rosiclare and in flat-lying caverns at Cave-in-Rock.

CARBONATES AND BORATES

The minerals described thus far have been compounds with metallic cations and anions of single elements: sulfur, oxygen, or a halogen. All the minerals remaining to

Fig. 7-65. Fluorite, Cumberland, England. (Ward's, Rochester, New York.)

be described are compounds with complex anions: the carbonates, borates, phosphates, sulfates, and silicates.

The anionic complexes of the carbonates and borates are similar and thus are considered together. In the carbonates the small C^{4+} ion is joined to three oxygen O^{2-} ions in a regular triangular arrangement to form the anionic complex $(CO_3)^{2-}$. In the borates, the B^{3+} ion is also bonded to three O^{2-} ions in a triangular arrangement, $(BO_3)^{3-}$, but these triangular groups are linked together in complex arrangements.

Because the anionic complexes govern the structural arrangement of the cations, a classification based on the ratio of anion to cation is not suitable. Instead, the carbonates are divided into three groups: calcite, aragonite, and hydroxyl. However, the important groups are the *calcite group* and the *aragonite group*. The members within each group are isostructural and many of them show appreciable solid solution between members. The hydroxycarbonates are represented by two copper minerals: malachite and azurite. The borates form a distinctly separate group because of the polymerization of the complex borate anion.

Carbonates

Calcite group	Aragonite group
Calcite, $CaCO_3$	Aragonite, $CaCO_3$
Magnesite, $MgCO_3$	Witherite, $BaCO_3$
Siderite, $FeCO_3$	Strontianite, $SrCO_3$
Rhodochrosite, $MnCO_3$	Cerussite, $PbCO_3$
Smithsonite, $ZnCO_3$	Hydroxycarbonates
Dolomite, $CaMg(CO_3)_2$	Malachite, $Cu_2CO_3(OH)_2$
	Azurite, $Cu_3(CO_3)_2(OH)_2$

Borates

Borax, $Na_2B_4O_5(OH)_4 \cdot 8H_2O$	Ulexite, $NaCaB_5O_6(OH)_6 \cdot 5H_2O$
Kernite, $Na_2B_4O_6(OH)_2 \cdot 3H_2O$	Colemanite, $CaB_3O_4(OH)_3 \cdot H_2O$

Carbonates

Calcite Group

Calcite, $CaCO_3$

Calcite gives its name to the group of rhombohedral carbonates, which includes magnesite, siderite, rhodochrosite, and smithsonite. All these minerals are isostructural with similar crystal forms and perfect rhombohedral cleavage. The angles between the cleavage faces of the different species vary from 72 to 75°. Although of slightly different crystal structure, dolomite is included here with the calcite group. Calcite has more varieties that occur in abundance than any other mineral except quartz. Each is described under the heading "Varieties."

Habit. Calcite is rhombohedral and crystallizes in a great variety and complexity of forms (Figs. 7-66 to 7-77 and 3-80 to 3-86). The fundamental rhombohedron (Fig. 7-67), the flat negative rhombohedron (Fig. 7-66), and the scalenohedron (Fig. 7-75) are common forms (see Plate III-1). There are other rhombohedrons lengthened in the vertical direction, as shown in Figs. 7-68 and 7-69. Figure 7-70 represents the hexagonal prism; Fig. 7-73 is the same with the obtuse rhombohedron *e* of Fig. 7-66 in combination. Figure 7-76 is a scalenohedron twinned on the base. Crystals forming in scalenohedrons or acute rhombohedrons are often called *dogtooth spar.* Besides the crystals illustrated, there are many other combinations of faces, some highly complicated, that can be deciphered only by one who has a thorough knowledge of crystallography. The types of calcite that do not show crystal forms are mentioned under "Varieties."

Physical Properties. $H = 3$, $G = 2.72$. The outstanding characteristic of calcite is its perfect rhombohedral cleavage. Whatever the crystal form, a mass, whether large or

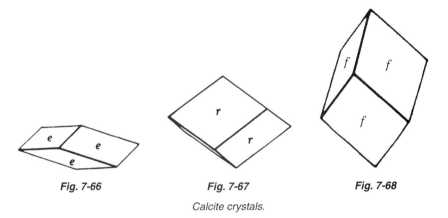

Fig. 7-66	Fig. 7-67	Fig. 7-68

Calcite crystals.

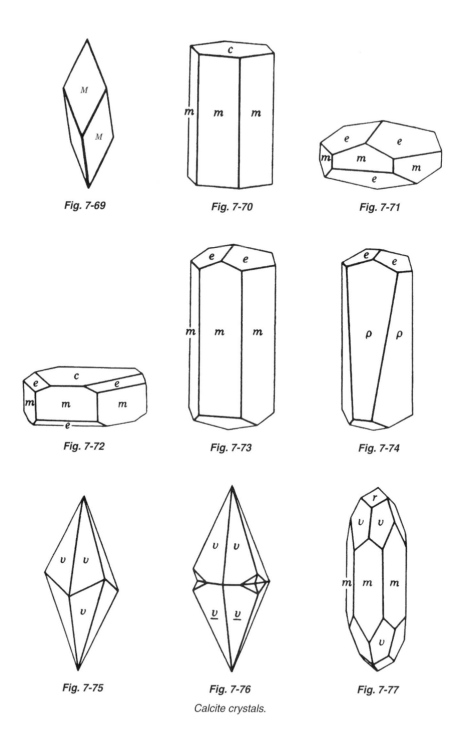

Fig. 7-69

Fig. 7-70

Fig. 7-71

Fig. 7-72

Fig. 7-73

Fig. 7-74

Fig. 7-75

Fig. 7-76

Fig. 7-77

Calcite crystals.

small, breaks easily under the blow of a hammer into fragments, all of which, as shown in Fig. 7-67, consist of rhombohedral blocks with an angle of 75° between the faces. Calcite is frequently twinned, and crystals may show a parting parallel to the twinning lamellae, that is, at the angle of the negative rhombohedron (Fig. 7-71). The twin plane in the remarkable experiment by which twinning may be imparted to a cleavage fragment is this negative rhombohedron (Fig. 3-126).

On most of its crystal faces and on the cleavage face, calcite has a hardness of 3, but on the basal pinacoid c (Fig. 7-70) it has a hardness of 2 and can be scratched by a fingernail. The luster is usually vitreous, but some varieties are dull to earthy. It is usually white or colorless but may be gray, red, yellow, green, or blue. Certain impurities may make it brown or black. One can see, therefore, that color cannot be used as a criterion for identification.

The clear colorless calcite, such as that originally brought from Iceland, called *Iceland spar*, is used for optical prisms because of its remarkable double refraction, or power of dividing a ray of light passing through it into two separate rays, so that a line seen through it appears double. This phenomenon has been described and is illustrated in Fig. 4-11.

Composition. Calcite is usually pure calcium carbonate, $CaCO_3$, but some varieties precipitated from seawater carry magnesium in solid solution. One of the best tests for calcite is to touch it with a drop of acetic acid or dilute (10%) hydrochloric acid, and note that it effervesces at once, giving off bubbles of carbon dioxide. The mineral should not be powdered, for similar-appearing dolomite will effervesce in the powdered form. When heated on charcoal, calcite does not fuse, but carbon dioxide is driven off, leaving calcium oxide. In this way, quicklime is made in the rotary kilns to be used in making mortar. The carbon dioxide driven off in the industrial process is utilized under pressure for many purposes; the most familiar is for charging carbonated drinks.

Varieties

LIMESTONE AND MARBLE. Calcite is an important rock-forming mineral, for it is almost the only constituent of the limestones that underlie thousands of square miles of the earth's surface and in places are thousands of feet thick. Through the remote geologic past as well as the present day, shellfish such as mollusks and brachiopods, as well as protozoans (foraminifera) and algae (coccolithophorids), secreted calcium carbonate from the seawater. Their shells accumulate on the seafloor and upon compaction build up great thickness of limestone. When limestones are subjected to great pressures within the earth, they may recrystallize, forming marble. Marble differs from limestone in texture, for in it one can see the cleavage faces on the coarser grains. Ordinary marble is white, but some varieties may be red, yellow, blue, or black and are then used for ornamental purposes. Very fine marble has been mined for centuries at Carrara, Italy.

Limestone finds its greatest use in the manufacture of cements and mortars and is thus the basic material of one of the great industries. It is also used as a building stone and as a flux in the smelting of iron ore. The lime used for agricultural purposes may be either quicklime or merely finely ground limestone.

Chalk, such as that found on the cliffs of Dover, England, is a fine-grained limestone made up of the tiny shells of marine organisms (coccoliths).

CAVE DEPOSITS. When water becomes charged with carbon dioxide, it has the power to dissolve limestones slowly as it works its way through them. In this way, most limestone caves and caverns have formed. After formation, conditions may change so that the water, as it drips into the cave, may evaporate and lose carbon dioxide and slowly deposit the dissolved calcium carbonate. Thus are built *stalactites,* which hang like icicles from the roof of the cavern, and *stalagmites,* which rise from the floor beneath. They are often very large and beautiful, with a great variety of shapes. These deposits are often banded and in some places occur on a large scale, so that the rock can be quarried and used for ornamental stone. *Mexican onyx* and the onyx of Pakistan is such a variety of calcite, delicate in coloring and beautifully translucent. Many caves with stalactites, stalagmites, and similar types of deposits are known in the United States. Probably the largest and most celebrated is Carlsbad Caverns, New Mexico.

SPRING DEPOSITS. Both hot and cold spring waters may deposit calcite on emerging from beneath the surface. Such deposits are known as *travertine* or *tufa.* The outstanding deposit of this type is at Mammoth Hot Springs, Yellowstone National Park, but small deposits of travertine are common in limestone regions.

SAND CRYSTALS. In certain places where water containing calcium carbonate in solution passes through sand, calcite may crystallize, incorporating the sand within it. Some crystals, such as those illustrated in Fig. 3-147, may be made up of as much as 60% quartz sand, but they still have the crystal form of calcite.

Occurrence. Calcite is such an abundant mineral that it is difficult in a limited space to list the outstanding occurrences. It commonly occurs in low-temperature hydrothermal veins intergrown with quartz, where it is a gangue mineral associated with galena, sphalerite, native silver, and acanthite. Fine specimens from such veins have been found in the copper mines of the Lake Superior region, at Derbyshire and elsewhere in England (Fig. 7-78), and in the Harz region of Germany. Low-grade metamorphism also has resulted in solutions depositing calcite in fractures and as amygdules in the vesicles of basalt, along with quartz, heulandite, and stilbite, as in the famous deposits on Iceland.

As described above, limestone rocks are formed in great quantities on the bottom of the sea and consequently found at many places throughout the world. Of special interest to the collector are openings in these rocks where crystallized forms occur as geodes or in cavities lined with beautiful crystals. The Tri-State mining district of Missouri, Kansas, and Oklahoma should be mentioned in particular for the caves of this region abound in beautiful calcite crystals, (Fig. 7-79) some of which are very large, weighing hundreds of pounds.

Magnesite, $MgCO_3$

Magnesite is a mineral that has greatly increased in importance during the twentieth century. At the end of the nineteenth century it was described as a rare mineral; by 1930 it was no longer considered rare and was used in the production of magnesium

Fig. 7-78. *Group of calcite crystals, Cumberland, England.*

Fig. 7-79. *Calcite crystal, Joplin, Missouri. (Ward's, Rochester, New York.)*

oxide for refractory bricks and insulating purposes. It was also used in the preparation of magnesium salts for medicines and paper manufacture; magnesium chloride is also recovered from the brines of salt wells at Midland, Michigan. Because carbon dioxide is given off at a lower temperature than it is from limestone, magnesite is used as a source of this gas. Along with brucite, magnesite has been an ore for magnesium, but today the entire production of the metal comes from seawater. The metal resembles aluminum but is even lighter (Al, G = 2.7; Mg, G = 1.74). Today it is alloyed primarily with aluminum, where it acts as a strengthener for use in structural aluminum for aircraft.

Habit. Magnesite is rhombohedral, but unlike calcite and dolomite, crystals are rarely seen. It usually occurs in compact masses and less frequently in granular masses showing definite cleavage.

Physical Properties. H = 3½–5, G = 3–3.2. Magnesite has perfect rhombohedral cleavage with a cleavage angle of 72°36'. With the compact massive variety having a hardness as high as 5. The luster is vitreous in crystals and dull in the massive variety. The color of pure magnesite is white, but impurities may color it gray, yellow, or brown.

Composition. Magnesite is magnesium carbonate, $MgCO_3$. It is almost insoluble in cold acid but dissolves with effervescence in hot hydrochloric acid.

Occurrence. Magnesite occurs in two distinct types, one fine-grained and compact, the other coarse and cleavable. The compact variety is found in irregular veins and masses in serpentinite, from which it has been derived through the action of groundwaters containing carbon dioxide. This type has been mined on the Island of Euboea, Greece, and in the serpentinites of the Coast Range of California. The cleavable variety occurs in sedimentary strata and appears to have formed by complete substitution of magnesium for the original calcium in limestone. The most famous deposit of this type is at Styria, Austria; in the United States large masses have been found in Stevens County, Washington.

Siderite, $FeCO_3$

Siderite, or *spathic ore* as it is frequently called, has been used as an ore of iron in certain places, as in Great Britain and Austria, but it is much less important than the three oxides.

Habit. Siderite is rhombohedral, and crystals usually show the unit rhombohedron with curved surfaces resembling the crystals of dolomite. It is usually found in cleavable aggregates but may also be concretionary in botryoidal or earthy masses, and as compact fine-grained masses.

Physical Properties. H = 3½–4, G = 3.85. Siderite has perfect rhombohedral cleavage with a cleavage angle of 73°. The luster is vitreous, and the color is light to dark brown. In the fine-grained varieties siderite can at times be mistaken for sphalerite.

Composition. Siderite is ferrous carbonate, $FeCO_3$, but both Mg and Mn may be present in solid solution. The 48.2% iron present is a much lower percentage than that in the iron oxides. When siderite is heated on charcoal, it becomes strongly magnetic and turns black. It dissolves with effervescence in hot hydrochloric acid.

Occurrence. Siderite occurs with quartz in high-temperature veins with galena, sphalerite, and tetrahedrite as at Coeur d'Alene, Idaho and Freiberg, Germany. Siderite may be formed by the action on limestone of solutions carrying iron. If replacement of calcium by iron is extensive, large bodies of ore may be formed, as at Styria, Austria or Bilbao, Spain. Concretionary siderite, known as *clay ironstone,* is often associated with coal sequences, as in Yorkshire, England and in western Pennsylvania. Extensive deposits of siderite were deposited during the Proterozoic as sedimentary iron formations consisting of interlayered chert, jasper, hematite, and siderite. Remnants of such siderite beds are found in the Steep Rock district, Ontario, where they are mined along with hematite and goethite as iron ore. It is quite likely that much of the original siderite in these banded iron formations has been converted by low-grade metamorphism to either hematite or magnetite, while much of the goethite has formed by the oxidation of siderite during weathering.

Rhodochrosite, $MnCO_3$

Habit. Rhodochrosite is rhombohedral, and crystals, although not common, are found in places in beautiful clear rhombohedrons (Plate III-2). Some crystals may show curved surfaces. It is usually in cleavable masses.

Physical Properties. H = 3½–4½, G = 3.45–3.6. Rhodochrosite has perfect rhombohedral cleavage with a cleavage angle of 73°. The luster is vitreous, and the color is a shade of rose-red. Some varieties may be very pale pink, and others, dark brown.

Composition. Rhodochrosite is manganese carbonate, $MnCO_3$. A small amount of iron may be present. Rhodochrosite dissolves in hot hydrochloric acid with effervescence. When it is fused in a sodium carbonate bead, the bead assumes a blue-green color, indicating manganese.

Occurrence. Rhodochrosite is found frequently as a gangue mineral in medium-temperature hydrothermal veins intergrown with quartz, barite, or fluorite and associated with sphalerite, galena, enargite, and tetrahedrite. In certain veins at Butte, Montana, it was so abundant that it was mined as an ore of manganese. Beautiful clear crystals of a deep rose-red color have been obtained from silver-bearing veins in Lake County, Colorado. The world's outstanding locality for rhodochrosite is at Capillites, Catamarka, Argentina. Here in an old silver mine, banded vein material is mined for ornamental purposes.

Smithsonite, $ZnCO_3$

Smithsonite is sometimes called *dry-bone ore* by miners because of the cellular texture of certain specimens that resemble dried bone. It is used as a zinc ore but is considerably subordinate to the sulfide, sphalerite. In some places smithsonite is found in translucent green or greenish-blue masses that are cut and polished for ornamental purposes.

Habit. Like the other members of the calcite group, smithsonite is rhombohedral but is rarely found in crystals. It usually forms botryoidal and stalactic masses (Fig. 7-80) or honeycombed masses, dry-bone ore.

Physical Properties. H = 5, G = 4.4. Smithsonite has perfect rhombohedral cleavage, but because of the nature of the masses in which it occurs, the cleavage is seldom seen. Both the hardness and the specific gravity are unusually high for a carbonate. The luster is vitreous, and the color is usually dirty brown but may be white, green, blue, or pink. A yellow variety known as *turkey-fat ore* contains cadmium.

Composition. Smithsonite is zinc carbonate, $ZnCO_3$, and resembles calcite in that it effervesces in cold dilute hydrochloric acid but can be distinguished from calcite by its high specific gravity.

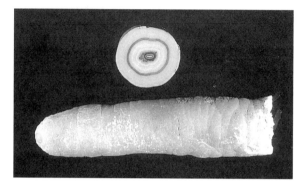

Fig. 7-80. *Smithsonite stalactite and cross section, Sardinia.*

Occurrence. Smithsonite is a secondary mineral formed by the alteration of sphalerite and is found in the upper portions of lead–zinc and copper–zinc deposits, where it is often associated with cerussite, anglesite, malachite, and pyromorphite, as at Tintic, Utah. Certain types, particularly *dry-bone ore,* are difficult to recognize, and it has not been uncommon in mining operations to overlook smithsonite as an ore mineral. Fine-grained massive specimens from Laurium, Greece, and Kelley, New Mexico, have been cut for ornamental purposes. Fine crystallized specimens have come from the Broken Hill mine, Zambia and from Tsumeb, Namibia.

Dolomite, $CaMg(CO_3)_2$

Dolomite is the rhombohedral carbonate intermediate between calcite and magnesite but with a different crystal structure. It is a common and important mineral but is less abundant than calcite in both crystalline and massive forms. Rock dolomite is used both as a building and as an ornamental stone, as well as in the manufacture of certain cements. When calcined it is used to precipitate $Mg(OH)_2$ from seawater and it has been used as a raw material for the production of metallic magnesium (see the section "Magnesite").

Habit. Dolomite is rhombohedral but crystallizes in a lower symmetry class than calcite. The unit rhombohedron is the only common crystal form (Fig. 7-81). The crystals have one peculiarity: The rhombohedron faces are almost always curved, giving a convex surface. The crystals of calcite are not curved, but those of siderite, the iron carbonate, are. Some crystal aggregates are formed of many small crystals and are so curved that they have a saddle-shaped form, as illustrated in Fig. 7-82.

Physical Properties. H = 3½–4, G = 2.85. Dolomite has perfect rhombohedral cleavage with angles so close to those of calcite that it is impossible to distinguish between them by inspection alone. It is a little harder than calcite and it is also a little denser. The luster is usually vitreous, but it is pearly in some varieties, known as *pearl spar.* The color is commonly pale pink but may be colorless, white, gray, green, brown, or black.

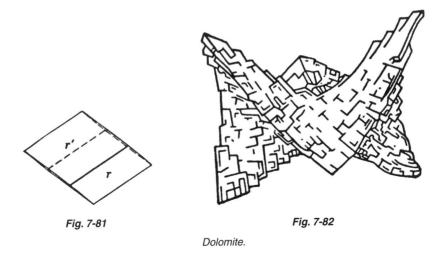

Fig. 7-81   **Fig. 7-82**

Dolomite.

Composition. Dolomite is the carbonate of calcium and magnesium, $CaMg(CO_3)_2$, with equal proportions of $CaCO_3$ and $MgCO_3$ in the pure mineral. Small amounts of iron commonly replace part of the magnesium; when considerable iron is present, the mineral is called *ankerite*. Unlike calcite, granular dolomite is not readily soluble in cold dilute (10%) hydrochloric acid or in acetic acid, but the powdered mineral will effervesce. This is the best way to distinguish between the massive varieties of the two minerals.

Occurrence. The principal occurrence of dolomite is as a rock mineral, and as such it makes up tremendous masses of dolomitic limestones and crystalline dolomitic marbles. Although dolomite can precipitate from high-salinity tropical lagoons, it is thought that most dolomitic rock has formed by magnesium either from seawater or in formation waters replacing half of the calcium in limestone. Crystals are common in veins and breccias of the Tri-State mining district centered around Joplin, Missouri. Some carbonatites, a rare igneous rock, are essentially dolomite.

Aragonite Group

Aragonite, CaCO₃

Aragonite is a polymorphic form of calcium carbonate but is much less important than calcite. It heads the list of the orthorhombic carbonates known as the *aragonite group*. This group includes aragonite, strontianite, witherite, and cerussite, all of which are closely related in their crystal and physical properties. The aragonite most commonly encountered is in pearls. When a small grain of sand or other object cannot be expelled by an oyster, it covers the grain with coatings of aragonite, which continues to grow until the oyster dies. In Japan, production of cultured pearls is a major industry. In this process, rounded beads made from clamshells, are inserted manually into oysters, which cover them with a coating of aragonite. The mother-of-pearl lining is also composed mostly of aragonite.

Habit. Aragonite is orthorhombic and frequently crystallizes in slender pointed crystals (Fig. 7-83) which are usually in radiating groups (Fig. 7-84). It also occurs in a tabular habit with a prominent side pinacoid modified by third-order and first-order prisms (Fig. 7-85). Twin crystals are common yielding a contact twin (Fig. 7-86) and a cyclic twin resembling a hexagonal prism with basal pinacoid (Fig. 7-87). In both types of twins the third-order prism is the twin plane. Such pseudohexagonal crystals are characteristic of all members of the aragonite group. Other types of aragonite also occur, such as the delicate coral-like *flos ferri* (Fig. 3-161) found in some iron mines. It also occurs in reniform and massive aggregates.

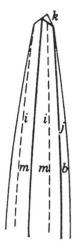

Fig. 7-83.
Aragonite crystal.

Fig. 7-84. *Aragonite crystals, Cumberland, England.*

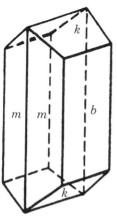

Fig. 7-85.
Aragonite crystal.

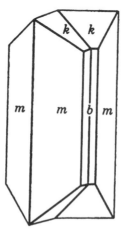

Fig. 7-86

Fig. 7-87

Aragonite twins.

Physical Properties. $H = 3\frac{1}{2}$–4, $G = 2.95$. There are two poor cleavages, one parallel to the side pinacoid and the other parallel to the third-order prism. In single crystals or columnar aggregates, therefore, all the cleavages are parallel to the length. This may be confused with columnar calcite, which also has cleavages parallel to the length of the individuals, but calcite has another cleavage plane at the end of the columns, whereas aragonite terminates in an irregular fracture.

Both hardness and specific gravity are slighter greater than those of calcite. The luster is vitreous. Aragonite may be colorless, white, or pale yellow.

Composition. Aragonite is calcium carbonate, $CaCO_3$, the same as calcite. It effervesces readily in cold dilute (10%) hydrochloric acid and gives to all other chemical tests the same reaction as calcite.

Occurrence. Aragonite is much less common than calcite and less stable once it is formed. It inverts sponatenously to calcite at temperatures above 470°C and if left in contact with water for an extended period, will slowly invert to calcite. Thus the aragonite deposited as mother-of-pearl, although common in shells, disappears rapidly and is rare in fossils. Thus pseudomorphs of calcite after aragonite are common.

Our understanding of the conditions needed for the formation of aragonite is still incomplete. It is known that aragonite is stable at higher pressures than calcite; thus in low-grade high-pressure metamorphic rocks such as in the glaucophane-bearing blue schists, aragonite is the common carbonate mineral. As aragonite is unstable at normal conditions, other special chemical conditions must be present to account for its occurrence in sinter deposits associated with hot springs and with malachite, smithsonite, and limonite in the oxidized zones of ore deposits. Mollusks secrete calcite, aragonite, or both in constructing their shells; however, the mechanism for this deposition is not fully understood.

Well-formed pseudohexagonal crystals (Fig. 7-87) are found embedded in a gypsum-rich clay at Aragon, Spain, whence the mineral derives its name. The best-developed pointed type of crystals is found in England at Alston Moor and Cumberland in veins with galena (Fig. 7-84). From the iron mines in Austria come the best specimens of flos ferri (Fig. 3-161). In the United States this variety has been found near Las Cruces, New Mexico, and at Bisbee, Arizona.

Witherite, $BaCO_3$

Witherite, barium carbonate, is a rare mineral compared to barite, barium sulfate. However, in certain places it has been abundant enough to become a minor source of barium, as at El Portal, Yosemite National Park, California.

Habit. Witherite is orthorhombic, and crystals resemble those of aragonite; that is, they are found in both acicular radiating groups and as pseudohexagonal dipyramidal twins that resemble quartz (Fig. 7-88). Massive witherite also occurs.

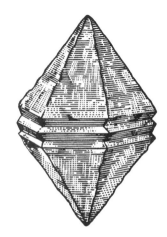

Fig. 7-88. *Witherite twin.*

Physical Properties. H = 3½, G = 4.4. As in aragonite, there is good cleavage parallel to the third-order prism and side pinacoid. Thus the cleavage fragments are parallel to the long dimension of the crystals. The specific gravity is unusually high for a nonmetallic mineral and easily detected by picking up a specimen. The luster is vitreous, and the color is white to gray.

Composition. Witherite is barium carbonate, $BaCO_3$. When a fragment is fused, it imparts a green color to the flame; this is the test for barium. Like calcite and aragonite, it is soluble in cold dilute hydrochloric acid with effervescence. It can be distinguished from these minerals by its high specific gravity.

Occurrence. Witherite occurs in low-temperature veins with barite and galena as at Cumberland and Durham, England, and Freiberg, Germany. It has also been found with fluorite in veins in dolomite at Rosiclare, Illinois.

Strontianite, $SrCO_3$

Strontianite, together with the strontium sulfate, celestite, is the source of the element strontium. Strontium has no great commercial significance but is used in fireworks and in the separation of sugar from molasses.

Habit. Strontianite is orthorhombic. Crystals are usually radiating acicular or in pseudohexagonal twins like those of aragonite. It may also be columnar, fibrous, or granular.

Physical Properties. H = 3½–4, G = 3.7. Strontianite has perfect third-order prismatic cleavage. The specific gravity is lower than witherite but still high for a nonmetallic mineral. The luster is vitreous, and the color is white, pinkish, gray, yellow, or green.

Composition. Strontianite is strontium carbonate, $SrCO_3$. A fragment will not fuse but will usually fly to pieces when heated, giving a red color to the flame. This is the best test to distinguish it from witherite, for like witherite, it is soluble in cold dilute hydrochloric acid.

Occurrence. Strontianite is a comparatively rare mineral, much less abundant than celestite. It is commonly found associated with celestite, barite, and calcite as low-temperature veins in limestone as at Westphalia, Germany, where it has been mined. It has also been found in limestones as concretionary masses, as at Clinton, New York.

Cerussite, $PbCO_3$

Habit. Cerussite is orthorhombic. Although it belongs to the aragonite group, its crystals are somewhat different from those of the other members. They are often in tabular plates (Fig. 7-89) or twinned with plates crossing each other at 60° angles (Fig. 7-90) to form reticular aggregates (Fig. 7-92). Another expression of the same type of twinning is pseudohexagonal pyramids as shown in Fig. 7-91. Cerussite is also found in granular masses.

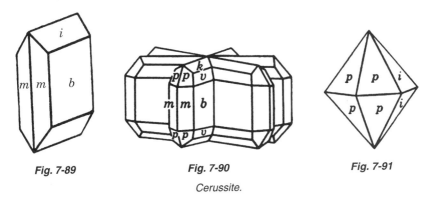

<table>
</table>

Fig. 7-89	Fig. 7-90	Fig. 7-91
	Cerussite.	

Fig. 7-92. *Cerussite, reticulated aggregate, Tsumeb, Namibia.*

Physical Properties. H = 3–3½, G = 6.55. Cerussite has third-order prismatic cleavage. The specific gravity is high for a nonmetallic mineral but to be expected in a lead compound. The luster is adamantine. Crystals are frequently colorless, but massive aggregates are usually white or gray. The clear and colorless crystals do not perhaps at first suggest to the eye that the mineral contains lead. A more careful examination, however, shows the adamantine luster possessed by crystals that strongly refract light. A high index of refraction, as this property is called, is usually found in very hard minerals (as in diamond) or in those that contain heavy atoms, such as lead. All the compounds of lead have an adamantine or resinous luster, as does the lead glass, called paste, of which imitation gems are made.

Composition. Cerussite is lead carbonate, $PbCO_3$. When mixed with sodium carbonate and heated on charcoal, it is reduced to a lead globule. Cerussite is the only member of the aragonite group that is not soluble in hydrochloric acid, but it will dissolve in warm dilute nitric acid with effervescence.

Occurrence. Next to galena, cerussite is the commonest ore of lead. The two minerals are frequently associated, for cerussite is produced in nature's laboratory by the action of carbonated water on galena. Not uncommonly, cerussite is found with a core of galena. Cerussite, associated with smithsonite and malachite, is a secondary mineral found in the upper portions of lead–zinc or lead–copper deposits, and mines that are worked for cerussite in the upper levels are usually worked for galena in depth. One of the best examples is at Broken Hill, New South Wales, Australia. In the United States, beautiful crystals came from the Stevenson–Bennett mine near Las Cruces, New Mexico. Here the cerussite was found in cavities in brecciated silicified dolomite.

Hydroxycarbonates

Malachite, $Cu_2CO_3(OH)_2$

Malachite, also known as *green copper carbonate,* is usually associated with the blue copper carbonate, azurite. Specimens containing the two minerals are frequently outstandingly beautiful because of their striking colors. Malachite has been an ore of copper, but more important today is its use for ornamental purposes. The Russians have been particularly skillful in making art objects of malachite. Thin slices are cut and used as a veneer on tabletops, vases, and pillars in churches. Most of the choice Russian malachite pieces have been brought together in the Hermitage in St. Petersburg and are on exhibit there.

Habit. Malachite is monoclinic, but crystals are rare and indistinct. It is usually in radiating fibers that form botryoidal masses (Fig. 7-93). Well-formed crystals of malachite are commonly pseudomorphs after azurite (Plate III-3).

Physical Properties. $H = 3\frac{1}{2}–4$, $G = 4$. The perfect platy cleavage of malachite is rarely seen because of the fibrous nature of the mineral. The luster is silky in fibrous varieties but vitreous in crystals. The color is a diagnostic bright green.

Fig. 7-93. Botryoidal malachite, Bisbee, Arizona.

Composition. Malachite is a basic carbonate of copper, $Cu_2CO_3(OH)_2$. Before a blowpipe, malachite fuses, giving a green flame. it dissolves with effervescence in cold hydrochloric acid and in this way can be distinguished from other green copper minerals.

Occurrence. Both malachite and azurite are copper ores of secondary origin; that is, they have formed near the surface by weathering and the action of carbonated water on the primary copper minerals, such as chalcopyrite and bornite. They are thus associated with minerals of similar origin, such as cuprite, native copper, and iron oxide. Fine specimens have come from the Ural Mountains, Russia (Fig. 3-157); Tsumeb, Namibia; Chessy, France; Broken Hill in New South Wales, Australia, and Katanga, Congo. In the United States the outstanding locality is Bisbee, Arizona.

Azurite, $Cu_3(CO_3)_2(OH)_2$

Habit. Azurite is monoclinic and, unlike malachite, is frequently found in well-formed crystals. It may also be in radiating groups.

Physical Properties. $H = 3\frac{1}{2}–4$, $G = 3.77$. The intense azure-blue color of azurite is its most striking property, but well-formed crystals may appear almost black. The luster is vitreous.

Composition. Azurite is a basic copper carbonate, $Cu_3(CO_3)_2(OH)_2$, very similar to malachite but containing slightly less water. It effervesces in cold hydrochloric acid, and other tests are the same as those for malachite.

Occurrence. It is commonly intimately associated with malachite in the highly weathered and oxidized upper portions of copper deposits.

Borates

There are over 100 borate minerals, but only four of them are abundant enough to warrant description in this book: borax, kernite, ulexite, and colemanite. Each has at one time been the chief source of the commercial products borax and boric acid.

Borax, $Na_2B_4O_5(OH)_4 \cdot 8H_2O$

The medicinal qualities of borax have been known for many centuries, and for this reason it has long been imported into Europe. The original source was Tibet. Later it was obtained by the evaporation of water from hot springs in Tuscany in northern Italy. In 1860 it was discovered in Lake County, California, and since that time, although the source has shifted from one place to another, the United States has been the chief producer. Borax is still used in medicine and as an antiseptic, but its chief use today is in borosilicate glass, washing powder, and in cleansing. It is an excellent solvent for metallic oxides and thus finds a use in certain smelting operations. Mineralogists use it in a similar manner when dissolving a mineral fragment in a borax bead to obtain the color imparted by certain elements. Boron is used in rocket fuels and as a neutron absorber in shields for atomic reactors.

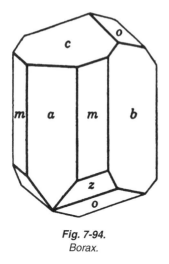

Fig. 7-94.
Borax.

Habit. Borax is monoclinic, and prismatic crystals (Fig. 7-94) have been common at certain localities. It is also found in porous masses.

Physical Properties. H = 2–2½, G = 1.7. Borax has perfect platy cleavage along the front pinacoid. The specific gravity is especially low. It is colorless in fresh crystals, but the crystals lose water and turn white on standing in a dry atmosphere. It is soluble in water and has a sweetish-alkaline taste.

Composition. Borax is hydrous sodium borate, $Na_2B_4O_5(OH)_4 \cdot 8H_2O$. It is easily fusible and imparts the strong yellow color of sodium to the flame. When moistened with sulfuric acid, it gives a green color to the flame, indicating boron. In the laboratory, colorless borax crystals lose five molecules of water and become the chalky white substance *tincalconite*, $Na_2B_4O_7(OH)_4 \cdot 3H_2O$.

Occurrence. Borax was a relatively rare mineral until the middle of the nineteenth century, when it was discovered in California. It was first found in Lake County and later in Death Valley, Inyo County. The famous twenty-mule team was used to haul borax from Death Valley, 135 miles to the railroad to the southwest. It is found dissolved in brines concentrated by evaporation, as in the waters at Searles Lake in California. It can also be found embedded in clay on the surfaces of dry lake beds (playas) with ulexite and as bedded deposits in Tertiary sediments that were ancient playas, as in Death Valley. Turkey and Argentina are also major producers (see Plate IV-1).

Kernite, $Na_2B_4O_6(OH)_2 \cdot 3H_2O$

Kernite was first described in 1926. Most new minerals that are described from time to time are usually found sparingly, some yielding barely enough material for an adequate description. Kernite was an exception to the rule, for it was found as thousands of tons and, moreover, became almost immediately a valuable commercial mineral. Today about half of the borax produced in the United States has its source in kernite. If one compares the formulas of borax and kernite, one finds that the only difference is the higher percentage of water in borax. Kernite is, therefore, a remarkable mineral, for 1 kilogram of it will yield about 1.4 kilograms of borax, the finished product!

Habit. Kernite is monoclinic and is usually found in coarse, cleavable aggregates.

Physical Properties. H = 3, G = 1.95. Perfect pseudoprismatic cleavage parallel to both the basal pinacoid and the front pinacoid gives rise to splintery fragments elongated parallel to the *b* crystallographic axis. The luster is vitreous to pearly. Kernite is usually colorless when fresh, but such specimens become coated with the chalky white mineral tincalconite on long exposure to the air.

Composition. Kernite is hydrous sodium borate, $Na_2B_4O_6(OH)_2 \cdot 3H_2O$. It fuses easily to a clear glass and like borax, is soluble in cold water.

Occurrence. The original occurrence, and at the present time the major occurrence is in the Mohave Desert at Boron, California, where it occurs nearly pure in massive quantities interbedded with borax, colemanite, and ulexite. The kernite is thought to have originated from bedded deposits of borax by partial dehydration upon burial. A similar kernite occurrence but much smaller is at Tincalayu, Salta, Argentina.

Ulexite, $NaCaB_5O_6(OH)_6 \cdot 5H_2O$

Ulexite has from time to time served as a source of borax. The boron for both borax and ulexite is thought to have originated in the waters of hot springs and fumeroles, associated with recent volcanism, and was concentrated and deposited by evaporation in enclosed desert basins.

Habit. Ulexite is triclinic. Crystals are found only as fine silky fibers, aggregates of which usually form rounded masses to which the name *cotton balls* is given.

Physical Properties. H = 1, G = 1.96. The luster is silky; the color, white.

Composition. Ulexite is a hydrous sodium and calcium borate, $NaCaB_5O_6(OH)_6 \cdot 5H_2O$. It fuses easily to clear glass, coloring the flame the deep yellow of sodium. If it is moistened with sulfuric acid and then introduced into the flame, a momentary flash of green may be seen, indicating boron.

Occurrence. Ulexite characteristically forms *cotton balls* embedded in clays on the surface of playa lakes, where it has crystallized by the complete evaporation of lake waters. In such a manner it has been found in California and Nevada, as well as in Chile and Argentina.

Colemanite, $CaB_3O_4(OH)_3 \cdot H_2O$

From 1886 when colemanite was found at Monte Blanco, on the rim of Death Valley, until the discovery of kernite in 1926, colemanite was a major source of borax, furnishing over half of the world's supply. It is still an important ore for making borosilicate glass.

Habit. Colemanite is monoclinic. Crystals are usually short prismatic. The mineral is most commonly found in cleavable masses.

Physical Properties. H = 4–4½, G = 2.42. A perfect platy cleavage parallel to the side pinacoid is colemanite's outstanding property. The luster is vitreous, particularly well seen on the cleavage face. Colemanite is colorless to white.

Composition. Colemanite is hydrous calcium borate, $CaB_3O_4(OH)_3 \cdot H_2O$. When a fragment is held in the flame, it fuses easily and crumbles, imparting a green color to the flame (boron).

Occurrence. Colemanite commonly occurs as large masses embedded in clay. Most likely these were originally beds of borax formed by evaporation which have been altered to colemanite by reaction with formation waters carrying calcium after burial. It has been mined at several locations in California, Turkey, and Kazakhstan.

PHOSPHATES

The phosphorus ion, P^{5+}, surrounds itself with four oxygens in a regular tetrahedral arrangement to form the complex anion $(PO_4)^{3-}$, the structural unit of the phosphates. Closely related to the phosphates and forming isostructural compounds with them are the arsenates and the vanadates with similar tetrahedral units as the complex anions $(AsO_4)^{3-}$ and $(VO_4)^{3-}$. Most of the arsenates and vanadates are very rare, and the only one considered here is vanadinite. Because so few phosphates are described, they are arranged in order of natural abundance.

Phosphates

Apatite, $Ca_5(PO_4)_3(F,Cl,OH)$	Amblygonite, $LiAlPO_4F$
Pyromorphite, $Pb_5(PO_4)_3Cl$	Wavellite, $Al_3(OH)_3(PO_4)_2 \cdot 5H_2O$
Vanadinite, $Pb_5(VO_4)_3Cl$	Turquois, $CuAl_6(PO_4)_4(OH)_8 \cdot 5H_2O$

Apatite, $Ca_5(PO_4)_3(F,Cl,OH)$

Although phosphorus is one of the 10 most abundant elements in the earth's crust and a large number of phosphate minerals are known, apatite is the only one that can be considered a common mineral; many of the others have resulted from its alteration or from the decay of organic matter. In certain rare instances apatite is found in beautifully colored transparent crystals and is cut as a gemstone. However, commercially much greater importance attaches to apatite as a source of phosphorus, an element necessary for plant life. It is largely from the chemical breakdown of apatite that phosphorus is contributed to the soil. Phosphorus is a constituent of most commercial fertilizers, and in some places apatite is mined as a source. Today, however, sedimentary beds known as phosphate rock or phosphorite are mined most extensively as a source of this element. After being mined, both apatite and rock phosphate are treated with sulfuric acid to make superphosphate, for in this form they are much more soluble in the dilute acids of the soil.

Habit. Apatite is hexagonal, and crystals are common, showing prism, base, and pyramids (see Figs. 7-95 to 7-97). It is also found in granular and compact masses.

Physical Properties. H = 5, G = 3.15–3.20. On some crystals a poor platy cleavage may be seen parallel to the basal pinacoid. At a hardness of 5, apatite can just be scratched by the knife. The color is usually green or brown, but it may be violet, blue, yellow, or colorless. The luster is vitreous.

Composition. Although apatite has traditionally been considered to be calcium fluophosphate, $Ca_5(PO_4)_3F$, its composition is often much more complex than this formula would suggest. There is extensive substitution of chlorine, hydroxyl, and even

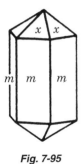

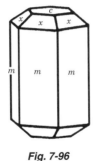

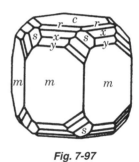

Fig. 7-95   *Fig. 7-96*   *Fig. 7-97*

Apatite crystals.

carbonate for the fluorine, and similarly, strontium, manganese, the lanthanides, uranium, and sodium for calcium.

Occurrence. Apatite is widespread and is found as an accessory mineral in tiny isolated grains in every kind of rock. In most igneous rocks it is present as a minor accessory mineral, but there are certain rare types in which apatite is a major constituent. Apatite with calcite is found in carbonatites as at Palabora, South Africa or the Alnö complex in Sweden, and one of the largest known masses occurs in granular form with nepheline and titanite on the Kola Penninsula, Russia. Gem apatite of a beautiful purple color has been found in pegmatites at Auburn, Maine. Apatite is also found in contact metamorphic limestones usually with chondrodite or scapolite, but also as crystals and masses in coarsely crystalline calcite (Plate IV-3) in Ontario, Canada, where it was mined. Apatite has been found in veins with quartz and adularia.

Phosphorus is an important element for sustaining life and is usually held as organic compounds in nucleotides and cell membranes of all living organisms. Some organisms deposit hydroxyapatite in the form of bone or teeth. When some of the phosphorus in soils is not taken up by plants, it is carried away in solution and eventually reaches the ocean, where it may be extracted by marine organisms. In certain places on the seafloor, organic remains rich in phosphorus may accumulate, building up a deposit of *rock phosphate*. Rock phosphate or *phosphorite* has as its principal constituent an amorphous material, *collophanite*. The formula for collophanite has been written as $Ca_3(PO_4)_2 \cdot H_2O$, but it usually contains varying amounts of other elements. In the case of the Eocene sedimentary deposits of Tunisia, Algeria, and Morocco, and the Permian Phosphoria formation of Wyoming and Idaho, the oölitic phosphorite is thought to have been deposited by microorganisms. However, in many places the rock phosphate is secondary, as in the "pebble" deposits of Florida and North Carolina, where concretions have been deposited by reaction of limestone with formation waters carrying phosphorus derived from the decomposition of accumulated organic matter in the sediments.

Pyromorphite, $Pb_5(PO_4)_3Cl$

Pyromorphite and vanadinite are isostructural with apatite and thus they are all closely related in both chemical composition and crystal form. However, because of the much larger size of the Pb ion to that of Ca or V, there is little solid solution between apatite and either pyromorphite or vanadinite.

Habit. Pyromorphite is hexagonal and is found in small hexagonal prisms which are frequently cavernous and may show rounded or barrel-shaped forms. The crystals may be clustered together, branching out from a slender stem, as shown in Fig. 7-98.

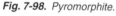

Fig. 7-98. Pyromorphite.

Fig. 7-99.
Pyromorphite crystals, Ward's, Rochester, Yew York

It occurs also as a thin crust or coating, which may be covered with tiny crystals or as globular masses.

Physical Properties. H = 3½–4, G = 6.5–7.1. Like all lead compounds, it has a high specific gravity. The luster is resinous, and the color is commonly green, varying from grass-green to both lighter and darker shades; it may also be pale brown. The streak is white even in the deep green varieties.

Composition. Pyromorphite is lead chlorophosphate, $Pb_5(PO_4)_3Cl$. When arsenic takes the place of half or more of the phosphorus, the mineral is called *mimetite*. In fact, mimetite is named from the Greek meaning *imitator* because of its close relationship to pyromorphite. Pyromorphite fuses easily on charcoal. If the fragment is examined after it is fused completely, it will be seen that it is nearly spherical and sparkles on the surfaces from the reflection of light from a multitude of facets. Hence the name of the mineral comes from the Greek words meaning *fire* and *form*. If the fused globule is heated further on charcoal with the addition of sodium carbonate, it yields a globule of metallic lead. If appreciable arsenic is present, fumes with the characteristic garlic odor will be detected.

Occurrence. Both pyromorphite and mimetite are secondary minerals formed with limonite, cerrusite, hemimorphite, and smithsonite in the upper oxidized portion of lead veins, as in the Coeur d'Alene district, Idaho or at Leadville, Colorado. In the early days of mining at Phoenixville, Pennsylvania, it was sufficiently abundant to have been mined as an ore of lead.

Vanadinite, Pb₅(VO₄)₃Cl

Vanadinite is closely related to pyromorphite in composition and form, although crystals are as a rule rarer and less distinct. Besides being a minor ore of lead, vanadinite was formerly a source of the rare element vanadium, which is now recovered along with uranium from *carnotite,* a uranyl vanadate. Vanadium is used chiefly for hardening steel, but is also used for fixing the colors in dyeing fabrics and as a yellow pigment in the form of metavanadic acid.

Habit. When crystals are sharp and clear it is one of the most beautiful minerals. Crystals are hexagonal prisms terminated by the base and more rarely by hexagonal pyramids (Fig. 7-100). Cavernous forms also occur as with pyromorphite.

Physical Properties. H = 3, G = 6.7–7.1. It is often a fine, deep red color. Less brilliant yellow and light brown varieties also occur.

Composition. The composition is Pb₅(VO₄)₃Cl; usually, small amounts of arsenic and phosphorus are present. The reactions for lead on charcoal are like those of pyromorphite.

Fig. 7-100.
Vanadinite.

Occurrence. As with pyromorphite, vanadinite is a secondary mineral formed with limonite, cerrusite, hemimorphite, and smithsonite in the upper oxidized portions of veins carrying galena, as in the Apache Mine near Globe, Arizona. It is found in fine crystals lining cavities in dolomite as at Jebel Mahser, Morocco. In a few places, such as Broken Hill, Zambia, it was mined as an ore of vanadium.

Amblygonite, LiAlPO₄F

Amblygonite is a rare mineral that in a few places is found abundantly enough to be mined as a source of lithium.

Habit. Amblygonite is triclinic, but crystals are small and rare. It is usually found in coarse cleavable masses.

Physical Properties. H = 6, G = 3.0–3.1. Amblygonite has prismatic cleavage, with a perfect cleavage parallel to the front pinacoid and a good cleavage parallel to a third-order pinacoid. The cleavages are nearly at right angles and thus resemble the cleavages of feldspar. The color is white, pale green, or blue. The luster is pearly on the base; elsewhere it is vitreous.

Composition. Amblygonite is lithium aluminum fluophosphate, LiAlPO₄F. Amblygonite fuses easily before a blowpipe, yielding a red flame. By this test one can distinguish it from feldspar, with which it may be confused.

Occurrence. Amblygonite is found in lithium-rich pegmatites such as Bikita, Zimbabwe and in the United States at Pala, California, the Black Hills, South Dakota, and several places in Maine. It occurs with microcline, lepidolite, and spodumene.

Wavellite, $Al_3(OH)_3(PO_4)_2 \cdot 5H_2O$

Habit. Wavellite is orthorhombic; crystals are rare; and it is usually in radiating globular aggregates (see Fig. 3-156).

Physical Properties. H = 3½–4, G = 2.33. Wavellite has good blocky cleavage formed parallel to the side pinacoid and *a*-axis prism. The luster is vitreous, and the color is green, yellow, white, or brown.

Composition. Wavellite is a hydrous basic aluminum phosphate, $Al_3(OH)_3(PO_4)_2 \cdot 5H_2O$. It is infusible, but when heated it swells and falls apart into fine particles.

Occurrence. Wavellite is a rare mineral usually found as crusts along joints and fractures, where it has been deposited by formation waters escaping from aluminous organic shales. Really fine specimens have been found growing in fractures in brecciated and altered novaculite in the districts around Hot Springs, Arkansas.

Turquois, $CuAl_6(PO_4)_4(OH)_8 \cdot 5H_2O$

Turquois is a mineral that has been used for ornamental purposes for centuries. The early material came from Persia and was carried through Turkey to reach Europe. The mineral derives its name from the French for *Turkish.*

Habit. Turquois is triclinic but is rarely found in minute crystals. It is usually cryptocrystalline, forming compact reniform masses. It also occurs in thin veins and as crusts.

Physical Properties. H = 6, G = 2.6–2.8. The bluish-green color is the outstanding property and the one that makes it desirable as a gem material. The luster is waxlike.

Composition. Turquois is a basic hydrous phosphate of aluminum, $CuAl_6(PO_4)_4(OH)_8 \cdot 5H_2O$. When touched with a drop of hydrochloric acid and heated, it gives a blue flame (test for copper).

Occurrence. Turquois is found in the distal portions of hydrothermal alteration zones around copper deposits, where igneous rocks such as trachyte carrying apatite and disseminated chalcopyrite are altered along fractures to kaolinite, the turquois being found as both fracture fillings or nodules in kaolin. The famous Persian deposits in the province of Khorasan, which are still producing, are of this type. In the United States it has been found around copper deposits in Arizona, Nevada, and California, but the traditional source of turquois for the Navajo indians was found in seams in brecciated trachyte just southwest of Sante Fe, New Mexico.

SULFATES

The sulfate anion is one in which the S^{6+} ion surrounds itself with four oxygens in a regular tetrahedral arrangement yielding the complex anion, $(SO_4)^{2-}$, the structural unit of all the sulfates. Although somewhat different structurally, the molybdates and tungstates are here included with the sulfates. For the cations Mo^{6+} and W^{6+} are also surrounded by four oxygens, although not in a regular tetrahedral arrangement, giving the complex anions $(MoO_4)^{2-}$ and $(WO_4)^{2-}$. The minerals included here are arranged according to abundance and crystal system.

Sulfates

Orthorhombic	Tetragonal
Barite, $BaSO_4$	Scheelite, $CaWO_4$
Celestite, $SrSO_4$	Wulfenite, $PbMoO_4$
Anglesite, $PbSO_4$	Monoclinic
Anhydrite, $CaSO_4$	Wolframite, $(Fe,Mn)WO_4$
	Gypsum, $CaSO_4 \cdot 2H_2O$

Barite, $BaSO_4$

The outstanding property of barite is its high density; because of its density it is often call *heavy spar*. As it is the most common and widespread barium mineral, it is the chief source of that element. The carbonate, witherite, is the only other important barium mineral. The principal use of barium is in the substance *lithopone*, produced by reacting a mixture of barium sulfide and zinc sulfate. This is used in paints and in the manufacture of linoleum and textiles. Most of the barite mined is ground to a fine powder and made into a sludge of high specific gravity for use in drilling deep wells.

Habit. Barite is orthorhombic. Crystals are common; the usual habit is tabular parallel to the base (Fig. 7-101). Some of the other modifying forms are shown in Figs. 7-102 to 7-105. Aggregates of divergent plates frequently form *crested barite* or

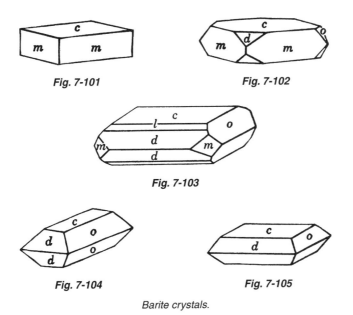

Fig. 7-101   Fig. 7-102

Fig. 7-103

Fig. 7-104   Fig. 7-105

Barite crystals.

barite roses, as shown in Fig. 7-106. Barite also occurs in massive or granular aggregates, and in such forms it is most difficult to recognize.

Fig. 7-106. *Crested barite.*

Physical Properties. H = 3–3½, G = 4.5. Barite has both basal and prismatic cleavage, yielding blocky fragments. At first glance these may resemble those of calcite, for the angles between the prism faces are close to the rhombohedron angle of calcite (Fig. 4-3). On comparing these two minerals, the student will find that the basal cleavage is at right angles to the prismatic cleavage in barite, whereas in calcite the third cleavage is not at right angles to the other two. The specific gravity of 4.5 is extremely high for a nonmetallic mineral. Barite may be colorless, white, or light shades of blue, yellow, and red. The luster is vitreous.

Composition. Barite is barium sulfate, $BaSO_4$. When heated before a blowpipe, it yields the yellowish-green flame of barium. It can be distinguished from witherite, which gives the same flame test, by its insolubility in hydrochloric acid.

Occurrence. Barite is often found as a gangue mineral, along with quartz and calcite in low-temperature veins carrying galena, chalcopyrite, and silver. In this way, beautiful crystals have come from several localities in England, chiefly Derbyshire, Westmoreland, Cornwall, and Cumberland (Fig. 7-107). It is also deposited as both replacements and fracture fillings by formation waters in limestone, such as around Athens, Kentucky, while residual masses from dissolution occur in clay overlying limestones, as in Washington County, Missouri.

Black barite, so called because of its high organic content, has been deposited in massive sedimentary layers up to 50 feet in thickness. Most likely, these are the result of exhalitive processes on the seafloor, in which hydrothermal solutions rich in barium have precipitated barite upon contact with seawater. Such sedimentary barites of Devonian age are found in Nye County, Nevada and near Magnet Cove, Arkansas,

Fig. 7-107. Barite crystals, Cumberland, England.

interbedded with chert and shale. Remobilization by formation waters yields nodules and rosettes.

Celestite, SrSO₄

Celestite is the most important source of strontium. The principal use of strontium is to desaccharize beet-sugar molasses. More familiar uses are in red pyrotechnics and in signal flares.

Habit. Celestite is orthorhombic, and the crystals resemble so closely those of barite that it is difficult to distinguish between them without careful angular measurements. Crystals are commonly tabular (Fig. 7-108) but may be elongated parallel to the *a* axis (Fig. 7-109). It also occurs in granular and fibrous aggregates.

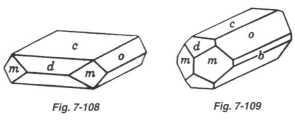

Fig. 7-108   **Fig. 7-109**

Celestite crystals.

Physical Properties. H = 3–3½, G = 3.95–3.97. Perfect cleavage parallel to the base and prism yields blocky fragments closely resembling those of barite. Although high for a nonmetallic mineral, the specific gravity is less than that of barite. The luster is vitreous, but it may be pearly on the base. The color is commonly white, but the crystals, as well as the fibrous forms, often show a tinge of blue, and although this is not an essential characteristic, the blue tinge is so common that it has given to the species the name, derived from the Latin *coelestis,* meaning celestial.

Composition. Celestite is strontium sulfate, $SrSO_4$. A fragment heated in forceps colors the flame the deep red of strontium. Celestite's insolubility in hydrochloric acid distinguishes it from the soluble strontium carbonate, strontianite.

Occurrence. Both as disseminations and as the linings of cavities, celestite has been deposited by formation waters derived from limestones and sometimes from beds of rock salt or gypsum. Much of this strontium may be released during the conversion of original aragonite to calcite. Masses of celestite with gypsum are found in Triassic marls of Gloucester and Somerset, England. A most striking occurrence is on the Island of Put-in-Bay in Lake Erie. Here one can descend into a cave that is completely lined with celestite crystals, some measuring more than a foot in length.

Anglesite, $PbSO_4$

Anglesite derives its name from the original locality on the island of Anglesey, Wales.

Habit. Anglesite is orthorhombic, and crystals resemble in form and angles those of barite and celestite but are usually of a more varied and complex development. It is found also in granular or compact masses and may show concentric banding around a core of galena.

Physical Properties. $H = 3$, $G = 6.2$–6.4. Anglesite has an imperfect cleavage parallel to the prism and basal pinacoid, yielding blocky fragments. The luster is adamantine when the specimen is crystalline, but it may be dull when the specimen is impure and earthy. The crystals are usually clear and colorless, but massive material may be gray, pale shades of yellow, or brown.

Composition. Anglesite is lead sulfate, $PbSO_4$. On charcoal before a blowpipe it decrepitates and fuses readily to a clear bead, which becomes milk-white on cooling. When heated on charcoal with soda, it yields a bead of metallic lead. It dissolves with difficulty in nitric acid and does not effervesce as does cerussite; hence the two minerals are easily distinguished.

Occurrence. Like cerussite, anglesite is a secondary mineral formed by the oxidation of galena. It is consequently found near the surface of lead veins and associated with other oxidized minerals, such as cerussite, smithsonite, and iron oxides. World-famous localities where this has occurred are Broken Hill, New South Wales, Australia; Monte Poni, Sardinia; and Otari, Namibia. Similar occurrences in the United States include Phoenixville, Pennsylvania; the Tintic district, Utah; and the Coeur d'Alene district, Idaho.

Anhydrite, $CaSO_4$

Anhydrite receives its name because although it is calcium sulfate, it does not, like gypsum, contain water. It is less common than gypsum and has little industrial use. Although it is orthorhombic as are barite, celestite, and anglesite, it has, because of the much smaller calcium ion, a different atomic arrangement. This is manifested in a completely different type of cleavage.

Habit. Anhydrite is orthorhombic, but crystals are rare and it usually occurs in cleavable masses. It may also be massive, fibrous, or granular.

Physical Properties. $H = 3$–$3\frac{1}{2}$, $G = 2.89$–2.98. Anhydrite has perfect cleavage parallel to the three pinacoids, which forms pseudocubic blocks. The color is usually white or gray but there may be a blue or red tinge.

Composition. Anhydrite is calcium sulfate, $CaSO_4$. Unlike gypsum, it contains no water.

Occurrence. Anhydrite is much less common than gypsum. It is found as a minor accessory mineral associated with orthoclase in the potassic alteration zone adjacent to magmatic intrusions. It is also one of the first minerals deposited around the vents of the deep-sea hot springs and may be found as bedded layers in exhalitive deposits with pyrite, chalcopyrite, galena, or sphalerite.

As with dolomite, anhydrite can be precipitated directly by evaporation from seawater at temperatures above 42°C. Massive bedded deposits are found interstratified with massive beds of salt, gypsum, and limestones, such as at Stassfurt, Germany. Anhydrite formed by the dehydration of gypsum is the major mineral in the cap rocks of salt domes in Texas and Louisiana.

Scheelite, $CaWO_4$

Scheelite is an ore of tungsten but is less important than wolframite. It is named after K. W. Scheele, the discoverer of tungsten.

Habit. Scheelite is tetragonal. Crystals are usually simple dipyramids (Fig. 7-110). It is also found in granular aggregates.

Physical Properties. $H = 4\frac{1}{2}$–5, $G = 5.9$–6.1. Scheelite has cleavage parallel to the second-order dipyramid, which breaks into blocks that closely resemble an octahedron. A clean cleavage surface will shown traces of cleavage lines in the form of triangles. The specific gravity is unusually high for a mineral with nonmetallic luster. The luster is adamantine, and the color is white, yellow, green, or brown. Scheelite is almost unique among minerals in that most specimens of it will fluoresce a pale blue in ultraviolet radiation. This property, therefore, can be used in prospecting for the mineral or in gaining an idea of the amount present during mining.

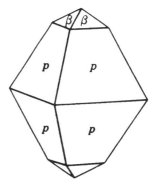

Fig. 7-110. Scheelite.

Composition. Scheelite is calcium tungstate, $CaWO_4$. It is decomposed by boiling in hydrochloric acid, yielding a yellow residue of tungstic oxide. If metallic tin or zinc is added and boiling is continued, the solution turns first blue and then brown.

Occurrence. Scheelite formed by hydrothermal processes is found in commercial quantities, where granites intruding limestones have produced scheelite-bearing skarns, as at Sangdong, Korea or Pine Creek, California. It also occurs in veins of quartz associated with cassiterite, molybdenite, and wolframite, as at Cornwall, England and Silver Dyke, Nevada, and in pegmatite dikes, as at Oreana, Nevada.

Wulfenite, PbMoO$_4$

Wulfenite is an ore of molybdenum but is considerably less important commercially than molybdenite. It is, however, of interest to the mineral collector because of the great variety and beauty of many of its specimens.

Habit. Wulfenite is tetragonal, and crystals are usually square tabular in habit with a prominent base (Fig. 7-111). Others are as thin as a knife-edge, with the flat table beveled by the faces of a low pyramid (Fig. 7-112 and Plate IV-2).

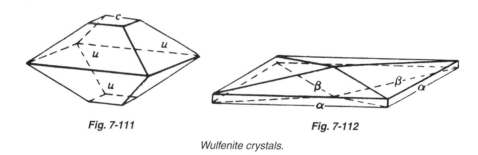

Fig. 7-111   **Fig. 7-112**

Wulfenite crystals.

Physical Properties. H = 3, G = 6.7–6.9. The color is most commonly a bright orange-yellow to reddish yellow, but it may be green or brown. The luster is resinous or adamantine. The square habit, bright color, and high luster make it a most striking mineral.

Composition. Wulfenite is lead molybdate, PbMoO$_4$. It fuses easily before a blow-pipe and gives a lead globule when fused with sodium carbonate.

Occurrence. Wulfenite is a secondary mineral found in the upper oxidized portion of lead veins associated with other secondary lead minerals, such as pyromorphite and vanadinite. Beautiful crystals in dissolution cavities have been found in several places in the western United States, including Red Cloud, Hamburg, and Mammoth mines in Arizona. Fine crystals were also obtained from the Stevenson Bennett mine near Las Cruces, New Mexico. Here they were found in cavities in silicified brecciated dolomite.

Wolframite, (Fe,Mn)WO$_4$

Wolframite is the chief ore of tungsten, a metal that has risen to great importance in our present civilization. Its most familiar use is in the filaments of incandescent electric light bulbs, where it is valuable because of its extremely high melting point, 3350°C. When the current is turned on, the filament becomes white hot without melting. It will, however, oxidize when hot; our present lamps, therefore, are filled with an inert gas that ensures the absence of oxygen.

The most important use of tungsten is in hardened steel for armor plate and metal-piercing projectiles. High-speed cutting tools made from tungsten steel will retain their temper even when red hot. Machines can thus be speeded up, and as a result, more work can be turned out in a given time. Tungsten carbide, harder than corundum, is used as an abrasive material for cutting glass and hard steel, and as tips on drill bits of all sizes.

Habit. Wolframite is monoclinic. Crystals are usually flattened parallel to the side pinacoid with a bladed habit. It is also found in massive granular aggregates.

Physical Properties. $H = 5–5\frac{1}{2}$, $G = 7.0–7.5$. Wolframite has perfect platy cleavage parallel to the side pinacoid. The specific gravity is high. The luster is submetallic; the color, brown to black. The streak is also brown to nearly black.

Composition. Wolframite is ferrous and manganous tungstate, $(Fe,Mn)WO_4$. Ferberite, $FeWO_4$, and huebnerite, $MnWO_4$, are isostructural and a complete solid solution series exists between them. Because these pure end members are rare, the name for the intermediate members, wolframite, is used. Since the physical properties of all members of the series are similar, the distinction between them must be made using advanced testing methods.

Occurrence. As with cassiterite, wolframite usually occurs in veins of quartz in or adjacent to granites in which the feldspars have been completely altered to muscovite (greisen). The wolframite is commonly associated with tourmaline, arsenopyrite, molybdenite, topaz, and fluorite. Deposits of this type are found in the Nanling district, China; Panansqueira, Portugal; and Nigeria.

As with cassiterite, minor amounts of wolframite are found with microcline, muscovite, and tourmaline in some pegmatites, as at Oreana, Nevada; Tavoy, Myanmar; and Black Hills of South Dakota. Ferberite has been found near Boulder, Colorado and huebnerite from Ouray, Colorado and in the Hamme district of North Carolina. Wolframite has also been recovered with columbite from alluvial deposits in Nigeria.

Gypsum, $CaSO_4 \cdot 2H_2O$

Gypsum is a common mineral and of considerable commercial importance because of its use in the production of plaster of Paris. This material is made by heating ground gypsum until about three-fourths of the water is driven off. When the plaster of Paris is mixed with water, it slowly crystallizes and hardens, assuming the shape of the confining surfaces. It is thus used to make casts and molds of all kinds. Its most extensive use is in the building industry in the form of gypsum lath, wallboard, and plaster for interior use.

Alabaster is a fine-grained massive variety of gypsum that is cut and polished for ornamental purposes.

Habit. Gypsum is monoclinic, and crystals often have the habit of those illustrated in Figs. 7-113 and 7-114 (see also Fig. 7-117). Twin crystals are also common, especially those of the "swallowtail" type, like that in Fig. 7-115. It is also found in cleavable and fine granular masses. A fibrous variety with a silky luster is known as *satin spar.* As mentioned above, the fine-grained massive variety is *alabaster.*

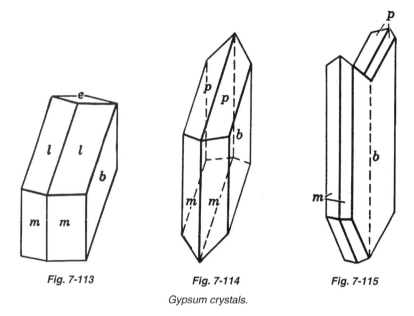

| Fig. 7-113 | Fig. 7-114 | Fig. 7-115 |

Gypsum crystals.

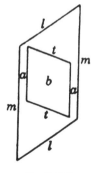

Fig. 7-116.
Gypsum cleavage.

Physical Properties. H = 2, G = 2.32. The crystals have very perfect cleavage parallel to the side pinacoid, and sometimes very large, thin, and perfectly transparent plates may be obtained. The variety yielding these is called *selenite.* The plates look a little like mica but are much softer and, though somewhat flexible, are quite inelastic. Careful inspection of the surfaces of these plates will reveal rhombic lines at angles of 66° and 114° from the two other cleavages parallel to the front pinacoid and the *a*-axis prism. When broken carefully, a plate will show these two other directions of cleavage. In one of these directions a plate breaks rather sharply, *snap cleavage,* with a conchoidal edge parallel to the front pinacoid (*a*). In direction *t* of Fig. 7-116, the plate is somewhat flexible and separates with a fibrous fracture, *bend cleavage.* Gypsum has a hardness of only 2, and hence it is easily scratched by a fingernail. The luster is usually vitreous but may be silky or pearly on the side of the best cleavage. The selenite variety is clear and colorless; the massive kinds are generally snowy white, as in alabaster. Impurities may give it various shades of yellow, red, and brown.

Composition. Gypsum is hydrous calcium sulfate, $CaSO_4 \cdot 2H_2O$. When a fragment is held in the flame, it becomes opaque white and exfoliates, fusing to a globule. Abundant water is given off in the closed tube.

Occurrence. Gypsum is most commonly found as a sedimentary rock interstratified with limestones and shales and usually underlying beds of rock salt. Such beds usually results from the evaporation of seawater, bringing about the crystallization and precipitation of gypsum and other dissolved salts. Commercial deposits of gypsum formed in this manner are known in many parts of the United States. The principal producing states are New York, Ohio, Michigan, Iowa, Texas, Nevada, and California. One of the most unusual occurrences of gypsum is in New Mexico at the White Sands National Monument. The gypsum, which has been deposited as the result of the evaporation of water in an enclosed basin, has been blown into great dunes. Here one can look for miles and see nothing but the rolling snow-white surface of the dunes.

Small crystals, such as those from Ellsworth, Ohio shown in Fig. 7-117, have formed by crystallization from formation waters in mud, which permitted them to

Fig. 7-117. *Gypsum crystals, Ellsworth, Ohio.*

grow freely in all directions. Solutions enriched in calcium sulfate sometimes form during the dissolution of limestones, and large fine crystals have come from caves in Wayne County, Utah, and Naica, Mexico (Fig. 3-5).

SILICATES

The silicates form the largest class of minerals. Relatively few of them are used as ores, but they are extremely important as rock-forming minerals. With only minor exceptions, all the minerals of the igneous rocks and most of the minerals in metamorphic rocks are silicates, and thus they make up the bulk of the earth's crust. Of the minerals that are either common enough or important enough to be described in this book, nearly 40% are silicates.

To nineteenth-century mineralogists, the composition of the silicates was a mystery and they attempted to arrange them on the basis of theoretical silicic acids derived from H_4SiO_4. It was only when crystal structure determinations were made in the 1920s using the methods of x-ray diffraction that the mineralogists learned that the fundamental structural unit of all silicates consists of a silicon atom surrounded

Fig. 7-118.
SiO₄ tetrahedron.

by four oxygen atoms in the form of a regular tetrahedron, $(SiO_4)^{4-}$ (Figs. 7-118 and 7-119a). In this way they resemble the phosphate, $(PO_4)^{3-}$, and sulfate, $(SO_4)^{2-}$, anion units. However, unlike the sulfates and phosphates, the structural studies revealed that the silica tetrahedra occur not only in isolated units but also join together (polymerize) in the form of doublets, rings, chains, sheets, and frameworks (Fig. 7-119b to g). The chemical formula of each silicate mineral is determined by the nature of the linkage.

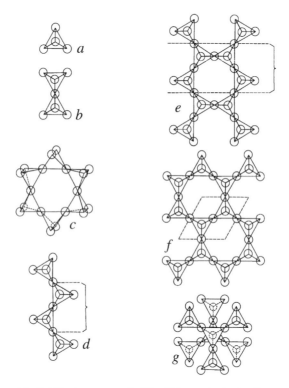

Fig. 7-119. Arrangement of the SiO₄ tetrahedra in silicates.

Structural Division of the Silicates		Figure
Framework or tektosilicates	SiO_2	7-119g
Sheet or phyllosilicates	$(Si_4O_{10})^{4-}$	7-119f
Chain or inosilicates	$(Si_2O_6)^{4-}$ or $(Si_8O_{22})^{12-}$	7-119d, e
Ring or cyclosilicates	$(Si_6O_{18})^{12-}$	7-119c
Isle or neso- and sorosilicates	$(SiO_4)^{4-}$ or $(Si_2O_7)^{6-}$	7-119a, b

The chemical formulas of the silicates are complicated not only by the linkage process of polymerization but also by solid solution. In some minerals aluminum replaces up to 50% of the silicon. Since the ionic charge of Al is only 3+ compared to the 4+ of silicon, cations must be added elsewhere to make up the charge difference. In silicates, aluminum occurs in two ways: in octahedral coordination with six oxygens surrounding each aluminum, or like silicon in tetrahedral coordination with four oxygens surrounding each aluminum. In the simple isle-silicates, there is only octahedral aluminum, while in the framework silicates, all of the aluminum is tetrahedral. For intermediate silicates such as the chain and sheet silicates, there can be both tetrahedral and octahedral aluminum. Generally, mineral formation at low temperature favors octahedral aluminum, as illustrated by kaolinite relative to phlogopite.

Framework (Tekto-) Silicates

Although the basic structural unit of framework silicates is the SiO_4 tetrahedron, these are fully polymerized in the framework structure such that each of the four corners of the tetrahedron is attached to the corner of another tetrahedron (Fig. 7-119g). Differences among the various silica minerals are the angular relationships among these linkages. Ideally, as a result of these complete linkages, only one composition should exist, SiO_2. However, because of the high degree of polymerization, aluminum regularly substitutes for silicon in ratios up to 1:1. To show this, it is convenient to write the formula for pure silica, SiO_2, as $4[SiO_2] = Si_4O_8$. Then if one aluminum substitutes for a silicon, this becomes $(AlSi_3O_8)^-$. The required charge-compensating ions Na^+ or K^+ yield the alkali feldspars. If two aluminums substitute, this becomes $(Al_2Si_2O_8)^{2-}$, and the compensating cations of Ca^{2+} yield anorthite.

Framework Silicates

Silica group	Scapolite series
Quartz, SiO_2	Scapolite, $(Na,Ca)_4(Al_2Si_2O_8)_3(Cl,CO_3)$
Tridymite, SiO_2	Feldspathoids
Cristobalite, SiO_2	Leucite, $KAlSi_2O_6$
Opal, $SiO_2 \cdot nH_2O$	Nepheline, $(Na,K)_2Al_2Si_2O_8$
Potassium feldspars	Sodalite, $Na_8(Al_2Si_2O_8)_3Cl_2$
Sanidine, $KAlSi_3O_8$	Lazurite, $Na_8(Al_2Si_2O_8)_3S$
Orthoclase, $KAlSi_3O_8$	Zeolites
Microcline, $KAlSi_3O_8$	Heulandite, $CaAl_2Si_7O_{18} \cdot 6H_2O$
Plagioclase feldspars	Stilbite, $NaCa_2Al_5Si_{13}O_{36} \cdot 14H_2O$
Albite, $NaAlSi_3O_8$	Chabazite, $CaAl_2Si_4O_{12} \cdot 6H_2O$
Anorthite, $CaAl_2Si_2O_8$	Analcime, $Na_2Al_2Si_4O_{12} \cdot 2H_2O$
	Natrolite, $Na_2Al_2Si_3O_{10} \cdot 2H_2O$

Silica Group

Quartz, SiO_2

Quartz is the most common mineral, and in some of its varieties one of the most beautiful. It makes up most of the sand of the seashore; it occurs as a rock in the forms of sandstone and quartzite and is an important constituent of many other rocks, such as granite and gneiss. It is a mineral that can usually be recognized by its form when crystallized, also by its hardness, conchoidal fracture, glassy luster, and infusibility. There are so many varieties, however, that it is only after long practice that one can be sure of always identifying it at once.

Habit. Quartz is rhombohedral, and the common habit of its crystals is a hexagonal prism showing horizontal striations terminated at each end by six pyramidal faces (two rhombohedrons), each having the shape of an acute isoceles triangle (Fig. 7-120). In some crystals the prism is not present and the shape is like that of Fig. 7-121, which appears to be a hexagonal dipyramid but is actually made up of two rhombohedrons. It is not uncommon to find the faces of the rhombohedron lettered *r* much larger than those of the other rhombohedron, or they may be present alone (Figs. 7-122 and 7-123).

On some crystals small modifying faces such as *x* and *s* in Figs. 7-124 and 7-125 may be present, revealing the true symmetry of quartz. They appear rather complicated as, indeed, they are, and the study of the structure and crystallography is a matter for the skilled mineralogist. Nevertheless, even the beginner can learn to recognize the difference between the crystals represented by Figs. 7-124 and 7-125. Figure 7-124 is called a *right-handed* crystal and has the little *x* face to the *right* above the prism face *m;* the other is a *left-handed* crystal and has a similar face to the *left* above *m.*

Some crystals may be so poorly formed that the hexagonal nature is not immediately apparent. These, such as Fig. 7-127, usually can be oriented by horizontal striations on the prism faces, and it should be remembered that in all cases the angles remain the same, despite the seeming irregularity in the form.

Although doubly terminated crystals, like those shown in Figs. 7-120 and 7-122, are found in some places, more commonly the prisms are attached at one end with only the other end free to develop (Fig. 3-3). Some crystals may be slender and tapering. Not infrequently, the crystals are so small that the forms can be distinguished only with a magnifying glass. A surface covered with many such tiny crystals is called *drusy.*

Twin crystals of quartz that show the reentrant angles of most twins are not often found (Fig. 3-128). Yet a careful study of most crystals will reveal that they are made up of two individuals in twin position interpenetrating each other irregularly. To prove such twinning it may be necessary to etch the crystal in hydrofluoric acid. One can then see that light is reflected differently from the etch pits formed on the two individuals. On some crystals natural etching has produced a difference in luster in various parts of a surface, thus outlining the twins or they may show interrupted striations.

The crystals illustrated have a 3-fold symmetry axis and three 2-fold symmetry axes but no symmetry center and are *low-temperature α-quartz*. When quartz is heated above 573°C, an internal rearrangement of the atoms forms *high-temperature β-quartz,* with a different structure and higher symmetry. Many massive varieties of quartz, some of which have been given names, will be mentioned in subsequent paragraphs.

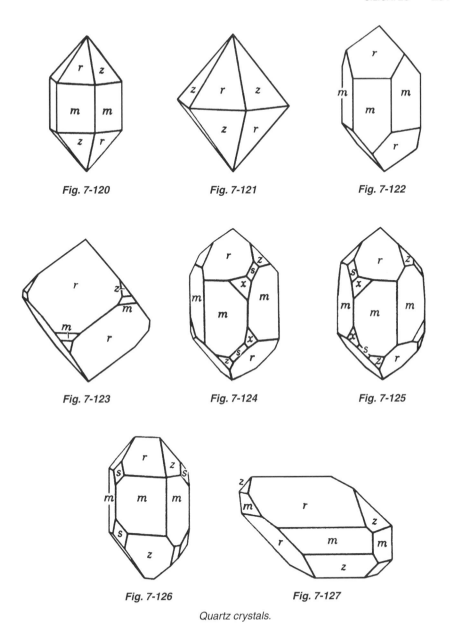

Fig. 7-120   Fig. 7-121   Fig. 7-122

Fig. 7-123   Fig. 7-124   Fig. 7-125

Fig. 7-126   Fig. 7-127

Quartz crystals.

Physical Properties. H = 7, G = 2.66. With a hardness of 7, quartz cannot be scratched by a knife, but it easily scratches glass; the specific gravity of pure crystals is 2.66. The luster is vitreous in crystals, but in some massive kinds it may be greasy or waxy; the impure varieties, like jasper, are dull. The color varies widely; crystals are usually colorless but may be purple, yellow, and brown to nearly black; pink, green, and red kinds usually do not occur in crystals. In the massive forms the color is often in bands or clouds, as described under the varieties.

Composition. Quartz is silicon dioxide, SiO_2. It is grouped with the silicates because its properties are more closely related to those minerals than to the oxides.

Quartz is found in a great number of varieties, differing particularly in color and state of aggregation, and as many have been used for ornamental purposes, the varieties have received a number of distinct names.

COARSELY CRYSTALLINE VARIETIES. Those occurring in distinct crystals are named chiefly according to their color, as outlined below.

Rock crystal is a clear colorless variety. If quite free from flaws, it is used for prisms for optical apparatus and cut into thin oriented plates for the control of radio frequencies. Most of this optical quartz comes from Brazil, and during the war years of 1942–1945, hundreds of tons of it were imported into the United States to be manufactured into radio oscillators. These oscillators were the forerunner of the tiny vibrating quartz plates used to control the time in today's watches. Quartz crystals for this purpose are now made synthetically.

Smoky quartz has a smoky brown color, which in some crystals may be very dark. It is cut into ornaments, as in Switzerland, where it is found in beautiful specimens. Such crystals were undoubtedly at one time clear and have darkened by exposure to emanations from radioactive materials. Some clear quartz can be darkened artificially by exposure to a strong x-ray beam (see Plate VI-1).

Amethyst is a fine purple variety cut as gemstones and used for ornamental purposes. The color is attributed to small amounts of iron.

Citrine is a light yellow–colored crystal sometimes called *false topaz.* Some jewelers sell citrine under the name of *topaz,* and one should be careful not to confuse it with true topaz. Much of the citrine on the market is heat-treated amethyst.

Milky quartz is milky white in color from the presence of small liquid inclusions (Fig. 7-128). It is the common type of quartz found in veins and in pegmatites. When two specimens of milky quartz are rubbed together, they luminesce, a phenomenon known as *triboluminescence.*

Rose quartz is a pale to deep pink variety occurring only in pegmatites and used for ornamental purposes. Rose quartz is usually massive, but crystals have been found, first at Newry, Maine and later in Brazil. The color is attributed to the presence of titanium (see Plates V-4 and VI-2).

Quartz with inclusions. Many minerals have been observed in coarsely crystalline quartz and only a few of the more common are mentioned here. *Rutilated quartz* is rock crystal that contains many randomly oriented slender needles of brown or golden rutile (see Plate V-2). It is frequently cut as a gemstone. Similar inclusions of green actinolite fibers and acicular black tourmaline crystals are also found. Quartz enclosing closely packed, parallel asbestos fibers when cut as a cabochon has a chatoyancy and is called *cat's eye quartz.* It should not be confused with the highly prized variety of chrysoberyl known as cat's eye. *Tiger's-eye* is a yellow to brown quartz somewhat like cat's eye quartz in effect. Its chatoyancy results from its fibrous nature, which is inherited from crocidolite asbestos, from which it has been formed

PLATE I

Photo by R. Jones.

(4) Sulfur crystals, Girgenti, Sicily.

(1) Perfect octahedral diamond, 84 carats. Kimberley, South Africa.

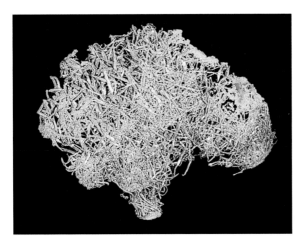

(2) Crystallized gold, Breckenridge, Colorado.

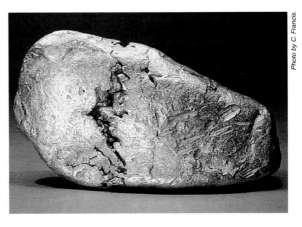

Photo by C. Francis.

(3) Gold nugget, 9.4 cm long, California.

PLATE II

(1) Ruby crystal, Myanmar (Burma).

Photo by C. Francis.

(2) Rutile cyclic twin, Parkesburg, Pennsylvania.

*(3) Iridescent limonite,
California.*

*(4) Washing gem
gravels, Sri Lanka.*

PLATE III

Photo by C. Francis.

(1) Calcite crystals coated with hematite, Cumberland, England.

Courtesy of Ward's, Rochester, New York.

(2) Rhodochrosite crystals, Alma, Colorado.

(3) Azurite and malachite, Bisbee, Arizona.

PLATE IV

(1) Rincon salar encrusted with borax and ulexite, Salta, Argentina.

(2) Wulfenite, Red Cloud Mine, Yuma, Arizona.

Photo by D. Cook.

(3) Apatite in calcite, Canada.

PLATE V

(1) Watermelon tourmaline (elbaite), Dunton mine, Newry, Maine.

Courtesy of Ward's, Rochester, New York.

(4) Rutilated quartz, Brazil.

Photo by D. Cook.

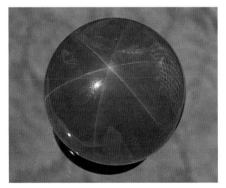

(2) Rose quartz sphere showing asterism, Brazil.

(3) Tourmaline crystal, Brazil.

PLATE VI

(1) Microcline feldspar
and smoky quartz,
Crystal Peak, Colorado.

(2) Rose quartz crystals, cut
stone, and snuff bottle, Brazil.

Photo by D. Cook.

(3) Agate, Brazil.

PLATE VII

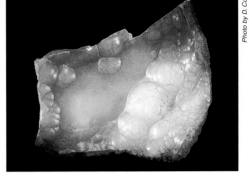

Photo by D. Cook.

(1) Opal, Australia.

(2) Chalcedony.

(3) Iridescent labradorite,
Labrador.

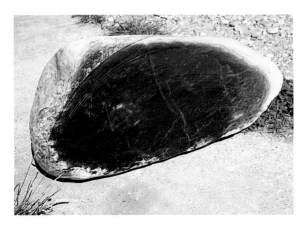

(4) Nephrite jade boulder,
about 1 m long, New Zealand.

PLATE VIII

Photo by J. Savickas.

(1) Garnet, Gore Mountain, New York.

(2) Datolite colored by native copper, Keweenaw Peninsula, Michigan.

(3) Rhodonite, Franklin, New Jersey.

Fig. 7-128. *Milky quartz coated with smaller crystals, Ouary, Colorado.*

by replacement. Although not found in crystals, *aventurine* should be mentioned here. It is quartzite spangled with scales of mica. The chrome mica fuchsite imparts a green color to the quartz, and a reddish brown is due to mica spangles of that color.

FINE-GRAINED VARIETIES. The fine-grained or microcrystalline varieties of quartz can be divided into two groups, depending on the nature of their microcrystalline or cryptocrystalline units, fibrous and granular. They will be separated in the following descriptions, but without the use of a high-powered microscope the student cannot be expected to distinguish between them.

FIBROUS VARIETIES. *Chalcedony,* when used alone, designates a light, honey yellow to gray, translucent fibrous quartz with a waxy luster (see Plate VII-2). It has a slight porosity, resulting in a lower specific gravity (2.60) than that of coarsely crystalline quartz. It commonly occurs in botryodial, mammillary, or stalactitic masses (Fig. 3-159) with the fiber lengths at right angles to the surface. Chalcedony is a general term, and specific names are given to some of the different-colored varieties.

Agate is a variegated chalcedony with the colors arranged in delicate concentric bands frequently alternating with bands of opal (as in Fig. 7-129 and Plate VI-3). These bands often follow the irregular outline of the cavity in which the silica was deposited. *Moss agate* is a kind of chalcedony containing brown or black mosslike or dendritic forms distributed rather thickly through the mass. These forms consist of some metallic oxide (manganese oxide is common) and have nothing more to do with vegetation than the frost figures on the windowpane in winter.

Carnelian is an orange-red or brownish-red chalcedony colored by the iron oxide hematite. It grades into *sard,* which is colored brownish red to brown to orange by the iron oxide goethite.

Onyx, like much agate, is made up of layers of different-colored chalcedony and opal, but the banding is straight and the layers are in parallel planes. Alternating lay-

Fig. 7-129. *Agate, Brazil.*

ers of white and black or white and brown are most common. Onyx is used for cameos, the head being cut from one layer and the background being formed by the other. Both agate and onyx are often artifically colored to make them more attractive for ornaments. Some varieties of calcite have been termed *Mexican onyx,* but they are very soft and should not be confused with true onyx.

Sardonyx is like onyx but has layers of sard or carnelian alternating with layers that are white or black. *Prase* is a translucent leek-green chalcedony; *chrysoprase* is an apple-green chalcedony. *Heliotrope,* or *bloodstone,* is a green chalcedony with small spots of red jasper scattered through it.

GRANULAR VARIETIES. In the granular varieties, the microscopic particles are equidimensional rather than fibrous.

Jasper is the major granular variety. It is usually impure and red from inclusions of hematite. It may also be brown, yellow, or dark green. The green and red colors may be present in the same specimen irregularly distributed or arranged in bands.

Flint is nearly opaque with a dull luster and is usually gray, smoky brown, or brownish black. The exterior is often white from a coating of lime or chalk in which it was originally embedded. It breaks with a conchoidal fracture, yielding sharp cutting edges, and hence was easily chipped by the American Indians into arrowheads and hatchets.

Chert is a compact silica rock resembling flint in most of its properties but is usually white or light gray in color. It frequently forms thin but extensive beds in limestone; flint is usually in isolated nodules. *Hornstone* is a name sometimes applied to chert.

Silicified wood consists largely of chalcedony or jasper that has replaced the woody structure of the tree. It may vary much in color and gives beautiful effects on the polished surface, like the specimens from the "petrified forest" near Holbrook, Arizona. Some silicified wood has been formed by replacement by opal and properly belongs under that species.

Occurrence. Quartz is one of the essential constituents of granite, granodiorite, gneiss, mica schist, and many related rocks; it is the principal constituent of quartzite and sandstone. In these rocks it is found in small glassy grains that rarely show crystal outline. From the weathering and disintegration of such rocks, quartz is set free to be washed into the streams and eventually into the ocean. Thus it is the principal material of the pebbles of gravel beds and the sands of the seashore. Such occurrences are familiar to all, but the mineralogist and crystallographer are more interested in the rarer localities where quartz is found well crystallized or in the cryptocrystalline varieties.

Well-formed crystals may be found in veins with ore minerals or, as from Hot Springs, Arkansas, in veins containing almost no other material. Some of the finest specimens have come from cavities in pegmatite and granite rocks. In cavities in volcanic flows and breccias as well as in some limestones, chalcedony, agate, carnelian, and so on, may be present filling all the available space; or they may line the cavity only partially and crystals of quartz may occupy the central part (Fig. 7-130).

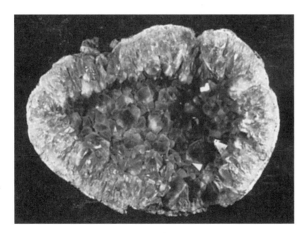

Fig. 7-130. *Geode with amethyst crystals, Uruguay.*

Embedded nodules or masses of fine-grained silica constitute the flint of the chalk formations, as in the chalk cliffs of Dover, England, and the chert of other limestones.

Tridymite and Cristobalite, SiO_2

Although beginners cannot be expected to recognize tridymite and cristobalite, they should know these minerals exist because they are of considerable interest to the mineralogist. Both minerals are high-temperature forms of silica and have the same composition as quartz, SiO_2. They are found, abundantly in some places, as constituents of silica-rich fine-grained volcanic rocks. Cristobalite is stable only above 1470°C; tridymite, only above 870°C.

Opal, $SiO_2 \cdot nH_2O$

Opal is essentially silica but contains a few percent water; thus it does not have the same properties as quartz. Its chief interest lies in its use as a gem; stones are usually cut *en cabochon* to exhibit to best advantage the play of colors in the variety called *precious opal.* Another variety of opal, called *diatomaceous earth,* is white and chalky and one would never suspect that it is related to the beautiful gemstones. It is used as an abrasive, as a substance for filtering solutions, and as an insulation material.

Habit. Opal is one of the few minerals that is not crystalline and is called amorphous. It is found frequently in botryoidal or stalactitic masses. The variety diatomaceous earth resembles chalk in appearance.

Physical Properties. H = 5–6, G = 1.9–2.2. Opal can usually be scratched by a knife and has a noticeably low specific gravity. It breaks with a smooth conchoidal fracture. The color may be white to yellow, red, brown, green, gray, and black. Several types of opal are recognized on the basis of their differing physical properties.

Precious opal is the most beautiful variety, much admired because of the delicate play of colors due to the optical effect of internal reflections. One kind of precious opal with a bright red flash of light is called *fire opal.* The beautiful opal found in Queensland, Australia, shows an iridescent blue like the effect of a peacock's feather. A great number of trade terms are used to describe the appearance of opal based on body color, transparency, and play of color (see Plate VII-1).

Common opal does not exhibit the play of colors. It is often white but varies widely in color and appearance and frequently has a resinous or waxy luster.

Hyalite is a clear glassy opal in globular or botryoidal masses that form crusts in rock cavities. In some specimens the glassy globules look like drops of gum.

Wood opal is silicified wood in which the mineral material is opal instead of quartz. Magnificent specimens of precious opal, faithfully preserving the external appearance of the branches they have replaced, are found in Humboldt County, Nevada. However, other wood opal breaks up into slender chalky-white splinters and shows none of the beauty of precious opal.

Geyserite or siliceous sinter is a type of opal deposited from hot springs, as by the geysers of Yellowstone National Park. It is usually white or gray and often has a pearly luster on the surface. It is soft and porous and is frequently built up into concretionary forms of varied and beautiful appearance.

Diatomaceous earth, sometimes called *infusorial earth* or *diatomite,* is a kind of opal-silica consisting of the microscopic shells of the minute organisms called diatoms. As these shells sink from near the ocean's surface to the seafloor, they build up beds that may be of great thickness and extent, as in the Monterey formation at Lompoc, California.

Composition. The formula for opal can be written $SiO_2 \cdot nH_2O$. The n indicates that it contains an indefinite amount of water, usually between 3 and 10% but may be as high as 20%. Unlike quartz, opal is soluble in alkalies. Thus, after agate has been immersed in an alkaline solution for some time, the layers containing opal are attacked and dissolved, leaving the layers of chalcedony unaffected.

Occurrence. Opal as a gemstone has been known for a long time. Most of the early material came from Cervenica in Hungry, but today Australia is the principal producer of precious opal. The Hungarian opal is found in seams in altered andesite. Opal is produced from several widely separated localities in Queensland, New South

Wales, South Australia, and Western Australia. The black opal from Lightning Ridge, New South Wales, occurs along joints in sandstone and conglomerates. In the United States it has been found at Opal Butte, Oregon and in the Virgin Valley, Nevada. In the Virgin Valley, wood embedded in ash and tuff beds has been opalized. There are several important precious opal localities in Mexico.

The Feldspars

The feldspars form the most important group of silicates, if not the most important of all the minerals. They are found as essential constituents of most crystalline rocks, such as granite, syenite, gabbro, basalt, gneiss, and many others, and thus make up a large percentage of the earth's crust. All of them are silicates of aluminum with potassium, sodium, and calcium, and rarely barium.

Although some feldspar is monoclinic and some triclinic, all occur in crystals that have a general resemblance to each other. All have cleavage in two directions, making angles of 90° or nearly 90° with each other. A careful examination will show that these two directions are unlike, that is, cleavage parallel to one face is easier than parallel to the other. The hardness of all the feldspars is about 6: they are not scratched by a knife. The specific gravity lies between 2.55 and 2.75, not far from that of quartz; the color is variable and may be white, pale yellow, reddish, greenish, or gray and have a chalky luster compared to the glassy luster of quartz.

Sanidine, Orthoclase, and Microcline, $KAlSi_3O_8$

While sanidine and orthoclase crystallize in the same crystal system, microcline crystallizes in a different crystal system. They all have the same chemical composition, physical properties, and occurrence. Together they are known as *potash feldspar*. It is frequently difficult to distinguish between them in the hand specimen, and thus in the past much feldspar has been called orthoclase that should be called microcline.

Potash feldspar is used extensively in the manufacture of porcelain, both in the body of the material and in the hard glaze with which much of it is finished. It is also a source of aluminum in the manufacture of glass. Green microcline known as *Amazon stone* is cut and polished and used for ornamental purposes.

Habit. Various orthoclase crystals are depicted in Figs. 7-131 to 7-134. Both Sanidine and orthoclase are monoclinic; microcline is triclinic. The crystals illustrated in Figs. 7-131 and 7-132 are orthoclase, but Fig. 7-132 represents equally well a crystal of microcline. The near identity of crystals of the two species is due to twinning. Microcline, being triclinic, has no symmetry plane parallel to the side pinacoid, and this plane can thus be a twin plane on which the crystal can repeatedly be twinned. This is *albite twinning* and is always present in microcline, giving it the apparent symmetry of a monoclinic cyrstal. Microcline is also twinned according to the *pericline law.* The lamellae of the albite and the pericline twins cross at about 90° on the basal pinacoid, giving a characteristic tartan pattern on that face. Other types of twinning found in both species are *Carlsbad* (Fig. 7-135), *Baveno* (Fig. 7-137), and *Maneback* (Fig. 7-136). Figures 7-131 to 7-133 represent the common forms. The

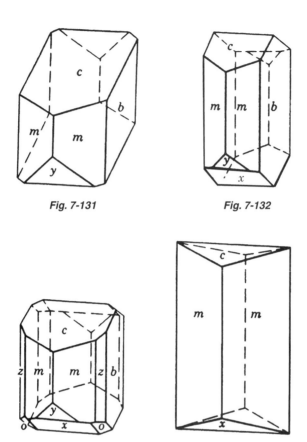

Fig. 7-131 **Fig. 7-132**

Fig. 7-133 **Fig. 7-134**

Orthoclase crystals.

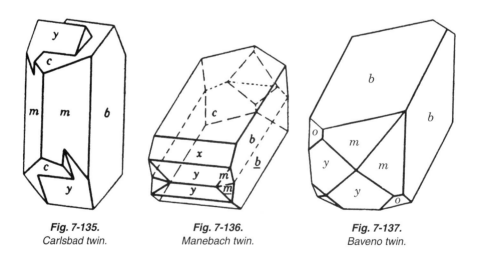

Fig. 7-135.
Carlsbad twin.

Fig. 7-136.
Manebach twin.

Fig. 7-137.
Baveno twin.

faces of the unit prism *m,* at angles of nearly 120° to each other, are short on some crystals (Fig. 7-131). The other most common forms are the basal pinacoid *c* and the side pinacoid *b.* There may also be other modifying planes.

Physical Properties. H = 6, G = 2.57. The name *orthoclase* is from two Greek words meaning *erect* and *fracture,* referring to the existence of the two prominent pseudoprismatic cleavages at right angles to each other. One of these is parallel to the basal pinacoid, and the other is parallel to the side pinacoid. Although microcline receives its name from two Greek words meaning *little* and *inclined* (referring to the fact that the cleavages are slightly inclined from a right angle), they appear to be perpendicular because of the albite twinning. This type of twinning is the outstanding physical property and is so characteristic of the triclinic feldspars that it enables one to distinguish them quickly from other minerals even when they are in tiny grains in a rock.

The hardness is the best test by which feldspar can be distinguished from calcite, barite, and other minerals showing good cleavage. The luster is vitreous, although it may be pearly on the basal cleavage surface. The color is commonly white, reddish, or pale yellow. Some crystals of orthoclase are clear and colorless, and others, extremely rare, are a transparent yellow and are cut as gemstones. A green variety of microcline, *amazonite,* is also cut and used for ornamental purposes (see Plate VI-1).

The name *adularia* is given to a glassy variety of orthoclase that is usually found in pseudoörthorhombic crystals (Fig. 7-134). Some adularia gives a beautiful bluish opalescence, especially when polished, and is called *moonstone.* Some moonstone belongs to the albite species.

Composition. Chemically, the only difference among the three potash feldspars is in the nature of the solid solution between aluminum and silicon. This difference can be shown by writing their chemical formulas as follows: microcline, $KAlSi_3O_8$; orthoclase, $K(AlSi)Si_2O_8$; sanidine, $K(AlSi_3)O_8$, with the state of the solid solution depending on the temperature history. Thus pseudomorphs of orthoclase after sanidine and microcline after orthoclase are common and can be determined only by advanced x-ray diffraction methods. Some sodium may replace potassium, and in sanidine as much as 50% of the potassium is replaced. In the variety *hyalophane,* barium replaces part of the potassium. An exact determination among these species is difficult and the beginner must be content to base an identification on observed crystal forms and general occurrence.

Occurrence. Sanidine is the species usually found as glassy crystals, *phenocrysts,* embedded in various volcanic rocks, as in trachyte from Rhineland, Germany. Orthoclase often showing distinct Carlsbad twinning is the common feldspar of granite, gneiss, and related rocks, while microcline is the usual feldspar found in pegmatites, where it may be present in large masses, as found in the New England states, North Carolina, South Dakota, Colorado, and elsewhere. In many of these pegmatites it is often possible to obtain feldspar in considerable quantity free from the associated quartz and mica. It is then mined and used in making porcelain and glass. Soda feldspar, albite, is commonly associated with the potash feldspar in pegmatites, as well as many interesting minerals, such as tourmaline, beryl, apatite, amblygonite, spodumene, and many others. In some pegmatites microcline is intimately intergrown with quartz (Fig. 7-138). The name *graphic granite* is given to this intergrowth.

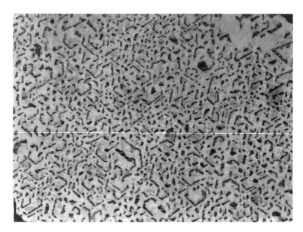

Fig. 7-138. Graphic granite, Bradbury Mountain, Maine (quartz, dark; microcline, light).

Albite, $NaAlSi_3O_8$, and Anorthite, $CaAl_2Si_2O_8$

Albite and anorthite are isostructural end members of a solid-solution series; that is, these two will mix together in all proportions so that crystals of any intermediate composition may be formed. This series is known collectively as the *plagioclase feldspars,* or since one end member contains sodium and the other calcium, the *soda-lime feldspars.* Although there is a complete gradation from one end to the other, various species names have been given on a completely arbitrary basis to feldspar of intermediate composition. They are:

	Percent Anorthite	Specific Gravity
Albite, $NaAlSi_3O_8$	0–10	2.62
Oligoclase	10–30	2.63–2.66
Andesine	30–50	2.67–2.69
Labradorite	50–70	2.70–2.72
Bytownite	70–90	2.72–2.75
Anorthite, $CaAl_2Si_2O_8$	90–100	2.76

Habit. The plagioclase feldspars are triclinic (Fig. 7-139). Crystals of albite and anorthite are more common than those of intermediate composition; all of them are usually small. Albite crystals are frequently flattened parallel to the side pinacoid and crowded together in parallel plates somewhat resembling barite. A snow-white albite in such platy aggregates is called *cleavelandite.*

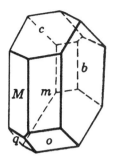

Fig. 7-139. *Albite.*

Twinning is so common in the plagioclase feldspars that it is the rare crystal that shows none. *Albite twinning,* with the side pinacoid as the twin plane, is almost universal. This is a repeated twinning that is evidenced by parallel lines or striations seen on the basal pinacoid. In fact, it is only by recognizing the presence of albite twinning that the elementary student can say that feldspar is triclinic rather that monoclinic. Carlsbad, Maneback, and Baveno twins (Figs. 7-135 to 7-137) may also be present in the triclinic feldspars.

Physical Properties. H = 6, G = 2.62–2.76. The plagioclase feldspars have good pseudoprismatic cleavage parallel to both the base and the side pinacoid, and although the angle between them is not 90°, the albite twinning makes it appear so. On the basal cleavage the evidence of this twinning can be seen by the fine lines which catch the light when the specimen is held so as to reflect it (Fig. 7-140).

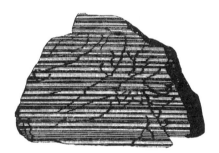

Fig. 7-140. *Albite twinning.*

The hardness is 6; the specific gravity varies continuously from 2.62 in albite to 2.76 in anorthite and is one method the beginner may use to distinguish among them. It may be colorless, white, or gray, and less frequently greenish, yellowish, or red. The luster is vitreous to pearly. A beautiful play of colors may be seen on some specimens of labradorite and andesine (see Plate VII-3).

Aventurine or *sunstone* is a variety of oligoclase that contains inclusions of hematite that give a golden shimmer to the mineral. Some of the material known as *moonstone* is a variety of albite showing an opalescent play of colors. As already noted, some moonstone is orthoclase.

Composition. The plagioclase feldspars form sodium and calcium aluminum silicates; complete solid solution exists between albite, $NaAlSi_3O_8$, and anorthite, $CaAl_2Si_2O_8$. Some potassium may be present particularly in the albite end of the series.

Occurrence. Like orthoclase and microcline, the plagioclase feldspars are rock-forming minerals and are even more widely distributed. They are present in most igneous rocks, especially gabbros, diorites, and anorthosites. Indeed, the classification of these rocks is based largely on the amount and kind of feldspar present. However, in general, it is true that the darker the rock, the more calcium is present in the feldspar. To classify them correctly, one must, for example, distinguish between oligoclase and labradorite, a task that must be left to the advanced mineralogist or petrographer. All we can hope to do at this stage is to make a rough division on the basis of specific gravity, and even for this classification, a fragment larger than the average grain of an igneous rock is necessary.

In addition to its occurrence as a rock-forming mineral, albite is frequently found in pegmatite dikes. It may be in small crystals or in larger masses replacing earlier microcline. The platy variety, cleavelandite, is usually confined to pegmatites.

Scapolite Series

Scapolite, $(Na,Ca)_4(Al_2Si_2O_8)_3(Cl,CO_3)$

Habit. Scapolite is tetragonal, and crystals are usually prismatic, showing prisms of both the first and second order and the dipyramid r (Fig. 7-141 and 3-59).

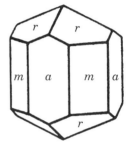

Physical Properties. H = 5–6, G = 2.65–2.74. Scapolite has prismatic cleavage parallel to both first- and second-order prisms: in all, a total of four directions making angles of 45° and 135° with each other. The luster is vitreous to pearly, and the color is commonly white to gray. There are also yellowish, reddish, greenish, and bluish varieties.

Fig. 7-141.
Scapolite.

Composition. Like plagioclase or olivine, scapolite is of varied composition and the name belongs more properly to a series rather than to a definite species with complete solid solution existing between the *marialite,* $Na_4(AlSi_3O_8)_3Cl$, and *meionite,* $Ca_4(Al_2Si_2O_8)_3CO_3$. Intermediate compositions also have been called wernerite. Varying amounts of SO_4 and F may also be present. Note the close compositional correspondence with plagioclase. Scapolite fuses easily before a blowpipe to a glass full of bubbles.

Occurrence. Characteristically, scapolite is formed in limestones as the result of igneous intrusion and is associated with diopside, garnet, apatite, and titanite. In this manner, large pink masses have been found at Bolton, Massachusetts, and yellow crystals of gem quality have come from Madagascar.

The Feldspathoids

As has already been mentioned, the feldspars are to be found, usually abundantly, in most igneous rocks. In a small group of igneous rocks, however, there is not only insufficient silica to form quartz but also insufficient silica to combine with the alkalies and aluminum to form feldspars. The rock minerals that form in their place with a lower percentage of silica are called *feldspathoids.* The most important of these are leucite, nepheline, sodalite, and lazurite.

Leucite, KAlSi$_2$O$_6$

Habit. Leucite is tetragonal. It is pseudoisometric, for it is nearly always found in trapezohedrons, as shown in Fig. 7-142. However, if one examines a crystal under the polarizing microscope, it will be seen that it is made up of many tiny tetragonal crystals. When these leucite crystals formed from hot molten magma, they were truly isometric both in outward form and in internal structure. However, on cooling below 605°C, an internal rearrangement took place but the external appearance remained unchanged.

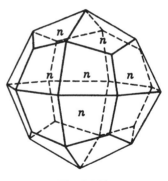

Fig. 7-142.
Leucite.

Physical Properties. H = 5½–6, G = 2.45–2.50. (The specific gravity of analcime, the only other mineral occurring in white trapezohedrons, is 2.27.) The color is white to gray. The name *leucite* comes from the Greek meaning *white.*

Composition. Leucite is potassium aluminum silicate, KAlSi$_2$O$_6$. It contains the same elements as orthoclase feldspar but has only 55% SiO$_2$, whereas orthoclase has 64.7% SiO$_2$. Upon melting, orthoclase forms leucite and liquid.

Occurrence. Leucite is a rare high-temperature mineral found in certain volcanic rocks. It is a major constituent of a rare type of rock enriched in potassium and deficient in silica known as lamproite. Such rocks are limited in extent but are found in the Leucite Hills, Wyoming and in the Highwood and Bear Paw Mountains, Montana. Major reserves of diamonds have been found in the lamproite pipes in Australia. The lavas of Mount Vesuvius in Italy contain abundant crystals of leucite.

Since leucite is a rock-forming mineral, it is found embedded in a fine-grained matrix; analcime, with which it may be confused, is found as crystals lining cavities in volcanic flows and tuffs.

Nepheline, (Na,K)$_2$Al$_2$Si$_2$O$_8$

Habit. Nepheline is hexagonal, but crystals are rare. It is usually massive and is common as small grains in certain igneous rocks.

Physical Properties. H = 5½–6, G = 2.55–2.65. Cleavage parallel to the prism faces is seen only in the larger crystalline masses. In crystals the luster is vitreous, but when nepheline is massive, the luster is greasy. For this reason the name *eleolite,* from the Greek word meaning *oil,* is sometimes given to the massive variety. The color is white to yellow in crystals, but the massive variety may be gray, greenish, or reddish.

Composition. Nepheline is sodium-potassium aluminum silicate, (Na,K)$_2$Al$_2$Si$_2$O$_8$. Although it has a formula closely analogous to that of plagioclase, the extra sodium requires a different arrangement of the atoms. It can be fused before a blowpipe, giving a strong yellow flame of sodium. It is soluble in hydrochloric acid and can thus be distinguished from feldspar.

Occurrence. Nepheline is a rock-forming mineral and is rarely found except in very silica-deficient and alkali-rich igneous rocks and pegmatite dikes associated with them. The granular rock, *nepheline syenite,* and the lava equivalent, *phonolite,* are the rocks in which it is found most abundantly. Masses of these rocks, however, are usually small. In the United States such nepheline-bearing rocks are found as a part of the carbonatite complex at Magnet Cove, Arkansas and in a couple of small syenite bodies at Beemerville, New Jersey. Larger masses of igneous rock with coarse crystalline nepheline are found near Bancroft, Ontario. The largest known mass of nepheline rocks is on the Kola Pennisula, Russia.

Sodalite and Lazurite, $Na_8(Al_2Si_2O_8)_3(Cl_2,S)$

Although both sodalite and lazurite are blue, lazurite is the mineral name of the gem and ornamental stone *lapis lazuli.* It has been known and highly prized for many centuries. When ground to a powder it was used as the blue paint pigment, *ultramarine.* This same pigment is now made synthetically.

Habit. These two isostructural minerals are isometric. They are usually massive and compact, but when rare crystals are found, they occur as dodecahedrons.

Physical Properties. Sodalite: H = 5½–6, G = 2.15–2.30; lazurite: H = 5–5½, G = 2.4–2.45. Cleavage is dodecahedral but is rarely observed. The color is usually blue, but sodalite may also be white, gray, yellow, or red. Lazurite is famous for its deep azure-blue to greenish-blue color.

Composition. Sodalite is sodium aluminum silicate with chlorine, $Na_8(Al_2Si_2O_8)_3$-Cl_2, while lazurite, $Na_8(Al_2Si_2O_8)_3S$, has sulfur. They are fusible before a blowpipe, giving a strong yellow flame of sodium. They are soluble in hydrochloric acid.

Occurrence. Sodalite is a rather rare rock-forming mineral, associated with other feldspathoids, particularly nepheline. It is found in crystals in the lavas of Vesuvius. The massive blue variety typically occurs with nepheline syenites as at Magnet Cove, Arkansas; the Kola Peninsula, Russia; and Bancroft, Ontario. Lazurite is also a rare mineral occurring in limestones that have been metamophosed by alkaline igneous rocks. It is usually associated with pyroxene and pyrite. Most of the properties of lazurite are similar to those of sodalite, and the beginner may have some difficulty in distinguishing them. However, the association of pyrite with lazurite is so common that it can be used as a determining criterion. The best lapis lazuli has come from Afghanistan and is a mixture of lazurite with calcite, pyrite, and pyroxene.

The Zeolites

The zeolite family includes a number of beautiful minerals having a close relation to each other both in manner of occurrence and in chemical composition. They are all hydrous silicates; that is, they contain water that is given off when a fragment is heated. Like other hydrous silicates, they are of inferior hardness, chiefly 3½–5½, and low specific gravity, ranging from 2.0 to 2.4. However, they react differently on heating from other hydrous silicates. When other hydrous silicates are heated and the water driven off, the structure collapses. The water in zeolites is weakly held in channelways and is driven off at relatively low temperatures without collapse of the structure. A dehydrated zeolite can be completely rehydrated by immersion in water. Because of their large interconnected cavities, zeolites are called *molecular sieves,* for they can, on dehydration, selectively absorb certain hydrocarbons and exclude others. In addition, they are used as water softeners, in which the Ca^{2+} of hardwater is exchanged for Na^+ of the zeolite and for the catalytic cracking of petroleum. Because the demand for zeolites for these purposes is great, they have been produced synthetically on a large scale. They are readily decomposed by hydrochloric acid. Many of them bubble up, or intumesce, when heated before a blowpipe, and this has given the name to the family from the Greek, *to boil.* Natural zeolites have been used as soil conditioners and as desiccants. Since 1950, zeolites analogous to chabasite have been manufactured by the hydrothermal crystallization of gels.

All the zeolites are said to be *secondary minerals,* which means that unlike the feldspar, quartz, and so on, which are an essential part of the rock, they crystallized subsequent to the time of the formation of the rock in which they occur. They have been formed in most cases out of glass, feldspar, or related minerals in the rock itself and hence usually occur in crevices, seams, or cavities instead of in the solid mass. They form upon burial of rocks at temperatures from 100 to 250°C and are indicators of the lowest degree of metamorphism.

All are silicates of aluminum with lime or soda or potash. Like the feldspars, they do not contain iron or magnesia; indeed, in the past they have been called *hydrous feldspars.*

The zeolites are commonly found associated with each other and with the minerals datolite, prehnite, apophyllite, and calcite. They may be found around hot springs such as Waikairi, New Zealand or along cracks in almost any kind of rock (Fig. 7-147). Most commonly they are found growing in cavities (Fig. 7-143) in volcanic

Fig. 7-143. Heulandite, Paterson, New Jersey.

breccia and tuffs or in amygdaloidal cavities of ancient basaltic lava flows, sometimes known as *traprock,* such as that which forms the Palisades of the Hudson River.
Famous localities have been developed where railroad cuts or tunnels have been cut
through ridges of traprock, as at Bergen Hill, New Jersey, Nova Scotia, and the Dec-
can Traps of India. Traditionally, this is the way in which all zeolites occur. Recently,
however, they have been found in large deposits in Nevada and in Tanzania as an
alteration of volcanic tuffs.

Heulandite, $CaAl_2Si_7O_{18} \cdot 6H_2O$

Habit. Heulandite is monoclinic and is characteristically
found in crystals somewhat flattened on the side pinacoid and
having a pseudo-örthorhombic appearance (Fig. 7-144).

Physical Properties. $H = 3\frac{1}{2}$–4, $G = 2.18$–2.2. Heulandite has
perfect platy cleavage parallel to the side pinacoid, and the luster, vitreous elsewhere, is pearly on this face. Heulandite is
most commonly colorless or white, but it may be red, gray, or
brown.

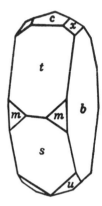

Fig. 7-144.
Heulandite.

Composition. Heulandite is essentially a hydrous calcium aluminum silicate, $CaAl_2Si_7O_{18} \cdot 6H_2O$. Some sodium and potassium may replace part of the calcium. Before a blowpipe it fuses
to a white glass and gives water in the closed tube.

Stilbite, $NaCa_2Al_5Si_{13}O_{36} \cdot 14H_2O$

Habit. Stilbite is monoclinic, and although crystals are common, well-formed single
individuals are rare. It is usually found in bundles of crystals, often looking like a
sheaf of wheat tied tightly about the center (Fig. 7-145). This habit has given it the
old name *desmine,* from the Greek for *bundle.*

Physical Properties. $H = 3\frac{1}{2}$–4, $G = 2.1$–2.2. Stilbite has perfect platy cleavage
parallel to the side pinacoid; a beautiful pearly luster is seen on the cleavage surface,
the side face of the bundles. This property gives the mineral its name from the Greek
word meaning *luster.* The color is usually white but may be yellow, red, or brown.

Composition. Stilbite is essentially a hydrous calcium-sodium aluminum silicate,
$NaCa_2Al_5Si_{13}O_{36} \cdot 14H_2O$. Before a blowpipe it swells up and fuses to a white enamel.
It yields water in the closed tube.

Chabazite, $CaAl_2Si_4O_{12} \cdot 6H_2O$

Habit. Chabazite is rhombohedral and is usually found in rhombohedral crystals
(Fig. 7-147) with nearly cubic angles. Penetration twins (Fig. 7-146) are common.

Physical Properties. $H = 4$–5, $G = 2.05$–2.15. Chabazite has a rhombohedral
cleavage, but it is much poorer than that shown in calcite. The luster is vitreous, and
the color is white, yellow, and red.

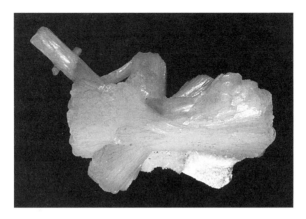

Fig. 7-145. *Stilbite, Nova Scotia.*

Fig. 7-146.
Chabazite.

Fig. 7-147. *Chabazite, West Paterson, New Jersey.*

Composition. Chabazite is essentially a hydrous calcium aluminum silicate, $CaAl_2Si_4O_{12}\cdot6H_2O$, usually with small amounts of potassium. Before a blowpipe it intumesces and fuses to a blebby, nearly opaque glass. It is decomposed by hydrochloric acid with the separation of silica.

Analcime, $Na_2Al_2Si_4O_{12}\cdot2H_2O$

Habit. Analcime is isometric and is usually found in trapezohedral crystals (Fig. 7-148), resembling one of the common forms of garnet. It occurs less frequently in cubes with three faces of the trapezohedron on each solid angle. In both form and color it resembles leucite, but its free-growing crystals, found in cavities in rocks, usually serve to distinguish it from leucite, which as a rock-forming

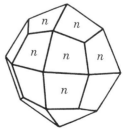

Fig. 7-148. *Analcime.*

mineral is embedded within a fine-grained rock mass. Sometime analcime is found as nodules in sedimentary clays, as at Calico, California.

Physical Properties. H = 5–5½; G = 2.27; the luster, vitreous; the color, colorless or white.

Composition. Analcime is a hydrous aluminum silicate, $Na_2Al_2Si_4O_{12} \cdot 2H_2O$. It fuses before a blowpipe to a colorless glass and gelatinizes with hydrochloric acid.

Natrolite, $Na_2Al_2Si_3O_{10} \cdot 2H_2O$

Habit. Natrolite is orthorhombic, usually occurring in fine acicular, or needlelike, crystals. For this reason it is sometimes called the *needle zeolite.* These crystals are often arranged in radiating tufts lining cavities in the enclosing rock. When crystals are larger, they show that the form is nearly a square prism with a low pyramid on the summit. There are also massive varieties having a fibrous or fine-columnar radiated habit.

Physical Properties. H = 5–5.5, G = 2.25. Although rarely seen, natrolite has a perfect prismatic cleavage. The luster is vitreous but may be pearly. Natrolite is commonly colorless or white but may be gray, yellow, or reddish.

Composition. Natrolite is a hydrous sodium aluminum silicate, $Na_2Al_2Si_3O_{10} \cdot 2H_2O$. The composition is partially reflected in the name, which comes from the Latin word meaning *sodium.* Natrolite fuses to a clear transparent glass, coloring the flame yellow. It is soluble in hydrochloric acid and gelantinizes upon evaporation.

Sheet (Phyllo-) Silicates

As with the framework silicates, the basic structural unit is the SiO_4 tetrahedron, but in the sheet silicates the linking of the tetrahedra is confined to a single plane (Fig. 7-119f), and this is what gives all the minerals in this division such perfect platy cleavage. The smallest unit that reproduces the sheet has the formula $(Si_4O_{10})^{4-}$, and this is best illustrated in kaolinite and serpentine. As with the framework silicates, aluminum substitutes for silicon. The substitution may either be for just one of the silicons $(AlSi_3O_{10})^{5-}$, as in the micas, or the substitution may be random $(Al,Si)_4O_{10}$, as in chlorite.

Sheet Silicates

Kaolinite, $Al_4Si_4O_{10}(OH)_8$	Serpentine, $Mg_6Si_4O_{10}(OH)_8$
Pyrophyllite, $Al_2Si_4O_{10}(OH)_2$	Talc, $Mg_3Si_4O_{10}(OH)_2$
Mica group	Chlorite group
Muscovite, $KAl_2(AlSi_3O_{10})(OH)_2$	Chlorite, $(Mg,Fe,Al)_6(Al,Si)_4O_{10}(OH)_8$
Biotite, $K(Mg,Fe)_3(AlSi_3O_{10})(OH)_2$	Other sheet silicates
Phlogopite, $KMg_3(AlSi_3O_{10})(OH)_2$	Apophyllite, $KCa_4(Si_4O_{10})_2F \cdot 8H_2O$
Lepidolite, $K(Li,Al)_3(Al,Si)_4$-	Prehnite, $Ca_2Al_2Si_3O_{10}(OH)_2$
$O_{10}(OH,F)_2$	Chrysocolla, $Cu_4H_4Si_4O_{10}(OH)_8$

Kaolinite, $Al_4Si_4O_{10}(OH)_8$

Kaolinite is an important mineral, for it makes up a large part of the soil mantle that covers a high percentage of the rocks of the earth's crust. It is the chief constituent of kaolin or clay. Other clay minerals of lesser importance are *dickite, nacrite, halloysite,* and *beidellite.* All these minerals are of secondary origin; that is, they have been derived by alteration of aluminum silicates, particularly feldspar.

Clay is one of the most important natural substances used in industry. Common brick, tile, pottery, sanitary ware, and porcelain are a few of the clay products. The chief value of clay lies in the ease with which it can be molded into any desired shape; and then, when the clay is heated, part of the combined water is driven off, producing a hard, inert substance.

Habit. Kaolinite is monoclinic, but its closest approach to crystals are thin, rhombic, or hexagonal-shaped plates. It is usually in earthy fine-grained masses.

Physical Properties. $H = 2-2\frac{1}{2}$, $G = 2.6$. Kaolinite has perfect platy cleavage parallel to the basal pinacoid. But because of the minuteness of the individual particles, they are seldom distinguishable. The luster is earthy. When pure, kaolinite is white, but it is often variously colored by impurities.

Composition. Kaolinite is a hydrous aluminum silicate, $Al_4Si_4O_{10}(OH)_8$. Although it is a chemical compound corresponding to the formula given, it cannot be distinguished from the other clay minerals without optical or x-ray tests.

Occurrence. As already mentioned, kaolinite is a widespread mineral in the soils of the earth's surface, but as such it is relatively impure. In certain places, large masses of feldspar have altered to pure kaolinite, which can be mined and manufactured into white porcelain or china. In Cornwall, England, such large masses were formed by low-temperature hydrothermal alteration of the feldspar in granite. Some of the kaolinite formed by the weathering of feldspar remains in place to form a mantle of soil over the rocks. Some of it is washed into the streams and may thus find its way into the ocean or a lake to be deposited in layers that, on compaction, are called clay. Pure layers of such clay are found around Augusta, Georgia. Usually, such sedimentary clay deposits are impure and thus cannot be used for high-quality china or porcelain but are used as fillers, as paper coatings, and in the manufacture of brick, tile, and other ceramic products.

Pyrophyllite, $Al_2Si_4O_{10}(OH)_2$

Pyrophyllite received its name from the Greek meaning *fire* and *leaf* because it exfoliates on heating. It is quarried and has much the same uses as talc and by inspection is difficult to distinguish from that mineral. A compact variety known as *wonderstone* from South Africa can be machined in a lathe into a variety of shapes so that it can be used as high-temperature electrical insulators, as foundry cores, and as high-pressure capsules in the manufacture of diamonds.

Habit. Pyrophyllite is monoclinic, but crystals are rare. It is most commonly found as radiating lamallar aggregates or as compact masses.

Physical Properties. H = 1–2, G = 2.8. Pyrophyllite has perfect platy cleavage parallel to the basal pinacoid. As with talc, the laminae are flexible but not elastic. The luster is pearly to greasy. Pure varieties are off-white, gray, or green.

Composition. Pyrophyllite is hydrous aluminum silicate, $Al_2Si_4O_{10}(OH)_2$. It can be distinguished from talc by a test for aluminum.

Occurrence. Pyrophyllite forms by the dehyration of kaolinite during the medium-grade metamorphism of aluminous shales. Fine specimens in the form of rosettes occur at the Brewer mine, South Carolina. Here solutions that formed quartz veins have reacted with aluminous host rocks. Other pyrophyllite-rich rocks occur at several localities in the Piedmont region of North and South Carolina.

Serpentine, $Mg_6Si_4O_{10}(OH)_8$

Serpentine is a remarkable mineral because it assumes a variety of forms (it is not known to occur in crystals). The crystals of serpentine that are found are pseudomorphs, having been derived from some other mineral by chemical change. Thus the magnesium-iron silicate, olivine, is often changed to serpentine, but the form of the olivine is retained.

Serpentine assumes commercial importance because of the fine fibrous variety, *chrysotile,* the most abundant type of asbestos (Fig. 7-149). Although two varieties of fibrous amphibole are used as asbestos, 95% of the asbestos produced is chrysotile, which is used in the manufacture of heat and fire-resistant fabrics. Because of the flexibility of the fibers, it is woven into cloth and gloves worn while tending hot furnaces and suits for fighting fire. The shorter fibers are used in products for insulating

Fig. 7-149. *Chrysotile asbestos in serpentinite, Thetford, Quebec.*

against heat and electricity. In recent years these uses have been curtailed because of the presumed carcenogenic nature of asbestos. However, the principal type of asbestos used in the United States is chrysotile, which has been shown to be far less hazardous than the amphiboles riebeckite ("blue asbestos") and amosite ("brown asbestos"). A massive variety of green serpentine quarried as a building stone and for ornamental purposes is called *verd antique* marble.

Habit. Serpentine is monoclinic, but as has been stated, crystals are unknown. The closest approach to a crystal is found in the fibrous variety, chrysotile. A platy type of serpentine is known as *antigorite,* while the massive polymorph is known as *lizardite.*

Physical Properties. $H = 4$, $G = 2.5$. The hardness varies from 2 to 5 but is usually about 4. The softer material is easily scratched and often has a smooth feel, sometimes almost greasy. The specific gravity is 2.2 in fibrous varieties but may be as high as 2.65 when serpentine is massive. Light and dark shades of green may be present on the same specimen, giving a mottled appearance. The luster in the massive varieties is greasy or waxlike, but in the fibrous material it is silky.

Composition. Serpentine is hydrous magnesium silicate, $Mg_6Si_4O_{10}(OH)_8$. Iron aluminum and nickel may be present in small amounts. When heated in a closed tube, it yields considerable water and can thus be distinguished from varieties of amphibole that appear similar. It is decomposed by hydrochloric acid.

Occurrence. Serpentine is probably formed by the moderate-temperature (under 400°C) alteration by seawater of rocks composed of olivine, enstatite, and other pyroxenes which underlie the floors of the oceans. As such rocks may later be exposed in the cores of mountain ranges, serpentine is a common and widely distributed mineral. Such examples are found along the Appalachians, the Coast Ranges of California, the Alps, and similar mountain ranges throughout the world. The name *serpentinite* is used for those rock masses made up largely of the varieties lizardite and antigorite. Chrysotile asbestos is formed by fluid-induced replacement recrystallization along tension fractures in serpentinite, as illustrated by the seams of asbestos in massive serpentine which have been mined extensively at Thetford, Quebec. Russia is the largest producer of chrysotile asbestos from deposits on the eastern slope of the Ural Mountains. Serpentine is also found as an alteration product of forsterite in contact-metamorphic dolomites.

The weathering of serpentine can result in a number of minor secondary minerals, of which *garnierite* is economically the most important. It is a hydrous silicate of nickel and magnesium having a bright green color. Garnierite appears to be amorphous and thus is found as incrustations or in massive form. It has served as an ore of nickel and been mined on the island of New Caledonia, at Riddle, Oregon, and at Nicaro, Cuba.

Talc, $Mg_3Si_4O_{10}(OH)_2$

Talc is the softest of the minerals and is thus placed at the beginning of the scale of hardness. It is easily scratched by a fingernail and has a slippery, soapy feel. For this reason, rock made up mostly of talc is called *soapstone* or *steatite.*

Talc has many uses. In powdered form it is used in the manufacture of paint, paper, roofing material, and rubber as well as in cosmetics in the form of face and talcum powder. Small parts fashioned from talc and then fired in a furnace to drive off the water become hard and durable and are used in electrical appliances. Slabs of soapstone are used for laboratory tabletops, washtubs, and sinks. Today, less talc is used in this way than formerly, for modern tubs and other items made of steel with an enamel coating are not only better looking but also more sanitary.

Habit. Talc is monoclinic. Crystals are rare, and those that are seen are usually thin plates with a rhombic or hexagonal outline. It is found most commonly in foliated or compact masses.

Physical Properties. H = 1, G = 2.7–2.8. Talc has perfect platy cleavage parallel to the basal pinacoid, but it can be observed in only certain types. Plates of talc are flexible but are not elastic and when bent will not return to their initial position. The luster is pearly, especially in the foliated kinds, and the color in the finest of these is a beautiful sea-green. There are also white foliated varieties. The massive material may be white, dark gray, or green.

Composition. Talc is hydrous magnesium silicate, $Mg_3Si_4O_{10}(OH)_2$. It is an inert mineral and, unlike serpentine, is unattacked by acids. It yields water on intense heating in a closed tube.

Occurrence. Like serpentine, talc is formed by the alteration by seawater of rocks composed of olivine and pyroxene underlying the ocean floors. Even though it is considered a higher-temperature (above 400°C) mineral, it occurs commonly as soapstone associated with serpentine, chlorite, and amphiboles such as tremolite or anthophyllite. It can also form directly from the thermal metamorphism of serpentine. High-grade talc usually forms by the thermal metamorphism of siliceous dolomites and has been mined at Murphy, North Carolina; Chatsworth, Georgia; at various places in New York and Vermont; the Inyo Range, California; and Llano, Texas.

Mica Group

The mica group is made up of several closely related members that have many properties in common. Chief among these properties is the perfect cleavage, by means of which the micas may be split into leaves much thinner than a sheet of paper. In fact, it is difficult to set a lower limit to the thickness of a cleavage plate. These leaves or sheets are usually very elastic and spring back when bent. Mica crystals are usually platy with an outline that is either rhombic (Fig. 4-8) with angles close to 120° and 60° or hexagonal with all the angles nearly 120°. The micas are complex aluminum silicates with potassium and hydroxyl and, in some varieties, sodium, lithium, magnesium, and iron.

Another group of minerals, the *brittle micas,* is similar to the micas in appearance. However, as the name implies, the folia are brittle and not elastic as the true micas. The most important brittle micas are *margarite, ottrelite,* and *chloritoid.*

Muscovite, $KAl_2(AlSi_3O_{10})(OH)_2$

Muscovite is variously called *white mica, common mica,* and *potash mica.* Plates of it, called *isinglass,* are clear and transparent and have been used for many purposes. Because it is not affected by heat, muscovite is used for openings in stoves and furnaces. In the early days when glass was difficult to obtain, it was even used for the windows of houses, particularly in Russia, where muscovite was abundant (the old name for the country, *Muscovy,* was given to the mineral).

Today, most uses of muscovite are based on its excellent properties as an electrical and heat insulator rather than on its transparency. Many of the small parts used for electrical insulation are punched from sheets of mica. Other parts are built up of thin sheets of mica cemented together and pressed into shape before the cement hardens. Because it was once a critical material for making condensers and radio tubes, many tons of muscovite were flown from India to the United States during World War II.

Much muscovite that is mined is too fine-grained to be used as sheet mica but is called "scrap mica." This is ground and used in the manufacture of wallpaper and roofing material. Mixed with oil it is used as a lubricant.

Habit. Muscovite is monoclinic, but crystals are rare. When occasionally seen, they appear orthorhombic or hexagonal (Fig. 7-150). The prisms usually taper sharply and are rough, making the crystals appear dark. However, more light can pass through the crystal at right angles to the prism than perpendicular to the basal cleavage. Even when the sheets have no regular shape, their structure conforms to the pseudohexagonal outline; this is illustrated when a blunt point is held against a sheet and struck a blow with a hammer. A six-rayed star with branches intersecting at angles of 60° results (see the center of Fig. 7-151). Two of these branches are parallel to the prism faces, and the third is parallel to the side pinacoid. That the structure is pseudohexagonal is shown by inclusions of magnetite in muscovite that form a network along lines having the same directions (Fig. 7-152).

Fig. 7-150. *Muscovite crystal.*

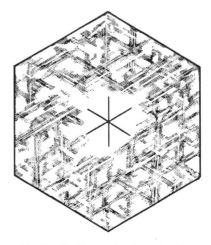

Fig. 7-151. *Percussion figure in mica.*

Fig. 7-152. *Oriented magnetite inclusions in muscovite.*

Muscovite most commonly occurs in scales or sheets without any regular form, its crystallization having been constrained by the surrounding minerals. Besides occurring in distinct plates it sometimes occurs in compact masses of minute scales known as sericite, other times with a featherlike habit, as in *plumose mica.* Often, the plates or scales are arranged in spherical aggregates.

Physical Properties. H = 2–2½, G = 2.76–2.88. As mentioned earlier, muscovite has highly perfect platy cleavage parallel to the basal pinacoid, permitting the mineral to be split into extremely thin sheets. The luster is vitreous to pearly. In thin sheets muscovite is colorless and transparent, but in thicker blocks it may be dark brown. Although smoky brown is the common color, it may be yellow, pink, or green.

Composition. Muscovite is a silicate of aluminum and potassium having essentially the formula $KAl_2(AlSi_3O_{10})(OH)_2$. It yields a small amount of water when heated very hot in the closed tube. It is fused with difficulty and usually melts only on very thin edges.

Occurrence. Muscovite is a common and abundant rock-forming mineral, usually found in scales or plates. It is formed during the metamorphism of sediments and felsic volcanic rocks and is found over the full metamorphic range, including phyllites, schists, and gneisses as well as water-rich granites. It forms within the phyllic zone during hydrothermal alteration of feldspars, andalusite, and cordierite as well as during fluorine alteration (greisenization) produced by the crystallization of some granites. In some pegmatite dikes, where the ordinary consituents of granite, feldspar, quartz, and mica are coarsely crystallized, large crystals of mica are found and may be mined with success. In certain pegmatite dikes, muscovite is found in immense sheets a yard or more across. These dikes are very interesting to the mineralogist, for in them he looks not only for well-crystallized specimens of the three principal minerals named but also for many rarer minerals. Here are found tourmaline, beryl, apatite, garnet, and even rarer minerals, such as columbite, microlite, and those of uranium.

Muscovite has been mined from pegmatite dikes in Maine, New Hampshire, Connecticut, North Carolina, South Dakota, and to a lesser extent in several other states. Much of the mica used in the United States is imported, India being a major supplier.

Biotite, $K(Mg,Fe)_3(AlSi_3O_{10})(OH)_2$

Unlike muscovite, biotite has few uses and is of little commercial significance except when it has undergone a slight alteration to *vermiculite*. When biotite is heated, there is no visible change in the mineral, but when vermiculite is heated, it swells perpendicular to the cleavage to many times its former size. It puffs up, leaving air space between the plates, making it an effective material for heat and sound insulation. It is an interesting experiment to heat several flakes of vermiculite and watch them apparently wiggle and squirm as they swell up. The name vermiculite comes from the Latin words meaning *to breed worms.*

Habit. Biotite is monoclinic. Although it may occur rarely in short prismatic crystals, it is usually found in scales with irregular boundaries or scaly aggregates.

Physical Properties. H = 2½–3, G = 2.8–3.2. Like all the micas, the perfect platy cleavage is the outstanding physical property. The luster is bright and pearly on the cleavage surface but vitreous on the edges. The color is usually black but may be dark green or brown. The wide range in color corresponds to the variation in composition.

Composition. Biotite is a potassium magnesium iron aluminum silicate, essentially $K(Mg,Fe)_3(AlSi_3O_{10})(OH)_2$. However, some kinds contain little iron, and other kinds contain much iron and almost no magnesium. The iron varieties have the darker color.

Occurrence. Biotite is the most abundant mica. It occurs as small scales in a wide range of igneous rocks but is especially common in feldspar-rich rocks such as granodiorites, quartz monzonites, syenites, and nepheline syenites and in related kinds of volcanic rocks such as andesites. In special cases such as some lamprophyres, the igneous rock may consist almost entirely of biotite as around Enoree, South Carolina and at Palabora, South Africa. Vermiculite formed by weathering has been mined at these localities as well as at Libby, Montana. Biotite is a product of low-grade metamorphism of clay-rich sedimentary rocks but is lacking in rocks of the highest-grade metamorphism. In some pegmatite dikes it is found in large sheets but is less common than muscovite.

Phlogopite, $KMg_3(AlSi_3O_{10})(OH)_2$

Phlogopite is a magnesium mica somewhat resembling biotite in color, but because it lacks iron it has desirable electrical properties. Sheets of it are therefore used for the same purposes as muscovite, chiefly as an electrical insulator.

Habit. Phlogopite is monoclinic and occurs in six-sided crystals which are often rough tapering prisms. It is also found in flakes and foliated masses.

Physical Properties. H = 2½–3, G = 2.86. The perfect platy cleavage is the outstanding physical property. The color may be yellow, brown, green, or white and often has a copper-red appearance on the cleavage surface. The "star mica" of northern New York and Canada is a brown phlogopite which shows a fine six-rayed star when a point light source is viewed through it (see Fig. 4-7). This is an example of asterism which results from the presence of minute needlelike crystals of rutile included in the mica. The needles have been constrained by the pseudohexagonal structure of the mica to grow in three directions at essentially 60° to each other. The needles lie in positions parallel to the rays of the six-rayed star obtained by percussion (Fig. 7-151).

Composition. Phlogopite is a hydrous potassium magnesium aluminum silicate, $KMg_3(AlSi_3O_{10})(OH)_2$. Unlike muscovite, it is decomposed by boiling concentrated sulfuric acid.

Occurrence. Phlogopite is characteristically found in metamorphosed limestones and dolomites as isolated crystals disseminated through the rocks, and this usually serves to distinguish it from muscovite and lighter-colored biotites. Phlogopite is also found with serpentine in kimberlites and in leucite-bearing rocks such as lamproites.

Lepidolite, $K(Li,Al)_3(Al,Si)_4O_{10}(OH,F)_2$

Lepidolite, frequently called *lithia mica,* is rare compared to the other micas and is found almost exclusively in pegmatites. In certain places it has been mined as a source of lithium for manufacture into various lithium compounds.

Habit. Lepidolite is monclinic; crystals are rare. It is often found in masses that seem to have a close granular structure. Individual scales are sometimes large, and occasionally, like other micas, lepidolite is found in plates.

Physical Properties. H = 2½–4, G = 2.8–3.0. Perfect platy cleavage parallel to the basal pinacoid is present as in the other micas but is not well shown in the granular compact varieties. The luster is pearly, and the color, commonly pink to lilac, is diagnostic when present, but it may also be grayish white or yellow.

Composition. Lepidolite is a potassium lithium aluminum fluorosilicate; the composition can be expressed by the formula $K(Li,Al)_3(Al,Si)_4O_{10}(OH,F)_2$. It fuses easily before a blowpipe, and because of the presence of the lithium, it yields a red flame. In this manner it can be distinguished from muscovite.

Occurrence. As mentioned above, lepidolite is found intergrown with quartz and feldspar in pegmatite dikes, where it is associated with other lithium-bearing minerals. Chief among its associated minerals are the pink and green varieties of tourmaline, spodumene, and amblygonite. It is found in various localities in Maine; Portland, Connecticut; the Black Hills, South Dakota; Pala, California; and at Bikita, Zimbabwe.

Chlorite Group

The chlorite group is made up of several minerals that are so closely related that it is impossible for the beginner to distinguish between them. The principal members of the group clinochlore, penninite, and prochlorite are collectively described below under the heading "Chlorite."

Chlorite, $(Mg,Fe,Al)_6(Al,Si)_4O_{10}(OH)_8$

Habit. The chlorites are monoclinic and crystallize in tabular crystals with pseudo-hexagonal outlines; their habit is similar to that of the micas. Distinct crystals are rare, and chlorite is most commonly found in massive aggregates and finely disseminated particles.

Physical Properties. H = 2–2½, G = 2.6–2.9. Chlorite has perfect platy cleavage parallel to the basal pinacoid. The thin cleavage plates can be distinguished from mica, for when they are bent they show no elasticity; that is, they will not return to the initial position as in the micas. The color is green of various shades. The name *chlorite* is from the Greek word for *green,* the word that has given the name to the yellow-green gas chlorine, but there is no other relation between the chlorites and chlorine. Some chlorite is pale yellow, white, or rose-red.

Composition. The chlorites are hydrous iron-magnesium aluminum silicates, $(Mg,Fe,Al)_6(Al,Si)_4O_{10}(OH)_8$. They fuse with difficulty. At high temperature, water is given off in a closed tube.

Occurrence. Chlorite is one of the defining minerals for the presence of greenschist metamorphism, where it is commonly associated with epidote, actinolite, and albite. It forms when clay-rich sediments and mafic igneous rocks are subjected to low-temperature metamorphism. Chlorite is also commonly found in the propylitic zone of hydrothermal alteration, where it forms by the alteration of pyroxenes, amphiboles, and biotite in quartz monzonites and similar granitic rocks as well as their volcanic equivalents, such as andesite. Chlorite is also found along fractures and in amygdaloidal cavities in basalts with epidote, quartz, calcite, and zeolites. Fine crystals of clinochlore have been found in a highly metamorphosed bedded magnetite carbonate body at the Tilly Foster mine, Brewster, New York.

Other Sheet Silicates

Apophyllite, $KCa_4(Si_4O_{10})_2F \cdot 8H_2O$

Habit. Apophyllite is tetragonal and occurs in square prismatic or pyramidal crystals but with a considerable variety in habit (Figs. 7-153 to 7-155). The crystals may be simple square prisms terminated by the basal pinacoid and may look like cubes. Other crystals with these forms may also have the faces of a dipyramid lying over the corners and thus appear similar to the combination of cube and octahedron. It should be noted, however, that the angle made by the pyramid (p) and the base (c) is 119°28′, not the same as that made by p and the prism a, which is 128°. Further, it is noticed on close examination that the faces a have a vitreous luster and are vertically striated, but the base c has a pearly luster and is often dull.

Physical Properties. H = 4½–5, G = 2.3–2.4. Apophyllite has perfect platy cleavage parallel to the basal pinacoid, and hence the pearly luster is noticed on this face.

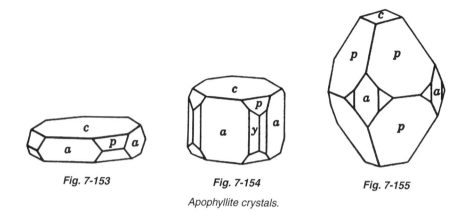

Fig. 7-153   Fig. 7-154   Fig. 7-155

Apophyllite crystals.

Elsewhere the luster is vitreous. The crystals are usually colorless or white but may show pale shades of green, yellow, or rose.

Composition. Apophyllite is hydrous calcium potassium silicate containing a small amount of fluorine, $KCa_4(Si_4O_{10})_2F \cdot 8H_2O$. It differs from the zeolites, with which it is associated, by containing no aluminum. Before a blowpipe it exfoliates, whitens, and fuses to a vesicular enamel. The name *apophyllite* refers to this property of exfoliating.

Occurrence. Apophyllite is found associated with zeolites, calcite, and datolite in amydaloidal cavities in basalts. In this setting, beautiful crystals have been found in Bergen Hill and Paterson, New Jersey, Nova Scotia, and India.

Prehnite, $Ca_2Al(AlSi_3O_{10})(OH)_2$

Habit. Prehnite is orthorhombic. It seldom occurs in distinct crystals but usually in crystalline masses with a botryoidal or mammillary surface, or in groups of tabular crystals showing a series of faces in parallel position.

Physical Properties. $H = 6-6\frac{1}{2}$, $G = 2.8-2.9$. The luster is vitreous, and although the common color is green, it is sometimes white or gray.

Composition. Prehnite is hydrous calcium aluminum silicate, $Ca_2Al(AlSi_3O_{10})$-$(OH)_2$. It fuses rather easily and yields a little water in a closed tube. Compared to the zeolites, it is slowly decomposed by hydrochloric acid.

Occurrence. Prehnite is associated with the zeolites and, like them, is a mineral of secondary origin. It is found lining cavities in basalts and related rocks. Beautiful specimens lining cavities have come from traprock at Bergen Hill and Paterson, New Jersey.

Chrysocolla, $Cu_4H_4Si_4O_{10}(OH)_8$

Chrysocolla ($H = 2-4$, $G = 2.0-2.4$) is a copper silicate formed by the weathering and oxidation of copper minerals and is associated with malachite, azurite, and cuprite in the surface cap over copper deposits. It occurs in cryptocrystalline or massive aggregates of a blue-green or sky-blue color with a composition of $Cu_4H_4Si_4O_{10}(OH)_8$. Chrysocolla resembles turquois but can be distinguished by its lower hardness. It has been mined as a minor ore of copper in New Mexico and Arizona; at Chuquicamata, Chile; and in the Congo.

Chain (Ino-) Silicates

In the chain silicates the SiO_4 units are linked together to form infinitely long strands. In this sense the silicates are in fact radicals: that is, molecules with an electrical charge. This charge is neutralized by cations that glue the strands together. Two types of strands are important, single chains (Fig. 7-119d), which form the pyroxenes, and double chains (Fig. 7-119e), which form the amphiboles. For single strands the basic repeat unit is $(Si_2O_6)^{4-}$, while for the double-chain silicates the basic repeat unit is $(Si_8O_{22})^{12-}$. In these silicates the substitution of aluminum for silicon is random and is written as $(Al,Si)_2O_6$ or $(Al,Si)_8O_{22}$. This random substitution requires random compensation in the compensating cations. Thus the relatively simple formula for diopside, $CaMgSi_2O_6$, becomes the complex formula $(Ca,Na)(Mg,Fe,Al) (Al,Si)_2O_6$ in augite.

Chain Silicates

Amphiboles	Pyroxenoids
Tremolite, $Ca_2Mg_5Si_8O_{22}(OH)_2$	Rhodonite, $MnSiO_3$
Hornblende, $NaCa_2(Mg,Fe,Al)_5(Si,Al)_8O_{22}(OH)_2$	Wollastonite, $CaSiO_3$
Pyroxenes	
Enstatite, $Mg_2Si_2O_6$	
Diopside, $CaMgSi_2O_6$	
Augite, $(Ca,Na)(Mg,Fe,Al)(Al,Si)_2O_6$	
Spodumene, $LiAlSi_2O_6$	
Jadeite, $NaAlSi_2O_6$	

Amphiboles

Included in the amphibole group of minerals are several species closely related to each other both chemically and structurally, although they crystallize in the orthorhombic, monoclinic, and triclinic systems. Furthermore, members of this group resemble corresponding members of the pyroxene group (p. 262). It is therefore important to know how to distinguish members of one group from members of the other.

The advanced mineralogist can make use of the optical properties to distinguish the amphiboles from the pyroxenes, but the beginner must rely on the physical properties, chiefly on the cleavage. Despite the fact that the amphiboles crystallize in three different crystal systems, they all have similar prismatic or pseudoprismatic cleavage. The intersection of these cleavage planes is at approximately 56° and 124°, as shown in Fig. 7-156. The pyroxenes also have prismatic cleavage, but the prism faces intersect at nearly right angles (87° and 93°), as shown in Fig. 7-157. Thus,

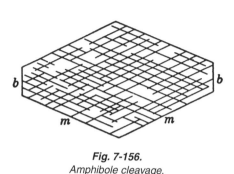

Fig. 7-156.
Amphibole cleavage.

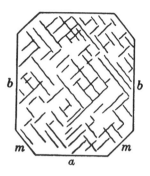

Fig. 7-157.
Pyroxene cleavage.

with a little practice, one can quickly tell whether the specimen in hand is an amphibole or pyroxene. The inherent reason for the differences in cleavage is in the nature of the silicate chains. The pyroxenes consist of strands of single chains of linked silica tetrahedra, and the equant nature of the chains yields a nearly 90° cleavage angle. In the case of the amphiboles, the strands consists of two conjoined chains, which broadens the angle to 124°. Although there are many exceptions, in general the amphiboles tend to be long and slender and the pyroxenes, short and stout.

Only two of the amphiboles, tremolite and hornblende, are common enough to warrant description here. Both are monoclinic.

Tremolite, $Ca_2Mg_5Si_8O_{22}(OH)_2$

Habit. Tremolite is monoclinic and commonly occurs in prismatic crystals. It is found also in bladed and columnar aggregates and in some places occurs in silky fibers that were once used as asbestos. However, the amphibole varieties have been shown to cause lung cancer and should be used no longer.

Physical Properties. H = 5–6, G = 3.0–3.3. Tremolite has perfect prismatic cleavage, with the two directions intersecting at angles of 124° and 56°. The color varies from white to green; the darker green varieties are called *actinolite.* A closely compact, tough vareity is called *nephrite,* one of the two types of *jade;* the other type is *jadeite,* a pyroxene. Nephrite jade is usually green, but some specimens are mottled with nearly white to dark green patches. A loosely felted variety of tremolite known as *mountain leather* and *mountain cork* has properties similar to the materials from which it is named.

Composition. The formula for pure tremolite is $Ca_2Mg_5Si_8O_{22}(OH)_2$. However, iron is usually present replacing some of the magnesium and giving the mineral a green color. The higher the percentage of iron, the darker the mineral. It is called *actinolite* when iron is present in amounts greater than 2%.

Occurrence. Tremolite is most commonly formed during the metamorphism of siliceous dolomites at intermediate metamorphic conditions, where it is commonly associated with calcite, talc, zoisite, and grossular. Along with epidote, chlorite, albite, and sometimes serpentine or talc, actinolite occurs in low-grade metamorphism of both basic igneous rocks and clay-rich sediments.

The finely fibrous asbestiform variety of tremolite has come from Switzerland. *Nephrite* has for centuries come from the Kuen Lun Mountains of southern Turkestan and been used by Chinese carvers. Today, it is found in many parts of the world. Important sources are near Lake Baikal, Russia; New Zealand; Taiwan; and British Columbia, Canada (see Plate VII-4). In the United States, nephrite is found in Alaska and near Lander, Wyoming. The other kind of jade, *jadeite,* resembles nephrite both in color and in the tough compact aggregates of fibers in which it occurs. It can be distinguished from nephrite by its higher specific gravity.

Hornblende, $NaCa_2(Mg,Fe,Al)_5(Si,Al)_8O_{22}(OH)_2$

Habit. Hornblende is monoclinic, and prismatic crystals terminated by a first-order prism (*r*) are common (Figs. 7-158 to 7-160). It is also found in columnar and bladed aggregates.

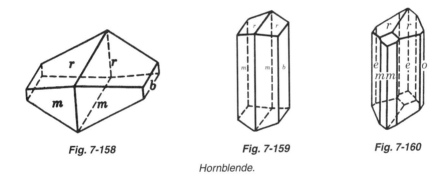

Fig. 7-158   Fig. 7-159   Fig. 7-160

Hornblende.

Physical Properties. H = 5–6, G = 3.2. As in all amphiboles, hornblende has perfect prismatic cleavage at angles of 56° and 124°. The luster is generally vitreous but is silky in fibrous varieties. The color is dark green to black, and although it may appear opaque, it will transmit light on thin edges.

Composition. Hornblende has a complex chemical composition containing calcium, sodium, magnesium, iron, fluorine, and water. In addition, it contains aluminum, an element not found in tremolite. Hornblende fuses with difficulty and yields water in a closed tube.

Occurrence. Hornblende is one of the common rock-forming minerals and is found in both igneous and metamorphic rocks. Although present in igneous rocks, it is often formed by hydrous alteration of pyroxene, but it also crystallizes directly from silicate melts crystallizing at moderate depths and is common as the dark grains disseminated through light-colored syenites and diorites. Hornblende is particularly common in dark-colored amphibolites formed by the medium-grade metamorphism of pyroxenes and feldspars in basic igneous rocks. It also forms at the higher grades of metamorphism of clay-rich sediments to form with quartz and feldspar hornblende schists and hornblende gneisses.

Pyroxenes

The pyroxenes form one of the most important of the silicate groups. As stated on page 259, they are similar to the amphiboles in many respects. It is important enough to restate here that the only way for the beginner to distinguish between members of the two groups is by the cleavage. The cleavage of the amphiboles is at 124° and 56°, and the cleavage of the pyroxenes is at 93° and 87°. The pyroxene diopside is similar to tremolite both in its appearance and in its association, and the pyroxene augite is similar to hornblende. Both of these minerals are monoclinic, but there is another pyroxene, enstatite, which is orthorhombic.

Enstatite, $Mg_2Si_2O_6$

Habit. Enstatite is orthorhombic; all the other pyroxenes are monoclinic. Crystals are rare, and it is usually massive or fibrous.

Physical Properties. $H = 5\frac{1}{2}$, $G = 3.2–3.5$. Enstatite has good prismatic cleavage at 87° and 93°. The color of pure enstatite is grayish white, but when iron is present it may be yellowish, olive-green, or brown. *Bronzite* is a variety with a bronzelike luster.

Composition. Enstatite is magnesium silicate, $Mg_2Si_2O_6$, when pure, but iron may be present replacing magnesium until the two are in equal amounts. When iron is present from 5 to 13%, the mineral is called *bronzite,* but when iron is between 13 and 20%, it is called *hypersthene* and is dark in color.

Occurrence. Enstatite is a common mineral along with olivine, augite, and plagioclase in the dark-colored ultramafic and mafic igneous rocks pyroxenite, peridotite, and gabbro; it also occurs in both iron and stony meteorites. It is a significant mineral in large layered igneous complexes such as the Bushveld, South Africa and Stillwater, Montana. Smaller peridotite masses are found at Edwards, New York and Webster, North Carolina.

Diopside, $CaMgSi_2O_6$

Habit. Diopside is monoclinic and is found in prismatic crystals showing a square or eight-sided cross section. It may also be granular.

Physical Properties. $H = 5–6$, $G = 3.2–3.3$. Diopside has imperfect prismatic cleavage at angles of 87° and 93°. A well-developed parting is present in some crystals parallel to the basal pinacoid. A less common parting parallel to the front pinacoid is characteristic of the variety *diallage.* The color is white to light green and in varieties rich in iron, may be deep green.

Composition. Diopside is a calcium magnesium silicate, $CaMgSi_2O_6$ while *hedenbergite* is calcium iron silicate, $CaFeSi_2O_6$. A complete solid-solution series exists between these two minerals. Chromium is present in the rare *chrome diopside,* a vivid green variety.

Occurrence. Diopside is a mineral characteristic of contact-metamorphic deposits, where it is found in limestone associated with forsterite, garnet, vesuvianite, tremolite, and titanite. It is also found in high-grade regionally metamorphosed rocks, and the variety diallage is present in some gabbros and peridotites. Pale green transparent crystals occur in calcite matrix in dolomite xenoliths in quartz monzonite in the Organ Mountains, New Mexico. Similar crystals with microcline in pegmatites from DeKalb Junction, New York have been cut as gemstones. Chrome diopside makes a highly prized gem.

Augite, (Ca,Na)(Mg,Fe,Al)(Al,Si)$_2$O$_6$

Habit. Augite is monoclinic, and crystals with a prismatic habit are common (Figs. 7-161 to 7-164). When the faces in the prism zone are the front pinacoid, side pinacoid, and prism, as in Fig. 7-161, they make angles of nearly 45° with each other.

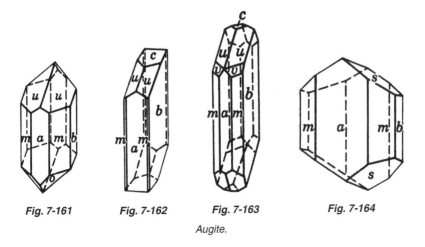

| Fig. 7-161 | Fig. 7-162 | Fig. 7-163 | Fig. 7-164 |

Augite.

Physical Properties. H = 5–6, G = 3.2–3.4. Prismatic cleavage yielding two directions at nearly right angles to each other is the most characteristic feature of augite, and by means of it one can distinguish it from hornblende. The color is dark green to black and in the average specimen appears opaque, for only on the thin edges will it transmit light.

Composition. The formula for augite is (Ca,Na)(Mg,Fe,Al)(Al,Si)$_2$O$_6$. It differs from the other pyroxenes in that it contains aluminum.

Occurrence. Augite is the commonest of the pyroxenes and is present as one of the chief dark constituents of many igneous rocks. It is particularly characteristic of dark mafic rocks such as basalt, gabbro, and peridotite and is found less commonly in the lighter-colored rocks, such as granite and syenite. Augite is also found with plagioclase in high-grade metamorphic rocks known as granulites, while the bright green variety *omphacite* is found with garnet in eclogites, a high-pressure rock with the composition of basalt consisting of pyroxene and garnet.

Spodumene, LiAlSi$_2$O$_6$

Spodumene is a rather rare mineral related to the pyroxenes but with a different type of origin and mineral association. In certain places where it occurs abundantly, it has been mined as a source of lithium. Spodumene is of particular interest because of the great size of some of the crystals, some of which measured over 47 feet long. Fine crystals of a lilac color, known as *kunzite,* and a deep green, known as *hiddenite,* are cut as gems.

Habit. Spodumene is monoclinic and is frequently found in prismatic crystals with rough faces striated parallel to the *c* axis. Twinning on the front pinacoid is common.

Physical Properties. H = 6.5–7, G = 3.15–3.20. Spodumene has perfect prismatic cleavage at angles of 87° and 93°. A parting, which in some specimens is more perfect than the cleavage, is usually present parallel to the front pinacoid. The color most commonly is white to gray but may be yellow and as already mentioned, lilac and green in the gem varieties.

Composition. Spodumene is a lithium aluminum silicate, LiAlSi$_2$O$_6$. Before a blowpipe it fuses to a clear glass coloring the flame red.

Occurrence. Spodumene is found exclusively in pegmatite dikes associated with feldspar, beryl, tourmaline, and other lithium-bearing minerals, such as lepidolite and amblygonite. The outstanding occurrence is at the Etta mine in the Black Hills of South Dakota, where the gigantic crystals mentioned have been mined. It is also found in various pegmatites in New England; at Kings Mountain, North Carolina; and Bikita, Zimbabwe. Kunzite is found in beautiful crystals at Pala, California and in Madagascar. Hiddenite has been found at Stony Point, North Carolina.

Jadeite, NaAlSi$_2$O$_6$

One might think that all jade would be jadeite, but it is only one variety. Most jade is nephrite, a hard, tough, compact variety of tremolite. To distinguish objects of jadeite from those made of the less valuable nephrite, the term *precious jade* is often used for jadeite. The jade of greatest value is a vivid green semitransparent jadeite sometimes referred to as Imperial jade. The most practical way to distinguish between nephrite and jadeite objects is to measure their specific gravity.

Habit. Jadeite is monoclinic and is rarely found as crystals. It is usually found as tough fine-grained masses.

Physical Properties. H = 6½, G = 3.2–3.4. Its specific gravity is greater than that of nephrite (G ≈ 3.0). Jadeite has good prismatic cleavage, but this is rarely observed in hand samples. It is usually a bright green in color but maybe white.

Composition. It is a sodium aluminum silicate, NaAlSi$_2$O$_6$, and gives a strong flame of sodium.

Occurrence. Jadeite is a rare mineral formed on the metamorphism of sodium-rich rocks at high pressures. The only important producer of this ornamental and gem material is Myanmar (Burma), where it is mined from dikes of metamorphosed rock. It is also mined as boulders and pebbles from the gravels of neighboring streams. Poorer-quality jadeite is found in Guatemala and in several places in the Coast Ranges of California associated with the blue-schist facies minerals glaucophane, lawsonite, and pumpellyite.

Pyroxenoids

The remaining members of the chain silicates have more complicated linkages among the chains which differ from the simple arrangements of both the pyroxenes and amphiboles and are grouped together as the pyroxenoids.

Rhodonite, $MnSiO_3$

Rhodonite is named from the Greek word for *rose* because of its beautiful rose-red color. It is not a common mineral but is found rather abundantly in some localities. One of the outstanding localities is in the Ural Mountains, where it has been mined and used as an ornamental stone. Here it is cut into thin slices for the veneering of tabletops and other objects.

Habit. Rhodonite is triclinic but closely related crystallographically to the pyroxenes. Crystals are usually flattened parallel to the basal pinacoid. Most commonly, the mineral occurs in cleavable masses.

Physical Properties. H = 5½–6, G = 3.4–3.7. Rhodonite has pseudoprismatic cleavage parallel to the third-order pinacoids at angles of about 88° and 92°. Rose-red is the most characteristic color, but it may also be pale pink or brown. The color of rhodonite is very similar to that of rhodochrosite, the manganese carbonate, but rhodonite can be distinguished from rhodochrosite by its greater hardness. Both these minerals may show a surface alteration of a black manganese oxide when exposed to atmospheric conditions for a short while.

Composition. The formula for rhodonite is $MnSiO_3$. *Fowlerite* is the name given to the zinc-bearing variety found at Franklin, New Jersey (see Plate VIII-3). Calcium also may be present, replacing part of the manganese. Rhodonite fuses before a blowpipe to a black glass. When a small fragment is fused in a soda bead, a turquoise blue color results, indicating manganese.

Occurrence. It forms as in the Coast Ranges of California during the low-grade metamorphism of sediments enriched in manganese either from the accumulation of manganese nodules or by exhalite activity, as at Broken Hill, New South Wales, Australia. A granular massive variety is found as lenses in mica schist at the Betts manganese mine near Cummington, Massachusetts, and well-crystallized specimens come from Franklin, New Jersey. It may also be deposited with rhodochrosite in some veins. As has been mentioned, the most important locality is at Sverdlovsk in the Ural Mountains, where massive rhodonite is found in a contact-metamorphic skarn.

Wollastonite, $CaSiO_3$

Wollastonite (H = 5–5½, G = 2.8–2.9) forms during the metamorphism of siliceous limestones and is associated with diopside, tremolite, vesuvianite, and epidote. It is triclinic, but crystals are uncommon; the usual occurrence is in coarsely cleavable aggregates. Pseudoprismatic cleavage parallel to the front and basal pinacoids yields fragments elongated parallel to the *b* crystallographic axis. The luster is vitreous but may be silky in fibrous varieties; the color is white to gray.

Wollastonite is calcium silicate, $CaSiO_3$. It is decomposed by hydrochloric acid, and because of this fact, can be distinguished from tremolite, which appears similar but is insoluble. In some places, as at Willsboro, New York, it is the chief constituent of the rock mass within bedded metamophosed limestones and is mined for use in the manufacture of tile.

Ring (Cyclo-) Silicates

In the ring silicates, the SiO_4 tetrahedra are linked together as discrete rings. In this way both the isle and ring silicates differ from the framework, sheet, and chain silicates, which are indefinite in extent. Three-, four-, and six-membered rings are known, but only the six-membered ring (Fig. 7-119*c*), $(Si_6O_{18})^{12-}$, is common, and only in a few minerals such as cordierite does any aluminum substitution for silicon occur.

Ring Silicates

Beryl, $Be_3Al_2Si_6O_{18}$

Tourmaline, $(Na,Ca)(Mg,Fe,Li)_3Al_6(BO_3)_3(Si_6O_{18})(OH)_4$

Cordierite, $(Mg,Fe)_2Al_3(AlSi_5O_{18})$

Beryl, $Be_3Al_2Si_6O_{18}$

Beryl is the most common mineral containing the rare element beryllium and can thus be considered an ore of that interesting metal. Beryllium is much lighter than aluminum or magnesium and would thus have many uses if only it could be found in larger amounts. Since the supply is limited, beryllium is used only sparingly in alloys, to which it gives many desirable properties. However, it is not as an ore of beryllium but as a gemstone that beryl has attracted attention through the centuries. The blue-green *aquamarine* and the deep green *emerald* are gem varieties. Aquamarine is a valued gemstone, but a fine-colored emerald may be of greater value than a diamond of comparable size.

Habit. Beryl is hexagonal and is one of the few species that is almost always in distinct and characteristic crystals, which aid in identifying it. The common forms are the hexagonal prism and the base (Fig. 7-165); more rarely, pyramid faces are present (Fig. 7-166 and 3-72). Some crystals may reach huge proportions, as in the pegmatite at Albany, Maine, where crystals 3 feet in diameter and 18 feet long have been found.

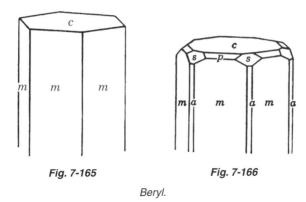

Fig. 7-165   Fig. 7-166

Beryl.

Physical Properties. H = 7½–8, G = 2.75–2.80. The hardness is just a little above that of quartz. Beryl has a poor platy cleavage parallel to the basal pinacoid, and it may not be observed in some specimens. The luster is vitreous, and in this respect beryl resembles quartz. The color is usually bluish green or light yellow, but it may be emerald green, golden yellow, pink, white, or colorless. When beryl is of gem quality, various names are given to the different-colored stones. The clear blue-green is *aquamarine;* the deep green, *emerald;* the pale pink to deep rose, *morganite;* and the clear yellow variety, *golden beryl.*

Composition. Beryl is a beryllium aluminum silicate, $Be_3Al_2Si_6O_{18}$. In pink and colorless beryl, small amounts of cesium are usually present replacing beryllium. The color of aquamarines and golden beryls is thought to be caused by small amounts of iron, while the color of emerald is attributed to minute amounts of chromium and sometimes vanadium.

Occurrence. Beryl is usually found with quartz, feldspar, and muscovite in pegmatite dikes associated with granitic rocks. The pegmatites of New England and North Carolina have afforded many beautiful specimens, while those in Brazil have served as a source of both beautiful specimens of gem beryls and many tons of rough crystals used as a source of beryllium. The finest emeralds come from Colombia, South America, where they occur with pyrite in calcite veins in black shales; the principal localities are Muso, El Chivor, and Cosquez. Good-quality emeralds are found in mica schist in the Transvaal, South Africa; Sverdlovsk, Russia; and Sandawana, Zimbabwe. Small pale emerald crystals, in biotite–phlogopite schist formed by contact metamorphism of serpentinites, come from North Carolina.

Tourmaline, $(Na,Ca)(Mg,Fe,Li)_3Al_6(BO_3)_3(Si_6O_{18})(OH)_4$

Tourmaline is one of the most attractive of the silicate minerals, and its varieties show a greater range of color than any other species. Fine-colored transparent crystals not only make superb mineral specimens but also have been cut and used as gemstones for many centuries (see Plate V-1,3).

Within recent years tourmaline has been used in certain scientific and industrial apparatus. This use, like one of the uses for quartz, is based on the property of piezoelectricity. Tourmaline is particularly suited for the construction of pressure gauges, and during World War II tourmaline gauges were used in measuring blast pressures from each atomic explosion.

Habit. Tourmaline is rhombohedral (Figs. 7-167 to 7-169). It is commonly in vertically striated prismatic crystals which show a triangular cross section. The three main

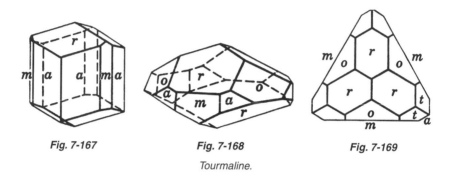

Fig. 7-167   Fig. 7-168   Fig. 7-169

Tourmaline.

faces are those of the trigonal prism, and the two smaller faces found at each of the points of the triangle are those of the second-order hexagonal prism. When crystals are doubly terminated, it is common to find a base at one end and a rhombohedron at the other. The commonest form is the obtuse rhombohedron (Fig. 7-167), which gives a terminal angle of about 133°. The different forms at the opposite ends of the crystal and the trigonal prism without parallel faces show that tourmaline belongs to a symmetry class lacking a center of symmetry. It should be remembered that only crystals without a center of symmetry show the property of piezoelectricity.

Although tourmaline commonly occurs in single crystals, it is also found in radiating groups and in massive aggregates (Figs. 7-170 and 7-171). The massive kinds

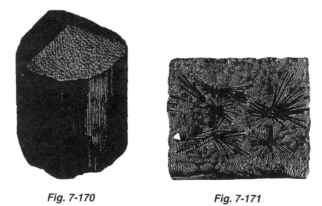

Fig. 7-170   Fig. 7-171

Tourmaline.

usually show a columnar habit, the mass appearing as if made up of a bundle of parallel crystals. There are also varieties that are compact.

Physical Properties. H = 7–7½, G = 3.0–3.25. Tourmaline has no cleavage but a conchoidal fracture. It is brittle, and the fracture of a black mass often resembles that of a piece of coal. Its characteristic fracture enables one to distinguish it from aggregates of amphibole and epidote. The luster is vitreous. The various colors, related to the chemistry, give rise to several varietal names. *Shorl,* the iron-rich black variety, is the commonest tourmaline. Magnesium is present in *dravite,* a brown tourmaline. The light-colored gem tourmalines, *elbaite* and *liddicoatite,* contain lithium and occur in several colors, to which the following names are given: red-pink, *rubellite;* blue, *indicolite;* green, *verdite;* and the rare colorless or white, *acrolite.* Not uncommonly, crystals differ in color in different parts. They may be pink at one end and green at the other, or there may be a pink center surrounded by a green exterior.

The property of piezoelectricity mentioned above has long been known in tourmaline, but only recently has any use been made of it. When a crystal is sequeezed at the ends of the *c* crystal axis, a positive electric charge is set up at one end and a negative charge at the other. The charge is proportional to the pressure exerted, and thus, with the proper recording equipment, it is possible to measure the pressure.

Composition. Tourmaline is a complex silicate of boron and aluminum with iron, magnesium, sodium, calcium, and lithium present in varying amounts. There is a rough correlation between color and composition. The common black tourmaline is rich in iron; the brown, in magnesium; and the pink, green, and blue, in lithium. The pink variety, rubellite, is often associated with the lithia mica lepidolite. The brown magnesian varieties fuse rather easily to a blebby enamel, the iron kinds fuse with difficulty, and the lithia variety is infusible.

Occurrence. Tourmaline occurs in a variety of rocks. It occurs with feldspar and quartz, as in the granites of southwest England. It is also common in quartz veins but is found especially well developed in pegmatite dikes. In these dikes the black iron-rich variety may be found near the walls, having formed at the same time as the enclosing quartz and feldspar. In many pegmatite dikes only the black tourmaline is found, but in the quartz core of some dikes, fine crystals of the bright-colored gem material may be present in cavities. World-famous localities for such gem tourmalines are in Brazil, Magagascar, and Russia; in the United States, pegmatites in Maine, California, and Connecticut have yielded many beautiful gemstones.

Tourmaline associated with quartz, orthoclase, cassiterite, and wolframite is common in high-temperature veins, as at Cornwall, England or Rooiberg, South Africa. The brown magnesian tourmaline occurs in contact-metamorphic limestones. It is also associated with galena, sphalerite, and pyrrhotite in exhalite mineral deposits such as the Sullivan mine, British Columbia, Canada.

Tourmaline may also be found in both schists and gneisses derived from the metamorphism of clay-rich rocks either continental or marine, which have been enriched in boron by either hot spring waters of by exhalite action. Tourmaline is highly resistant to weathering and may be found as a detrital mineral in sands of streams or beaches.

Cordierite, $(Mg,Fe)_2Al_3(AlSi_5O_{18})$

Cordierite, sometimes called *iolite* or *dichroite,* is a relatively rare mineral that is occasionally used as a gem. The name *dichroite* comes from the strong pleochroism shown by some specimens, which present different colors when viewed in different directions.

Habit. Cordierite is orthorhombic, but crystals usually appear hexagonal because of twinning. It is most common in massive embedded grains.

Physical Properties. $H = 7–7\frac{1}{2}$, $G = 2.6–2.66$. Cordierite has a poor cleavage parallel to the side pinacoid, but in many specimens cleavage is not observed. It has a vitreous luster, and the color is various shades of blue. When viewed in transmitted light, some specimens show a deep blue color in one direction and a very pale blue color in another. Cordierite resembles quartz in many of its physical properties and can be distinguished from it with difficulty.

Composition. Cordierite is a silicate of magnesium and aluminum, $(Mg,Fe)_2Al_3$-$(AlSi_5O_{18})$. There are no easy diagnostic tests for cordierite.

Occurrence. Cordierite forms either during contact metamorphism or during high-grade metamorphism of aluminous sediments. It is thus found in schists with biotite and andalusite or in hornfels and gneisses with sillimanite, corundum, orthoclase, and muscovite. In this way, cordierite is particularly abundant in the granulite terrane of Sri Lanka. More rarely, it occurs in certain pegmatites with andalusite or sillimanite, where crystals of gem quality may be found, as at Haddam, Connecticut; Micanite, Colorado; or Ankaditany, Madagascar.

Isle (Soro- and Neso-) Silicates

In the isle silicates, the SiO_4 tetrahedra are found as doublets (Fig. 7-119*b*), $(Si_2O_7)^{6-}$, as singlets (Fig. 7-119*a*), $(SiO_4)^{4-}$, or as in epidote, mixtures of both. Chemical formulas for most species are simple because there is no substitution of aluminum for silicon.

Isle Silicates

Double-isle silicates	
Hemimorphite, $Zn_4Si_2O_7(OH)_2 \cdot H_2O$	Zoisite, $Ca_2AlAl_2O(SiO_4)(Si_2O_7)(OH)$
Vesuvianite, $Ca_{10}Mg_2Al_4(Si_2O_7)_2$-	Epidote, $Ca_2(Al,Fe)Al_2O(SiO_4)(Si_2O_7)$-
$(SiO_4)_5(OH)_4$	(OH)
Single-isle silicates	
Olivine, $(Mg,Fe)_2SiO_4$	Topaz, $Al_2SiO_4(F,OH)_2$
Chondrodite, $Mg_5(SiO_4)_2(F,OH)_2$	Andalusite, Al_2OSiO_4
Willemite, Zn_2SiO_4	Sillimanite, Al_2OSiO_4
Zircon, $ZrSiO_4$	Kyanite, Al_2OSiO_4
Garnet, $R_3''R_2'''(SiO_4)_3$	Titanite, $CaTiOSiO_4$
Datolite, $CaBSiO_4(OH)$	Staurolite, $Fe_2Al_9O_6(SiO_4)_4(OH)_2$

Double-Isle (Soro-) Silicates

Hemimorphite, $Zn_4Si_2O_7(OH)_2 \cdot H_2O$

Hemimorphite (H = 4½–5, G = 3.4–3.5) is named from the hemimorphic character of the crystals (Fig. 7-172). In the United States it was formerly called *calamine*. It is a zinc silicate, a secondary mineral, found in the oxidized zone of zinc deposits, having formed by the alteration of sphalerite. It is orthorhombic but is found rarely in isolated crystals. The common habit is in mammillary or botryoidal masses with crystalline surfaces (Fig. 7-173). Occasionally, the surface is made up of flat tabular

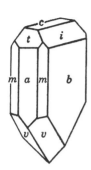

Fig. 7-172. *Hemimorphite.*

Fig. 7-173. *Hemimorphite, Sterling Hill, New Jersey.*

crystals projecting from the mass. It is usually white or slightly yellowish but may be tinged blue from a little copper. The luster is vitreous.

The composition is $Zn_4Si_2O_7(OH)_2 \cdot H_2O$, which gives 67.5% ZnO. Thus where hemimorphite is abundant, it is a valuable ore of zinc.

Vesuvianite, $Ca_{10}Mg_2Al_4(Si_2O_7)_2(SiO_4)_5(OH)_4$

Vesuvianite is named for the famous volcano Mount Vesuvious, an important locality for the mineral. The official name has alternated back and forth between *idocrase* and *vesuvianite,* and both names are in common use. Vesuvianite is of little economic importance, but a compact green variety found in California known as *californite* has been used as a jade substitute.

Habit. Vesuvianite, also called idocrase, is tetragonal (Fig. 7-174) and occurs in crystals of varied habit, some highly complex. A simple crystal is shown in Fig. 7-175 with a dipyramid, basal pinacoid, and first- and second-order tetragonal prisms. It is found also in massive form and in striated columnar aggregates.

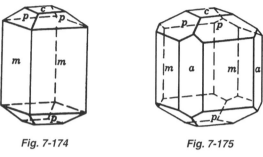

Fig. 7-174   **Fig. 7-175**

Vesuvianite.

Physical Properties. H = 6½, G = 3.35–3.45. The luster is vitreous, inclining to resinous, and the color is commonly brown to green. It may also be sulfur-yellow and bright blue. Transparent crystals have been found but are rare.

Composition. The composition of vesuvianite is complex but can be expressed by the formula $Ca_{10}Mg_2Al_4(Si_2O_7)_2(SiO_4)_5(OH)_4$. Boron or fluorine is present in some varieties, and beryllium has also been reported. It is actually a mixed double- and single-isle silicate possessing both (SiO_4) and (Si_2O_7) groups and has close structural affinities with grossular. Before a blowpipe it fuses with intumescence.

Occurrence. Vesuvianite is found most commonly in both contact metamorphosed and regionally metamorphosed siliceous limestones, where it is associated with grossular garnet, diopside, epidote, and titanite. It resembles some brown garnet and brown tourmaline but is more fusible, and if in crystals, is easily distinguishable.

Zoisite, $Ca_2AlAl_2O(SiO_4)(Si_2O_7)(OH)$

Zoisite (H = 6½, G = 3.1–3.3), named after Baron von Zois, is usually pale green to gray. The pink manganese-bearing variety is known as *thulite*. In 1967, gem-quality blue-colored crystals were found in Tanzania, and almost immediately the mineral became a popular gem sold under the name *tanzanite*. Crystals are elongate orthorhombic prisms parallel to the *b* axis with striations parallel to *b*. It has perfect platy cleavage parallel to the basal pinacoid. Its composition, $Ca_2AlAl_2O(SiO_4)(Si_2O_7)(OH)$, is the same as that of clinozoisite. Zoisite is a minor mineral associated with albite, biotite, and hornblende in medium-grade schists. Collectible specimens are usually found with vesuvianite and grossular in metamorphosed siliceous limestones.

Epidote, $Ca_2(Al,Fe)Al_2O(SiO_4)(Si_2O_7)(OH)$

Habit. Epidote is monoclinic and frequently found in crystals elongated parallel to the *b* crystallographic axis (Figs. 7-176 and 7-177). If the long dimension is placed in a vertical position, it is difficult to find the symmetry plane that shows the monoclinic character. Epidote also occurs in fibrous or columnar aggregates and others that are compact and granular.

Physical Properties. H = 6–7, G = 3.35–3.45. Epidote has a perfect platy cleavage parallel to the basal pinacoid and imperfect cleavage parallel to the front pinacoid.

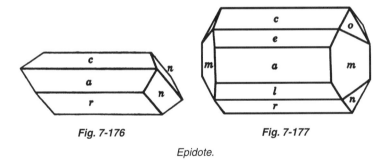

Fig. 7-176   Fig. 7-177

Epidote.

Crystals therefore break into fragments with the long dimension parallel to the *b* crystal axis. The luster is vitreous, and the color is commonly green. The peculiar yellow-green or pistachio-green of ordinary epidote is so characteristic that in the majority of specimens, color alone distinguishes it from other minerals that it may resemble. Well-formed crystals may be black, but if they have been bruised or cracked, the green color is usually visible.

Composition. Epidote is a silicate of aluminum, iron, and calcium, $Ca_2(Al,Fe)$-$Al_2O(SiO_4)(Si_2O_7)(OH)$. As with vesuvianite, it possesses both (SiO_4) and (Si_2O_7) groups. The ratio of iron to aluminum varies widely from $Fe/Al = 1:2$ in epidote to $Fe/Al = 0.3$ in clinozoisite, the iron-free member of a solid-solution series. It fuses with intumescence rather easily, and because of the iron it contains, gives a magnetic globule.

Occurrence. Epidote is a common mineral in many metamorphosed rocks, such as gneiss, mica schist, and amphibole schist, where it occurs with chlorite, albite, actin-olite, biotite, and quartz. Some rocks rich in epidote can resemble massive olivene but can usually be distinguished from it by the presence of quartz with the epidote. Epidote is also formed by the hydrothermal alteration of plagioclase in the propylitic zone. It also is found in contact-metamorphic deposits in limestone, where fine crys-tals have formed as at Knappenwand, Austria and Prince of Wales Island, Alaska (Fig. 7-178).

   Allanite and *clinozoisite* are closely related to epidote and belong to the epidote group of minerals. Allanite is a black radioactive mineral containing cerium and other rare elements which is found in syenites, granites, and pegmatites. Clinozoisite is a white to gray, greenish, or pale red mineral, like epidote in composition, except that it contains almost no iron. As with epidote, it is found in metamorphosed sedi-ments deficient in iron.

Single-Isle (Neso-) Silicates

Olivine, $(Mg,Fe)_2SiO_4$

Olivine, sometimes called *chrysolite,* is a common rock-forming mineral found most abundantly in dark-colored mafic rocks. The clear yellow-green variety is cut as a gem-stone and goes by the name *peridot.*

Fig. 7-178. *Epidote with quartz, Prince of Wales Island, Alaska.*

Habit. Olivine is orthorhombic. Crystals with good faces are rare and usually small. It is found most commonly in small grains disseminated through certain igneous rocks. Some rocks are made up almost exclusively of granular masses of olivine.

Physical Properties. H = 6½–7, G = 3.27–3.37. Although it is hard, granular varieties readily disaggregate, giving a false scratch. There is no cleavage but a pronounced conchoidal fracture. The luster is vitreous, and the color is yellow to olive-green and brown.

Composition. The formula for olivine, $(Mg,Fe)_2SiO_4$, represents an intermediate member in a solid-solution series between the magnesium-rich *forsterite,* Mg_2SiO_4, and the iron-rich *fayalite,* Fe_2SiO_4. The most common olivines are richer in magnesium. It is infusible before a blowpipe.

Occurrence. Olivine is a rock-forming mineral found in some rocks only as scattered grains, but in others it may be the principal constituent. It is most abundant in the dark mafic and ultramafic igneous rocks, such as gabbro, basalt, and peridotite, a rock composed of olivine and pyroxene. It is also found occasionally in contact-metamorphosed dolomites. In a certain rare type of rock known as *dunite,* it is the only important mineral present. Olivine readily alters to serpentine, so olivine grains may be coated with serpentine. In addition to its presence in rocks, it is also a common mineral in some meteorites. One celebrated meteorite found in Siberia in 1749 consists of a spongy mass of metallic iron containing bright yellow grains of olivine. The transparent yellow-green gem material, peridot, is found in Myanmar and on Zebirget, an island in the Red Sea.

Chondrodite, $Mg_5(SiO_4)_2(F,OH)_2$

Chondrodite (H = 6–6½, G = 3.1–3.2) is a silicate of magnesium and iron containing fluorine. To show its compositional relationship with olivine, the formula is commonly written as $Mg(OH,F)_2$ (brucite)·$2Mg_2SiO_4$ (olivine). It usually occurs in yellow grains embedded in crystalline limestone but has also been found as deep red crystals with magnetite in ore lenses in gneiss at the Tilly Foster iron mine near Brewster, New York. It is a member of the humite group, of which *norbergite, humite,* and *clinohumite* are the other members. All these minerals are similar in appearance and composition, varying only in the percentages of brucite.

Willemite, Zn_2SiO_4

Willemite ($H = 5\frac{1}{2}$, $G = 3.9$–4.2), a valuable ore of zinc, is commonly found in bright yellow, apple-green, or brown masses and more rarely in small colorless or pale green hexagonal crystals. The variety known as *troostite* is sometimes found in large flesh-red crystals with a resinous luster. It is a zinc silicate, Zn_2SiO_4. In troostite, manganese is present, replacing part of the zinc.

The most notable locality of willemite is Franklin, New Jersey, where it was mined as a zinc ore associated with calcite, franklinite, and zincite. However, unlike franklinite and zincite, which are peculiar to Franklin, willemite is formed by the weathering of sphalerite in the oxidized zone of other zinc deposits, as at Altenberg, Belgium and Tsumeb, Namibia. One of the most interesting characteristics of willemite is its strong fluorescence in ultraviolet radiation, and some specimens show a marked phosphorescence as well (p. 78).

Zircon, $ZrSiO_4$

Zircon is of interest because of the beautiful gems that have been cut from it. The common zircon gem is colorless, but yellowish-red and orange-red stones called *hyacinth* are particularly prized. The blue gems sold under the name *starlite* are zircons that have been colored artificially by heat treatment. Because of its high index of refraction, a well-cut colorless zircon resembles a diamond.

Zircon is the chief zirconium mineral and can be considered an ore of that rare metal, which is used in certain alloys. Since 1945, the chief use of metallic zirconium is in the construction of nuclear reactors. A more recent use of zirconium is in the manufacture of cubic zirconia, ZrO_2, a synthetic material used as a diamond substitute. Because of its high hardness, $8\frac{1}{2}$, and a refractive index approaching that of diamond, it makes a far more convincing diamond substitute than does zircon itself. Several hundred tons of cubic zirconia are cut each year to be set in jewelry. The rare mineral *baddeleyite,* which is a monoclinic form of ZrO_2 that occurs in nepheline syenites as in Pocas-de-Caldas, Brazil, is also an ore of zirconium and a source of zirconium oxide, which is used in making refractory bricks.

Habit. Zircon is tetragonal and is almost always found in small prismatic crystals. The common forms are the first-order tetragonal prism and pyramid (Fig. 7-179). More highly modified crystals also occur occasionally (Figs. 7-180, 7-181, 3-53 and 3-55).

Physical Properties. $H = 7\frac{1}{2}$, $G = 4.7$. The luster is brilliant adamantine, and the color varies from colorless through various shades of reddish brown and yellow.

Composition. Zircon is zirconium silicate, $ZrSiO_4$. It is infusible and is not attacked by acids.

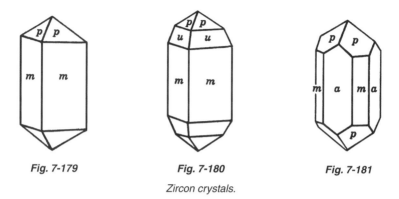

Fig. 7-179   **Fig. 7-180**   **Fig. 7-181**

Zircon crystals.

Occurrence. Zircon is a common accessory mineral in all types of igneous rocks but is particularly abundant in granites and syenites, where it is often enclosed in biotite, amphiboles, and feldspar. It usually occurs in these rocks in well-formed crystals, so tiny that it is difficult to see them well without a microscope. Large crystals may be found in granite and syenite pegmatites and occasionally in contact-metamorphic limestones. Fine specimens are obtained from pegmatites in New York, North Carolina, and Colorado, and very large ones from the nepheline–syenite pegmatites in Renfrew County, Ontario, Canada (Fig. 7-182) and in the Ural Mountains.

Fig. 7-182. *Zircon crystal in calcite, Renfrew County, Ontario.*

Because zircon is mechanically strong, chemically inert, and has a high specific gravity, it collects in placers. In places large accumulations of tiny crystals, weathered from their host rocks, are found in stream gravels and beach sands. Zircon is recovered from concentrations in beach sands in Florida, India, and Brazil, but the major world production is from beach sands on an island in Queensland, Australia.

Garnet, $R_3''R_2'''(SiO_4)_3$

Garnet is the name given to a mineral group made up of several subspecies, which have similar crystal habits, although their other properties vary widely.

Garnet is known chiefly through its use as an inexpensive red gemstone. However, the green andradite garnet known as *demantoid* is highly prized and is cut into beautiful gems. Garnet also has an industrial use as an abrasive, and the crushed mineral is used for sawing and grinding stones. It is also made into garnet paper, which is preferred to sandpaper for certain purposes.

Habit. Garnet is isometric and almost always occurs in distinct crystals, and since the crystals are commonly isolated and disseminated through the rock (Fig. 7-183), it is not difficult to recognize them (see Plate VIII-1). There are, however, massive kinds that require some skill for their identification.

Fig. 7-183.
Garnet crystals in mica schist.

The common crystal forms are the dodecahedron (Fig. 7-184) and the trapezohedron (Fig. 7-185), or a combination of these (Fig. 7-186). More rarely, the dodecahedron may be modified by the hexoctahedron (Fig. 7-187). It should be remembered that the dodecahedron has angles of 120° between two adjacent faces and that these faces themselves are diamond-shaped, with plane angles of 60° and 120°.

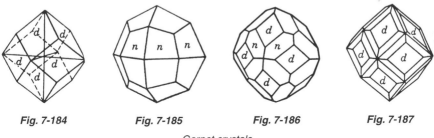

| *Fig. 7-184* | *Fig. 7-185* | *Fig. 7-186* | *Fig. 7-187* |

Garnet crystals.

Physical Properties. H = 6½–7½, G = 3.5–4.3. The luster is vitreous, and the color, although most commonly red, varies from colorless to yellow, brown, black, and green.

Composition. Although the compositions of the different garnets vary greatly, they can be expressed by the general formula $R_3''R_2'''(SiO_4)_3$. In this formula R″ may be calcium, magnesium, iron, or manganese, and R‴ may be aluminum, iron, titanium, or chromium. The individual subspecies are listed below with their chemical formulas and specific gravities. Garnets can be divided into two major groups: the *pyralspite* (pyrope, almandine, spessartine) and the *urgandite* (grossular, andradite, uvarovite). There is extensive solid solution between members within a group but relatively little between members of the different groups.

Group	Name	Composition	Specific Gravity
Pyralspites	Pyrope	$Mg_3Al_2(SiO_4)_3$	3.58
	Almandine	$Fe_3Al_2(SiO_4)_3$	4.37
	Spessartine	$Mn_3Al_2(SiO_4)_3$	4.19
Urgandites	Grossular	$Ca_3Al_2(SiO_4)_3$	3.59
	Andradite	$Ca_3Fe_2(SiO_4)_3$	3.86
	Uvarovite	$Ca_3Cr_2(SiO_4)_3$	3.90

Pyrope, or magnesia garnet, often has a deep red color. When perfectly clear, it is used as a gem and called *precious garnet. Rhodolite,* a purple to rose-red garnet, has a composition of about 2 parts pyrope and 1 part almandine.

Almandine includes a large part of the common garnet; the rest of the common garnet is andradite. It is a silicate of iron and aluminum and is ordinarily red but sometimes black. When clear, it is cut into gems and, like pyrope, is called precious garnet.

Spessartine, or manganese garnet, is comparatively rare. It is brownish red or hyacinth red in color. One type found in Virginia is perfectly transparent, with a peculiar red shade, and is used as a gem.

Grossular, the calcium aluminum garnet, may be colorless, white, pale yellow, green, brownish yellow, or cinnamon brown, and occasionally, rose-red. The commonest kind, called *cinnamon stone* or *essonite* is brownish red. The original grossular was green and received its name from the botanical name for the gooseberry.

Andradite is a calcium iron silicate and makes up much of the common garnet. The color is various shades of yellow, green, and brown to black. *Topazolite* is a topaz-yellow variety; *melanite* is black; *demantoid,* a highly prized green gem variety with a brilliant luster, is found in the Ural Mountains.

Uvarovite, or chrome garnet, is a rare type that has a fine emerald-green color. The finest specimens come from the Ural Mountains, where it is associated with chromite. Uvarovite is infusible, but all the other garnets fuse easily; if the fused mass is magnetic, it indicates that much iron is present.

Occurrence. Garnet is a common mineral found as an accessory constituent in granitic rocks but is more characteristically and abundantly associated with metamorphic rocks. Pyrope is formed only at high pressure deep in the earth and is a major mineral in the beautiful red-green rock known as eclogite. Pyrope is commonly found in kimberlites and has been used as a guide mineral in prospecting for diamonds. Almandine is the common garnet, occurring with biotite and feldspar in most schists and gneisses. Spessartine, which occurs with chlorite and actinolite, is also found in schists but usually those of lower grade. It is also the garnet commonly found in pegmatites. Grossular, often occurring with vesuvianite, wollastonite, and calcite, is the garnet typical of metamorphosed siliceous limestones. Andradite is the garnet formed by iron-rich solutions during contact metamophism. Uvarovite, the rarest variety, forms in fractures by the alteration of chrome-rich periodoties. As mentioned in the first pages of this book, garnet is found in rounded grains in streams and may make up a high proportion of the black and red sand of the seashore.

Datolite, CaBSiO₄(OH)

Habit. Datolite is monoclinic and commonly occurs in crystals that are complex and difficult to decipher even for the experienced crystallographer. The fact that many crystals are nearly equidimensional in the three axial directions (Fig. 7-188) adds to the difficulty of orientation.

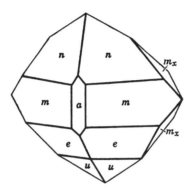

Fig. 7-188. *Datolite.*

A kind of datolite that occurs in massive aggregates resembles porcelain. Some of this type, found in the Lake Superior copper-mining region, is colored pink by finely disseminated copper enclosed within it (see Plate VIII-2).

Physical Properties. $H = 5-5\frac{1}{2}$, $G = 2.8-3.0$. Crystals are usually clear and glassy with a faint greenish tinge.

Composition. Datolite is a calcium borosilicate, $CaBSiO_4(OH)$. It yields a little water when heated in a closed tube and gives the green flame of boron when held in a blowpipe flame, at the same time fusing easily.

Occurrence. Datolite is a mineral of secondary origin found in cavities and cracks in basaltic lavas and related rocks. It has the same origin as and is frequently associated with the zeolites prehnite, apophyllite, and calcite. Well-formed crystals of this origin are found in trap rock in Massachusetts, Connecticut, and New Jersey. More rarely, datolite may be found in contact-metamorphic limestones.

Topaz, Al₂SiO₄(F,OH)₂

Topaz is highly prized as a gem mineral, and although gem-quality material occurs in several colors, the wine-yellow stones from Brazil are of most value and are called *precious topaz.* The pink color of some gem topaz is usually obtained by gently heating the dark yellow stones. Today, conspicuous in jewelers' windows, is jewelry set with deep blue topaz. The color of this popular stone is not natural but results from a double treatment: radiation followed by heating. Unlike the temporary color induced by treating some gems, the blue of topaz is permanent.

Habit. Topaz is orthorhombic, and prismatic crystals such as those shown in Figs. 7-189 to 7-191 are common. Frequently, beautiful etch figures are found on certain

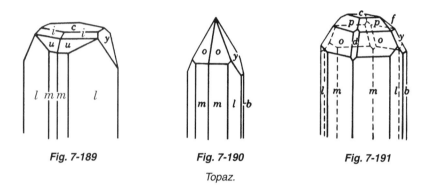

Fig. 7-189   Fig. 7-190   Fig. 7-191

Topaz.

faces but are lacking on others; thus in a striking manner the difference in crystal structure parallel to different forms is shown. Although it is usually in crystals, it also occurs in massive crystalline aggregates. Clear fragments and pebbles often resemble fragments of quartz but are much harder. Most crystals are small, but some clear and well-formed crystals weighing as much as 600 pounds have been found in Brazil.

Physical Properties. $H = 8$, $G = 3.4–3.6$. Topaz has perfect platy cleavage parallel to the basal pinacoid. The hardness is 8, greater than that of any other common mineral with the exception of corundum. The luster is vitreous, and the color varies from colorless to white, straw-yellow, wine-yellow, pink, bluish, and greenish.

Composition. Topaz is an aluminum fluosilicate, $Al_2SiO_4(F,OH)_2$.

Occurrence. Topaz is associated with granites and related felsic rocks. It is also found in cavities in rhyolite lavas, as in the Thomas Range, Utah and Guanajuato State, Mexico. It is often associated with high-temperature hydrothermal alteration either with quartz and orthoclase in the potassic zone or with quartz and muscovite in the greisen zone. It also occurs in quartz veins associated with tourmaline and cassiterite, as at Cornwall, England or in pegmatites with quartz, feldspar, and beryl. In this way fine wine-yellow and pale blue crystals have been found at Adun-Cholon, Siberia. Other outstanding pegmatite localities are Ouro Petro, Brazil (extremely large crystals), Sverdlorsk in the Ural Mountains, and the Tanokyami Yuma district, Japan. In the United States well-formed crystals have been found at the Little Three mine, Ramona, California; Pikes Peak district, Colorado; and Lord's Hill, Maine.

Andalusite, Al_2OSiO_4

Andalusite is named from the locality where it was identified, Andalusia in Spain. The most interesting variety is that called *chiastolite*. In this, parts of the crystals are white, and others contain carbonaceous impurities and are black. These inclusions are arranged in a regular manner to give a cruciform design in cross section (Fig. 7-192). The form on the cross section is a little like the Greek letter χ, and this resemblance has given it the name *chiastolite*.

Fig. 7-192. *Chiastolite.*

Andalusite has been mined in large quantities in the White Mountains of California for use in the manufacture of mullite, a highly refractory porcelain, such as that used in spark plugs. The compound Al_2OSiO_4 is polymorphous; that is, there is more than one substance with this composition; sillimanite and kyanite have the same chemical formula as andalusite.

Habit. Andalusite is orthorhombic and usually occurs in coarse prismatic crystals. The crystals are usually embedded and may be seen only when the surrounding rock is weathered away.

Physical Properties. H = 7.5, G = 3.16–3.20. The luster is vitreous, and the color varies from white or gray to pink, brown, or green.

Composition. Andalusite is an aluminum silicate, Al_2OSiO_4.

Occurrence. Andalusite in a low-pressure moderate-temperature metamorphic mineral usually formed during the contact metamorphism of aluminous shales, where it may be associated with cordierite, chlorite, or biotite. The crystals are often surrounded by fine-grained micaceous rock that weathers easily and leaves the andalusite protruding. Much andalusite, such as that found in New England, has been altered to fine-grained muscovite and therefore does not have the high hardness of the original mineral. Some andalusite is formed when aluminous rocks are dissolved in granitic or pegmatitic magma. Large prismatic crystals occur at the Champion mine in the White Mountains of California. Here a hydrothermally altered rhyolite has been metamorphosed. Chiastolite is found in fine-grained schists at Lancaster and Sterling, Massachusetts.

Sillimanite, Al_2OSiO_4

Sillimanite (H = 6–7, G = 3.23) occurs in orthorhombic prisms that are usually long and slender. Sometimes they are aggregated into parallel groups that may be fibrous. Hence the name *fibrolite* has been used for the mineral. Sillimanite has a perfect platy cleavage parallel to the side pinacoid. The luster is vitreous; the color, pale brown to gray and green. The chemical composition is the same as that of andalusite. Sillimanite is the high-temperature polymorph and is usually found in aluminous gneisses and schists, often with corundum, as at Pella, South Africa; Nongstein, India; and Williamstown, South Australia.

Kyanite, Al_2OSiO_4

Kyanite is named from the Greek word for *blue* because of its blue color. When it is transparent and of a rich color, fine gems can be cut from it. Like andalusite, it is used in the manufacture of high-quality refractories.

Habit. Kyanite is triclinic and usually occurs in long tabular crystals or in bladed aggregates.

Physical Properties. H = 5–7, G = 3.56–3.66. Kyanite tends to break into tabular blocks. This results from a perfect platy cleavage parallel to the front pinacoid, a good platy cleavage parallel to the side pinacoid, and a parting parallel to the basal pinacoid. It shows better than any other common mineral the variation in hardness with crystallographic direction, for parallel to the length of the crystals it is 5, but across the length it is 7. The luster is vitreous to pearly. Some bladed crystals may be a fine blue throughout, but others show a central blue strip between paler or even colorless sides. It may also be white, gray, or green.

Composition. Like andalusite, kyanite is aluminum silicate, Al_2OSiO_4.

Occurrence. Kyanite forms during the high-pressure metamorphism of aluminous shales and is typically found as an accessory mineral with feldspar, quartz, and biotite in schists and gneisses (Fig. 7-193). In North Carolina, South Carolina, and

Fig. 7-193. *Kyanite and staurolite (dark) in mica schist, St. Gotthard, Switzerland.*

Georgia, where it has been mined for use as a refractory, it is formed by the metamorphism of alumina-enriched hydrothermal alteration zones. This kyanite is found to occur with quartz and pyrite or with quartz, rutile, and ilmenite, as at Graves Mountain, Georgia. Kyanite is found in gneisses and schists, usually as an accessory mineral, but in some places it makes up a high percentage of the rock. Fine specimens occurring in mica schist have come from St. Gotthard, Switzerland.

Titanite, CaTiOSiO$_4$

Titanite takes its name from its titanium content. The official name has alternated back and forth between *sphene* and titanite, and both names are in common use. Clear crystals can be cut into lovely gems, but such high quality is rare. In certain places in the world, as on the Kola Peninsula in Russia, where titanite is abundant, it is mined as an ore of titanium.

Habit. Titanite is monoclinic and frequently occurs in well-formed wedge-shaped crystals (Figs. 7-194 and 7-195). It may also be lamellar with a well-defined parting plane parallel to the front pinacoid.

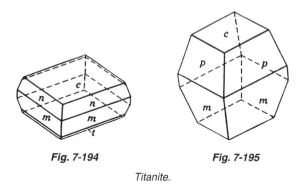

Fig. 7-194   **Fig. 7-195**

Titanite.

Physical Properties. H = 5–5½, G = 3.4–3.55. There is prismatic cleavage as well as the parting mentioned above. The color is usually various shades of yellow to brown but may also be gray, green, or black. The luster is resinous to adamantine.

Composition. Titanite is calcium titanosilicate, $CaTiOSiO_4$. It fuses with difficulty (scale value of 4).

Occurrence. Titanite is a common accessory mineral in igneous rocks and is found in tiny crystals disseminated among the grains of feldspar in granites, syenites, and nepheline syenites. It is also found in skarns, such as Mt. Linsay, Tasmania, where it occurs with vesuvianite, garnet, and calcite.

   As mentioned above, titanite associated with nepheline is found in large quantities in the nepheline syenites of the Kola Peninsula, Russia, where it is mined as a granular aggregate of crystals. Fine crystals associated with adularia and epidote are found in quartz veins at St. Gotthard, Switzerland and Zillertal, Austria. In the United States, crystals have been found in marble at Rossie, New York and as crystals with apatite in vugs in a perthite–quartz pegmatite near Eagle, Colorado.

Staurolite, $Fe_2Al_9O_6(SiO_4)_4(OH)_2$

Staurolite is named from the Greek words meaning *cross* and *stone* because of the remarkable crosses formed by its twinned crystals (Figs. 7-196 and 7-197). It is not unusual in a Christian country for superstition to attach great importance to these "cross stones" or "fairy stones." As a result, the right-angle twins have been made up into amulets and sold to tourists. In certain places the demand has exceeded the supply, and the natives carve crosses out of fine-grained mica schist to sell to unsuspecting customers.

Habit. Staurolite is monoclinic but often appears as crystals of orthorhombic form (Fig. 3-97). Single crystals are not rare, but it is more common to find twins with two crystals crossing each other nearly at right angles (Fig. 7-196) or at an angle of nearly 60° (Fig. 7-197). More complex twins, of the same interpenetration type, also occur where three or even four crystals are grouped together.

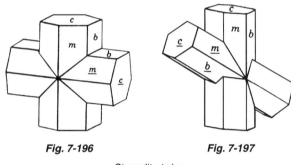

Fig. 7-196   **Fig. 7-197**

Staurolite twins.

Physical Properties. H = 7–7½, G = 3.65–3.75. Poor platy cleavage is present on some crystals parallel to the side pinacoid. The luster is vitreous, and the color is reddish, yellowish brown, brownish black, or gray.

Composition. Staurolite is a silicate of iron and aluminum, $Fe_2Al_9O_6(SiO_4)_4(OH)_2$, but is often impure. It is infusible and insoluble.

Occurrence. Staurolite forms during the medium-grade metamorphism of aluminous shales, where it frequently occurs with garnet, tourmaline, kyanite, or sillimanite in mica schist and gneiss. Examples of this association are known from several localities in New England, and very perfect crystals are found in Fannin County, Georgia and in the Sange de Cristo Mountains, New Mexico.

8

On the Determination
of Minerals

To the beginner it will seem difficult to become so familiar with the many mineral species as to be able to recognize each of them at sight. It is difficult, in fact impossible, even for trained mineralogists always to be prompt and sure in their determinations. For there are over 3000 distinct species, many of them very rare, and not a few appear in a great number of varieties. The varieties sometimes depend on fundamental differences of chemical composition, as among the garnets, and sometimes on less essential distinctions of habit, color, and state of aggregation, as with quartz, calcite, and fluorite. Hence it is obvious that the properties observed without aid of careful tests are often insufficient to determine a mineral positively. Although the student will soon learn to know the common minerals so well that they are considered old friends and most of them can be named on sight, the experienced mineralogist will often hesitate to name a specimen without a confirmation test of its hardness, for example. Confidence and hasty judgment belong to those who have little experience and scanty knowledge of the subject.

On the other hand, to recognize most of the minerals that are likely to be collected on a mineralogical excursion or obtained by exchange with other collectors is generally easy even for the beginner. For the number of common species is small, and quartz, feldspar, mica, calcite, barite, and the metallic species galena, sphalerite, pyrite, and chalcopyrite are constantly presenting themselves. Although their properties vary in different specimens, they are usually distinct, and in most cases a simple test will suffice for identification.

First of all, then, the mineralogist should know the common species well, for the chances are many times greater that an unknown specimen is one of them rather than a rare and little-known species. But a mineral may be rare, even a new one not described in any book, but this is unlikely. A real difficulty that experience does not entirely remove lies in the fact that at any major mineral locality, there are likely to be many nondescript specimens that show few distinct properties. These may be mixtures of several species and often are the result of the chemical decomposition of well-known minerals. Such specimens have no recognizable properties, and exhaustive microscopic, x-ray, and chemical work may be needed to prove their identity. In such cases the beginner may well turn for counsel to someone more experienced.

Ideally, the process of identification begins when the mineral is being removed at the collecting site. For all mineral species have a host matrix in which they occur. The student should carefully note the common minerals and the textural character of this matrix. From this the student should be able to infer the probable genetic host of the mineral: igneous (mafic, felsic, pegmatitic), hydrothermal (high or low temperature), skarn, metamorphic (slate, schist, gneiss), or sedimentary (sandstone, shale, limestone, evaporite, exhalite). Knowledge of the genetic association will greatly restrict the possible mineral species of an unknown mineral.

With a specimen of an unknown mineral in hand, the beginner should think first of the common species and mentally compare their properties with those of the unknown. If the properties appear to fit those of a common mineral, a confirmatory test should be made before making a final judgment. Next, one should ask if the proposed species fits the genetic association suggested by its matrix. If the properties observed do not add up to a provisional identification, determinative tables such as those that follow should be consulted. Care should be exercised not to decide too hastily but to give each property full weight. Do not give the name *albite* to a specimen of barite because it occurs as tabular glassy crystals and overlook the fact that it is much too heavy as well as too soft. Do not give the name *beryl* to a crystal of apatite because it is a green hexagonal prism and overlook the fact that it is much too soft. Do not call a green mineral olivine if it is in direct contact with quartz, an incompatible genetic matrix. Finally, do not hesitate to confess ignorance—because experienced mineralogists are always ready to do so, they are enabled from time to time to identify rare and interesting species and occasionally, a species new to science.

In the systematic determination of an unknown specimen, the first step, as emphasized in Chapter 2, is to learn everything possible about it through observation and handling. It has already been shown that in this way one can tell something about its habit, cleavage, luster, color, and degree of transparency. But at the same time, the other senses must be kept on the alert—to determine, for example, if a specimen is particularly heavy or light, or greasy to the touch. All these points will be quickly perceived and mentally recorded. Then a touch with the point of a knife blade will show something as to the hardness. This, it must be repeated, should be done carefully so as not to injure the specimen. If the mineral is not scratched by a knife, it will be well to determine whether it is scratched by or will scratch the smooth surface of a quartz crystal. After the hardness test, the mineral should be observed carefully with a hand lens (Fig. 8-1) for the presence of cleavage. If a cleavage surface can be observed, it should be studied carefully for the presence of cleavage traces from other directions, including the linear traces of a prismatic

Fig. 8-1. *Loupe. (Bausch and Lomb, Rochester, New York.)*

cleavage or the rectangular, rhombic, or triangular traces of a blocky cleavage. If the mineral has a metallic luster, the streak should be observed. Also, if the blade of the knife is magnetized, the magnetic character of magnetite and pyrrhotite will show itself. Additional tests are possible once the specimen has been returned to the field camp or laboratory.

Although a rough estimate of weight should already have been made by the hand, careful determination of the specific gravity requires more time and proper equipment and must be made back in the field camp or laboratory, Further, when all the properties mentioned have been noted, it will often be necessary to make blowpipe or chemical tests (review carefully pp. 104 to 113). A fragment for examination can generally be obtained without injury to the specimen by a careful blow with a light hammer. The blowpipe tests (see Chapter 5) may come before or after the other chemical tests. Careful study of the material that follows will enable the student to determine which test to make. For example, if the mineral has metallic luster, it should first be tested for sulfur in an open tube. With the same test, oxides of arsenic, antimony, and mercury will show themselves (p. 111). If necessary, a closed-tube test may be made next and then tests on charcoal (p. 110), which may confirm the results obtained earlier and show by the coating the presence of lead and molybdenum or, perhaps, yield a metallic globule of lead or copper. Magnetic residue will indicate iron. Further, after roasting off (p. 109) the sulfur, arsenic, or antimony, the residue may be tested for copper, cobalt, and so on, with borax bead. In the case of a nonmetallic mineral, the student should start by placing a small amount of powdered mineral on a watch glass with a few drops of hydrochloric (or nitric) acid (p. 112), which will effervesce, yielding an odorless gas (CO_2) if it is a carbonate. If the specimen is insoluable, even when finely powdered and heated in acid, that is also an important bit of information. Then the mineral may be tested in the forceps for the flame coloration and the fusiblility noted (p. 107), and if necessary, an examination can be made on the platinum wire in the borax bead.

Note, finally, that to obtain correct and concordant results the pure mineral must be experimented upon. In many specimens two or more species are so closely mixed that it needs sharp eyes, aided by a magnifying glass,* to separate them; this is particularly true of metallic minerals. Many species commonly occur in an earthy mass, so that it is difficult to obtain pure material. In that event an impurity of quartz or clay will often do no harm if its presence is noted. A fragment of cinnabar is entirely volatile on charcoal or in the open tube, but frequently it is associated with a gangue of clay, which will of course be left behind. Such a fragment heated in the glass tube often yields water, which comes from the nonessential gangue.

Even if at the beginning it seems as if very little is known about a specimen, careful use of the eyes and hands, and the various tests that may be made in a few minutes, will give a pretty complete table of its properties, and these may be used to fill out the blank list, as suggested on page 134. Usually, unless the specimen is rare and unusual, the observed properties will suggest the name of a species with which it can be compared. When this method of attack yields no definite result, the determinative tables on the following pages may be employed.

It should be pointed out that one should not rely completely on determinative tables, for in using them there is a tendency merely to make the observation and look it up in the table rather than to sum up mentally the properties that should indicate a given species. Instead, one should think about minerals in various ways. How many green minerals do I know? What other minerals that I know are present? What is the genetic association? What is the hardness? What is the cleavage? Is it unusually heavy or light? For these mental exercises to develop into a skill, the beginner must see as many different examples of a mineral as possible and to construct the student's own tables of diagnostic characteristics.

DETERMINATIVE TABLES

In the following tables the minerals are grouped according to their physical properties: physical tests made quickly on an unknown specimen give the necessary data to use them. The first division is on the basis of luster: (1) *metallic* and *submetallic,* and (2) *nonmetallic.* In the first group are placed those minerals that have the appearance of a metal and are quite opaque even on their thin edges. Most of these minerals have a colored streak that may be quite different from the color of the mineral, and this property complements the color. The second group includes all those minerals with a nonmetallic luster, those that will transmit light on their thin edges. These in general give a colorless or light-colored streak which is diagnostic in only a few cases. Therefore, streak is not given for the nonmetallic minerals.

The tables are next subdivided according to hardness. For metallic minerals the divisions are: (1) less than 2½; (2) greater than 2½, less than 5½; (3) greater than 5½.

---

*Every mineralogist should have a pocket magnifying glass, also called a hand lens or loupe, for even good eyes often need assistance, especially in examining small crystals.

For nonmetallic minerals the divisions are: (1) less than 2½; (2) greater than 2½, less than 3; (3) greater than 3, less than 5½; (4) greater than 5½, less than 7; (5) greater than 7. These limits are used because they are easy to determine. For example, 2½ is the hardness of a fingernail, and minerals scratched by a fingernail are in the first division—under 2½. A copper coin has a hardness of 3; a knife blade, 5½; and a quartz crystal, 7. Thus without special equipment it is possible to determine in which subdivision to look for the name of the unknown specimen. As stated earlier in this book, caution should be used in making a hardness test. Make sure that the scratch made by the knife blade is truly a scratch and is not merely steel that has been taken off by a hard mineral. It is well to try the scratch both ways; that is, try to scratch the knife with the mineral after scratching the mineral with the knife. Correct observation of hardness beyond 5½ is difficult and requires a good deal of practice. For these minerals the student should repeat the test several times. Finally, it should be pointed out that the state of aggregation may influence the hardness determination. For example, crystals of hematite have a hardness of 6, but when the mineral is compact and in a pulverulent aggregate it has a hardness of only 1–2. Minerals of this type have been listed in the tables under both hardness groupings.

Within each hardness division a further subdivision is made on the basis of cleavage, one that the student should be able to perceive easily. Minerals that show good cleavage are arranged in the order blocky, prismatic, and platy. Within each cleavage the minerals are ordered according to increasing hardness. However, a finely aggregated mineral may show no cleavage even though coarse crystals have perfect cleavage. When such a condition is common, the mineral is also listed with those showing no prominent cleavage.

It should be remembered that the tables are far from complete, although all the common minerals are listed. Thus it is possible, although unlikely, that the unknown specimen may be a rare mineral that is not listed. Further, since only the physical properties are used, it is quite impossible by use of the tables alone to identify all the minerals. Sometimes one will find that the data apply to two or three species. These minerals should then be looked up in Chapter 7 and additional tests such as chemical or blowpipe should should be made to confirm the identification more definitely.

To make the tables as compact as possible, the following abbreviations may be used:

H	hardness	Blk	black	Grn	green	Bss	brass
G	specific gravity	Yel	yellow	Brn	grown	Gld	golden
x1	crystal	Gry	gray	Blu	blue	Clr	clear
Drk	dark	Sil	silver	Cop	copper	Irid	iridescent
Lgt	light	Wht	white	Stl	steel		

On the next page is given a general outline of the classification. After determining the luster, hardness, and presence or absence of cleavage, one should consult the outline to find on which page in the tables the name of the unknown specimen appears.

Outline of Determinative Tables

Metallic or Submetallic Luster

1. Hardness less that 2½. Can be scratched by a fingernail and in general will leave a mark on paper.

   a. Prominent cleavage. Page 291.

   b. No prominent cleavage. Page 291.

2. Hardness greater than 2½, less than 5½. Can be scratched by a knife blade but cannot be scratched by a fingernail.

   a. Prominent cleavage. Page 292.

   b. No prominent cleavage. Page 293.

3. Hardness greater than 5½. Cannot be scratched by a knife blade.

   a. Prominent cleavage. Page 294.

   b. No prominent cleavage. Page 294.

Nonmetallic Luster

1. Hardness less that 2½. Can be scratched by a fingernail.

   a. Prominent cleavage. Page 295.

   b. No prominent cleavage. Page 295.

2. Hardness greater than 2½, less than 3. Cannot be scratched by a fingernail but can be scratched by a copper coin.

   a. Prominent cleavage. Page 296.

   b. No prominent cleavage. Page 297.

3. Hardness greater than 3, less than 5½. Cannot be scratched by a copper coin but can be scratched by a knife blade.

   a. Prominent cleavage. Page 298.

   b. No prominent cleavage. Page 299.

4. Hardness greater than 5½, less than 7. Cannot be scratched by a knife blade but can be scratched by quartz.

   a. Prominent cleavage. Page 301.

   b. No prominent cleavage. Page 302.

5. Hardness graeter than 7. Cannot be scratched by quartz.

   a. Prominent cleavage. Page 304.

   b. No prominent cleavage. Page 304.

Metallic or Submetallic Luster: Hardness Less Than 2½

Habit	Color Streak	H	Remarks and Occurrence	Name, Page
a. Shows good cleavage				
Cubic or octahedral xls	Black Gry-blk	2½	Cubic cleavage; G = 7.6; in quartz or calcite veins with sphalerite	Galena 152
Massive or as coatings	Gry-blk Gry-blk	2–2½	Easily sectile, blocky cleavage; G = 7.3; in veins with calcite, galena, and pyrite	Acanthite 150
Fibrous aggregates; surface dendrites	Black Black	1–2	Square prismatic cleavage; soft earthy alteration of rhodochro-site and rhodonite	Pyrolusite 184
Fine granular masses	Red Red	2–2½	Prismatic cleavage; G = 8.1; hot springs; in veins with quartz and stibnite	Cinnabar 161
Foliated massive; hexagonal plates	Black Black	1–1½	Platy cleavage, greasy feel; G = 2.2; in veins, as scales in lime-stones and schists	Graphite 148
Foliated masses; embedded scales	Blu-blk Black	1–1½	Platy cleavage, greenish streak on glazed porcelain; in quartz veins and pegmatites	Molybdenite 168
Platy masses or coatings	Blue Black	1½–2	Platy cleavage, indigo-blue color; as alteration coatings on born-ite, etc.	Covelite 151
Vertically striated xls; radiating	Blu-blk Gry-blk	2	Platy cleavage, fuses in a match flame; in quartz–calcite veins	Stibnite 162
b. Shows no prominent cleavage				
Earthy masses	Red Red-brn	1+	Weathering of iron-rich sediments	Hematite 177
Earthy masses	Yel-brn Yel-brn	1+	Weathering of pyrite or siderite	Limonite 187

Metallic or Submetallic Luster: Hardness 2½ –5½

Habit	Color Streak	H	Remarks and Occurrence	Name, Page
		a. Shows good cleavage		
Usually massive or coatings	Stl-gry Black	2–2½	Easily sectile, blocky cleavage; G = 7.3; in veins with calcite, galena and pyrite	Acanthite 150
Cubic or octahedral xls	Gry, blk Black	2½	Cubic cleavage; G = 7.6; in quartz or calcite veins with sphalerite	Galena 152
Prismatic xls, massive granular	Red, blk Red	2½	Rhombohedral cleavage, fuses in candle flame; G = 5.85; in calcite veins with galena	Pyrargyrite 170
Prismatic hexagonal xls	Red Red	2–2½	Rhombohedral cleavage, fuses in candle flame; in calcite veins with galena	Proustite 170
Usually in radiating hairlike xls	Bss-yel Black	3–3½	Rhombohedral cleavage, fuses readily; in cavities in limestone or serpentinite	Millerite 160
Granular cleavable masses	Brn-blk Brown	3½–4	Dodecahedral cleavage; in veins with quartz, calcite, barite galena, and pyrite	Sphalerite 154
Massive or in isometric xls	Red Red	3½–4	Poor octahedral cleavage; secondary with malachite over copper ores	Cuprite 172
Earthy, granular	Red Red	2½	Prismatic cleavage; G = 8.1; in veins with quartz and stibnite	Cinnabar 161
Fine-grained massive	Gry-blk Gry-blk	2½–3	Imperfect sectile and square prismatic; supergene zone of copper deposits	Chalcocite 151
Prismatic xls or bladed aggregates	Gry-blk Black	3	Square prismatic cleavage; in quartz veins with pyrite	Enargite 157
Skeletal or bladed forms	Sil-wht Gry-blk	2	Platy cleavage, fuses easily; G = 8.0–8.2; in quartz veins with gold and pyrite	Sylvanite 169
Granular aggregate	Sil-wht Sil-wht	2–2½	Platy cleavage, sectile, easily fusible; in calcite veins with stibnite; G = 9.8	Bismuth 144
Reniform or stalactitic masses	Tin-wht Gry-blk	3½	Platy cleavage; in calcite veins with realgar and stibnite	Arsenic 144

Metallic or Submetallic Luster: Hardness 2½–5½ (cont.)

Habit	Color Streak	H	Remarks and Occurrence	Name, Page
In fibrous or crystalline masses	Black Gry-blk	4	With pyrolusite in sedimentary beds; in barite or calcite veins	Manganite 186
Radiating fibers reniform	Brn-blk Yel-brn	5–5½	Platy cleavage; from oxidation of iron-rich minerals	Goethite 187
Flattened or bladed xls, massive	Brn-blk Brn-blk	5–5½	Platy cleavage; G = 7.0–7.5; with tourmaline in pegmatites, greisens, and quartz veins	Wolframite 224
			b. Shows no prominent cleavage	
Massive	Yel, wht Gry-blk	2½	Fuses in candle flame; G = 9.3; in quartz veins with gold	Calavarite 169
Nuggets; embedded plates or wire	Gld-yel Gld-yel	2½–3	Malleable, extremely heavy; G = 19.3; in quartz veins and stream gravels	Gold 137
Arborescent masses or as wire	Cop-red Cop-red	2½–3	G = 8.9; secondary with cuprite and malachite over copper ores	Copper 140
Arborescent masses or as wire	Sil-wht Sil-wht	2½–3	Malleable, tarnishes black; G = 10.5; in calcite veins	Silver 139
Massive or irregular grains	Bronze Black	3	Tarnishes purple; as globules in gabbro and in quartz veins	Bornite 152
Usually massive sphenoidal xls	Bss-yel Grn-blk	3½–4	Greenish-black streak; in quartz veins, also with pyrrhotite in magmatic or exhalitive ores	Chalcopyrite 156
Irregular massive	Bronze Black	4	Magnetic; as masses with chalcopyrite in gabbros, with pyrite in exhalites	Pyrrhotite 159
Nuggets or embedded scales	Sil-wht Gray	4–4½	In dunites with chromite or pyrrhotite; stream gravels	Platinum 142
Massive or in tetrahedral xls	Gry-blk Black	3–4½	In veins with quartz, calcite, chalcopyrite, and galena	Tetrahedrite 158
Massive or reniform columnar	Cop-red Black	5–5½	G = 7.8; in calcite veins with cobaltite and pyrrhotite	Nickeline 159
Massive	Brn-blk Yel-brn	5–5½	Compact masses from alteration of pyrite or siderite	Limonite 187
Usually granular massive	Black Brn-blk	5½	Pitchy luster; in bands or masses in peridotites and serpentinites	Chromite 176
Botryoidal, massive	Black Black	5–6	Impure masses with pyrolusite or manganite	Romanechite 190
Massive, reniform, or micaceous	Black Red	5½–6½	With quartz in veins, or in skarns and schists	Hematite 177

Metallic or Submetallic Luster: Hardness Greater Than 5½

Habit	Color Streak	H	Remarks and Occurrence	Name, Page
		a. Shows good cleavage		
Cubes or pyrito-hedrons	Tin-wht Black	5½	Blocky rectangular cleavage, py-ritelike crystals; in calcite veins with silver	Cobaltite 166
Usually massive, granular	Tin-wht Black	5½–6	Prismatic cleavage, easily fusible with garlic odor; in quartz veins with gold	Arsenopyrite 167
Radiating fibers	Brn-blk Yel-brn	5–5½	Platy cleavage; alteration of pyrite or siderite	Goethite 187
Massive or bladed xls	Brn-blk Brown	5–5½	Platy cleavage, G = 7.0–7.5; with tourmaline in pegmatites, greisens, and quartz veins	Wolframite 224
Platy xls, massive	Black Brn-blk	5½–6	Platy parting; with feldspar in magmatic segregations or pegmatites	Ilmenite 179
Radiating reniform rarely as xls	Black Red	5½–6	Parting with up to four directions; with quartz in veins or in skarns and schists	Hematite 177
		b. Shows no prominent cleavage		
Massive	Brn-blk Yel-brn	5–5½	Compact masses from alteration of pyrite or siderite	Limonite 187
Usually massive	Tin-wht Black	5½–6	Garlic odor when hot; in calcite veins with silver	Skutterudite 166
Massive or reniform columnar	Cop-red Black	5–5½	G = 7.8; in calcite veins with cobaltite, with pyrrhotite	Nickeline 159
Massive, granular	Black Brn-blk	5½	Pitchy luster; G = 9.0–9.7; yellow alteration	Uraninite 184
Granular aggregate	Brn, blk Brown	5½	Pitchy luster; as bands in peridotites and serpeninites	Chromite 176
Massive, stalactic, botryoidal	Black Black	5–6	Impure masses with pyrolusite or manganite	Romanechite 190
Cubes, octahedrons, pyritohedrons	Bss-yel Black	6–6½	In most rocks; masses in veins, shales, and exhalites	Pyrite 163
"Cock's comb" xls	Lgt yel Black	6–6½	In veins with chalcedony, calcite, galena, sphalerite; as nodules in coal beds	Marcasite 166
Granular aggregates octahedral xls	Black Black	6	Very magnetic; in most rocks; skarns; quartz veins	Magnetite 174
Disseminated gran-ular; octahedral xls	Black Drk brn	6	Associated with willemite; slightly magnetic	Franklinite 175

Nonmetallic Luster: Hardness Less Than 2½

Habit	Color	H	Remarks and Occurrence	Name, Page
a. Shows good cleavage				
Granular cleavable masses	Clear, white	2	Cubic cleavage, soluble, bitter taste; in salt beds with halite	Sylvite 192
In xls, fibrous or massive	Clear, white	2	Unequal blocky cleavage; embedded in clay or shale, with limestone or evaporites	Gypsum 225
Foliated or compact masses	White, drk grn	1	Platy cleavage, greasy feel; alteration of olivine, in metadolomites	Talc 251
Radiating lamallar aggregates	White, gray	1–2	Platy cleavage, pearly luster, greasy feel; in aluminous schists	Pyrophyllite 249
Foliated or columnar masses	Lemon-yellow	1½–2	Flexible platy cleavage; hot springs, in veins with chalcedony, calcite, and stibnite	Orpiment 162
Fine-grain claylike masses	White	2–2½	Platy cleavage; weathering of feldspar, as clay beds in sandstone	Kaolinite 249
Tabular xls, scales, or sheets	White	2–2½	Elastic platy cleavage; with quartz and feldspar in veins, pegmatites, and granites	Muscovite 253
Embedded scales or aggregates	Black	2½–3	Elastic platy cleavage; with quartz and feldspar in syenites, gneisses, and schists	Biotite 255
Flakes, scales, or foliated masses	Yellow brown	2½–3	Platy cleavage; in metadolomites and pegmatites	Phlogopite 255
Flakes or massive aggregates	Green	2–2½	Flexible platy cleavage; in skarns and schists, an alteration of biotite and pyroxene	Chlorite 257
Foliated or massive aggregates	White, gray	2½	Flexible platy cleavage, sectile, pearly luster; along fractures in serpentinite	Brucite 186
b. Shows no prominent cleavage				
Fibrous masses, "cotton balls"	White	1	With borax on dry lake beds; embedded in clay	Ulexite 213
Earthy, small xls, aggregates	Red to orange	1½–2	Hot springs, with stibnite, orpiment, calcite, and chalcedony veins	Realgar 162
Rhombic xls, fine aggregates	Pale yellow	1½–2½	Burns; coatings around volcanic vents; in beds with anhydrite and gypsum	Sulfur 145
Scales, plates, or waxy masses	Pearl-gray	2–3	Sectile; surface alteration over silver deposits	Chlorargyrite 193
Earthy, may be in rounded grains	Yellow, brown	1–3	Residual soils in tropical regions	Bauxite 189

Nonmetallic Luster: Hardness Greater Than 2½ Less Than 3

Habit	Color	H	Remarks and Occurrence	Name, Page
			a. Shows good cleavage	
Granular cleavable masses	Clear, white	2	Cubic cleavage, soluble, bitter taste; in beds with halite	Sylvite 192
Cubic xls and cleavable masses	Clear, white	2½	Cubic cleavage, soluble, salty taste; in beds with gypsum or limestone	Halite 191
Usually massive granular	Clear, gray	3–3½	Pseudocubic cleavage; as beds with limestone or salt	Anhydrite 222
Toothlike xls granular masses	Clear, white	3	Rhombohedral cleavage, effervesces; in veins, as limestone or marble	Calcite 197
Curved xls, massive	Clear, pink	3½–4	Rhombohedral cleavage; in veins and as dolomitized limestone	Dolomite 204
Tabular or platy xls, granular	Clear, white	3–3½	Blocky cleavage; G = 4.5; pearly luster; in veins, as sedimentary beds	Barite 219
Platy like barite, granular masses	Clear blue	3–3½	Blocky cleavage, crimson flame; as cavity fillings in limestones	Celestite 221
Massive or tabular xls	Clear, gray	3	Blocky cleavage; G = 6.2–6.5; alteration of galena	Anglesite 222
Pseudohexagonal twins, acicular	Clear, white	3½	Prismatic cleavage, effervesces; G = 4.3; in veins with barite and galena	Witherite 207
Splintery cleavable masses	Clear, white	3	Pseudoprismatic cleavage; with borax and ulexite in clay	Kernite 212
Small scales and irregular sheets	Lilac	2½–4	Platy cleavage, red flame; in pegmatites with quartz, feldspar, and spodumene	Lepidolite 256

Nonmetallic Luster: Hardness Greater Than 2½ Less Than 3 (cont.)

Habit	Color	H	Remarks and Occurrence	Name, Page
		b. Shows no prominent cleavage		
Rounded grains and earthy masses	Yellow, brown	1–3	Residual soils in tropical regions	Bauxite 189
Compact masses	Clear, white	2–2½	Weathering of feldspar; sedimentary beds	Kaolinite 249
Prismatic xls, porous masses	Clear, white	2–2½	Soluble, fuses easily; with ulexite in clay	Borax 211
Usually massive	Clear, white	2½	Appears like wet snow or paraffin; pseudocubic parting	Cryolite 193
Massive usually mottled, fibrous	Drk grn, yellow	2–5	Alteration of olivine and pyroxene	Serpentine 250
Massive compact	Turquois-blu, grn	2–4	With malachite in secondary zone over copper ores	Chrysocolla 258
Tabular plates, granular masses	Clear, brown	3–3½	Very heavy; G = 6.5; alteration of galena	Cerussite 208
Massive or in radiating masses	Clear, white	3.5	Effervesces; G = 4.3; green flame; in veins with barite and galena	Witherite 207
Radiating or globular aggregates	Yellow, green	3.5–4	Cracks and cavities in phosphatic-aluminous shales	Wavellite 218

Nonmetallic Luster: Hardness Greater Than 3, Less Than 5½

Habit	Color	H	Remarks and Occurrence	Name, Page
		a. Shows good cleavage		
Toothlike xls, granular masses	Clear, white	3	Rhombohedral cleavage, effervesces; in veins and as limestone	Calcite 197
Xls are curved rhombohedrons	Clear, white	3½–4	Rhombohedral cleavage, pearly luster; in veins and as dolomitized limestone	Dolomite 204
Dense compact masses, aggregates	White, gray	3½–5	Rhombohedral cleavage; compact masses in serpentinite replacement of limestone	Magnesite 200
Cleavable masses	Lgt brn, drk brn	3½–4	Rhombohedral cleavage, magnetic after heating; as veins or as concretions	Siderite 202
Cleavable masses	Pink, red	3½–4½	Rhombohedral cleavage; as veins with galena and sphalerite	Rhodochrosite 203
Botryoidal aggregates	Green, white	5	Rhombohedral cleavage rare, effervesces; alteration of sphalerite	Smithsonite 203
Rhombohedral xls	Pink, white	4½	Rhombohedral cleavage; in fractures and cavities of volcanic rocks	Chabazite 246
Frequently in platy aggregates	Clear, yellow	3–3½	Blocky cleavage, pearly luster, green flame; G = 4.5; as veins and sedimentary beds	Barite 219
Platy like barite, granular masses	Clear, lgt blue	3–3½	Blocky cleavage, crimson flame; cavity linings in limestones	Celestite 221
Usually in fine aggregates	Clear, gray	3–3½	Pseudocubic cleavage, as beds with gypsum, limestone, and salt	Anhydrite 222
Cubic xls	Clear, grn, violet	4	Octahedral cleavage; in veins with quartz or calcite	Fluorite 194
Granular cleavable masses	Yel-brn, black	3½–4	Dodecahedral cleavage, resinous luster; in veins with quartz, calcite, or barite	Sphalerite 154
Massive or embedded grains	Blue, gray	5½–6	Dodecahedral cleavage; in syenites with feldspar and nepheline	Sodalite 244
Bladed aggregates	Lgt blue, gry-wht	5 alg 7 acr	Platy cleavage; with quartz in meta-aluminous rocks	Kyanite 281
Pseudohexagonal twins; acicular	Clear, white	3½	Prismatic cleavage, green flame; G = 4.3; in veins with barite and with barite and galena	Witherite 207
Pseudohexagonal twins; acicular	Clear, white	3½–4	Prismatic cleavage, effervesces; hot springs, secondary zone over sulfide ores	Aragonite 205
Prismatic xls	Clear, white	3½–4	Prismatic cleavage, effervesces, red flame; G = 3.7; lining cavities in limestone	Strontianite 208

Nonmetallic Luster: Hardness Greater Than 3, Less Than 5½ (cont.)

Habit	Color	H	Remarks and Occurrence	Name, Page
			a. Shows good cleavage	
Radiating groups	White, lgt grn	4½–5	Prismatic cleavage; secondary mineral; alteration of sphalerite	Hemimorphite 271
Usually in cleavable masses	Clear, white	5–5½	Prismatic cleavage; in silicous marbles, in skarns with diopside, garnet, and scapolite	Wollastonite 266
Xls usually slender	White, grn, blk	5–6	Prismatic cleavage of 55°; with feldspar in schists, gneisses, granites, and skarns	Amphiboles 259
Xls usually short and stout	Green, black	5–6	Prismatic cleavage of 90°; with feldspar in gneisses, granites, and skarns	Pyroxenes 262
Usually massive	Rose-red	5½–6	Pseudo-prismatic cleavage; meta-maganese beds, skarns	Rhodonite 265
Prismatic xls	Pink, gray	5–6	Prismatic cleavage; in skarns with diopside and epidote	Scapolite 242
Sheaflike aggregates	White	3½–4	Platy cleavage; in cavities and fractures of volcanics	Stilbite 246
Xls tabular	White	3½–4	Platy cleavage, pearly luster; in fractures and cavities, especially volcanics	Heulandite 246
Pseudocubic xls	Clear, lgt grn	4½–5	Platy cleavage, pearly luster; in fractures and cavities with zeolites	Apophyllite 257
Prismatic xls, cleavable masses	Clear, white	4–4½	Platy cleavage, decrepitates; with ulexite, embedded in clay	Colemanite 213
Slender prismatic xls, radiating	Clear, white	5–5½	Platy cleavage; in cavities and fractures, especially volcanic rocks	Natrolite 248
			b. Shows no prominent cleavage	
Rounded grains and earthy masses	Yellow, brown	1–3	Residual soils in tropical regions	Bauxite 189
Mottled massive, fibrous (asbestos)	Blk-grn, yel-grn	2–5	Alteration of olivine and pyroxenes	Serpentine 250
Fine granular, massive	White	3	Effervesces in cold acid; as chalk beds, marbles	Calcite 197
Square tabular xls	Orange-white	3	Adamantine; G = 6.7–6.9; alteration of galena	Wulfenite 224
Radiating masses	Clear, white	3½	Effervesces, green flame; in veins with barite and galena	Witherite 207
Compact concretions	Lgt brn, drk brn	3½–4	Becomes magnetic on heating; as concretions with coal	Siderite 202

Nonmetallic Luster: Hardness Greater Than 3, Less Than 5½ (cont.)

Habit	Color	H	Remarks and Occurrence	Name, Page
		b. Shows no prominent cleavage		
Radiating xls	Clear, white	3½–4	Effervesces; with iron ore and gypsum	Aragonite 205
Prismatic xls	Clear, white	3½–4	Effervesces, crimson flame; cavities in limestone	Strontianite 208
Mammillary, fibrous	Bright green	3½–4	Effervesces; alteration of copper ore	Malachite 210
Small xls or massive	Bright blue	3½–4	Effervesces; alteration of copper ore	Azurite 211
Hexagonal xls, cavernous	Green, red, brown	3½–4	In secondary zone over galena deposits	Pyromorphite Vanadinite 215, 217
Mostly in radiating aggregates	Lgt grn, brown	3½–4	In fractures in phosphatic-aluminous shales	Wavellite 218
Commonly in dense compact masses	Clear, white	3½–5	Compact masses in serpentine; limestone replacements	Magnesite 200
Usually in square pyramidal xls	Yellow, brown	4½–5	Fluorescent; G = 5.9–6.1; in skarns, pegmatites, and quartz veins	Scheelite 223
Usually in radiating xl groups	White, grn-blu	4½–5	Secondary alteration of sphalerite	Hemimorphite 271
Hexagonal prisms, also massive	Green, blue	5	As concretions in sedimentary beds	Apatite 214
Botryoidal masses	Lgt grn, lgt blu	5	Effervesces; secondary alteration of sphalerite	Smithsonite 203
Usually in brilliant xls	Clear, lgt grn	5–5½	With zeolites in cracks and cavities of volcanic rocks	Datolite 279
Usually in trapezohedrons	Clear, white	5–5½	In cracks and cavities, especially of volcanic rocks	Analcime 247
Slender prismatic xls	Clear, white	5–5½	In cracks and cavities, especially of volcanic rocks	Natrolite 248
Wedge-shaped xls	Brown	5–5½	Adamantine; accessory mineral in granites-syenites; skarns	Titanite 282
Massive with embedded pyrite	Azure-blue	5–5½	In skarns and metamorphic aluminous rocks	Lazurite 244
Massive	Green, white	5½	Fluorescent; in marbles, alteration of zinc ores	Willemite 275
Botryoidal or stalactitic masses	White, irid.	5–6	Light; G = 2.0–2.3; hot springs, cavities in volcanic ash	Opal 235
Prismatic xls, also massive	Pink, gray	5–6	In skarns with epidote, garnet, and diopside	Scapolite 242
Massive or embedded grains	Blue, green	5½–6	In syenites with nepheline and feldspar	Sodalite 244

Nonmetallic Luster: Hardness Greater than 5½ Less Than 7

Habit	Color	H	Remarks and Occurrence	Name, Page
			a. Shows good cleavage	
Massive or embedded grains	Blue	5½–6	Poor dodecahedral cleavage; in syenites with feldspar and nepheline	Sodalite 244
Bladed aggregates	Blue-white	5 alg 7 acr	Platy cleavage parallel to xl length; with quartz in meta-aluminous rocks	Kyanite 281
Usually in cleavable masses	Clear, white	5–5½	Prismatic cleavage; in silicous marbles, in skarns with garnet and epidote	Wollastonite 266
Slender prismatic xls	White, clear	5–5½	Prismatic cleavage; in cavities and fractures, especially in volcanic rocks	Natrolite 248
Prismatic xls, fibrous or massive	Brown, bronze	5½	Square prismatic cleavage; with feldspar or olivine in gabbros or pyroxenites	Enstatite 262
Stout xls, irregular grains	Green, black	5–6	Square prismatic cleavage; in gabbros with feldspar or in skarns	Pyroxenes 262
Usually massive, cleavable–compact	Red or pink	5½–6	Square prismatic cleavage; in manganiferous meta-sediments, and skarns	Rhodonite 265
Xls slender, rarely fibrous	White, grn, blk	5–6	Prismatic cleavage at 55°; with feldspar in amphibolites, in skarns with diopside	Amphiboles 259
Cleavable masses resembling feldspar	White	6	Unequal prismatic cleavage, red flame; in pegmatites with microcline and spodumene	Amblygonite 217
Cleavable masses, irregular grains	White, red	6	Square prismatic cleavage; in granites, gneisses, and pegmatites	Orthoclase Microcline 237
Cleavable masses, irregular grains	White-gray	6	Square prismatic cleavage with striations; with pyroxene in gabbros and granites	Plagioclase 240
Compact or granular masses	Yel-grn, grn-blk	6–7	Unequal prismatic cleavage; with quartz in schists, in skarns with garnet	Epidote 272
Flattened striated xls	White, gry-grn	6½–7	Square prismatic cleavage, red flame; in pegmatites with quartz and microcline	Spodumene 264
Striated prismatic xls	Gray to pink	6½	Platy cleavage; with albite in schists or skarns with garnet	Zoisite 272
Thin, tabular xls	White, gray	6½–7	Platy cleavage; alteration of corundum or emery	Diaspore 189
Long, slender xls	Gry-grn, brown	6–7	Platy cleavage; as fibers in schists and gneisses	Sillimanite 281

Nonmetallic Luster: Hardness Greater than 5½ Less Than 7 (cont.)

Habit	Color	H	Remarks and Occurrence	Name, Page
			b. Shows no prominent cleavage	
Wedge-shaped xls	Brown-green	5–5½	Adamantine; with feldspar in granites and syenites	Titanite 282
Trapezohedral xls	Clear, white	5–5½	In cracks and cavities of volcanics; embedded in clay	Analcime 247
Trapezohedral xls	Gray, white	5½–6	In volcanics rich in potash and low in silica	Leucite 243
Massive with pyrite	Azure-blue	5–5½	In skarns or metamorphosed aluminous rocks	Lazurite 244
Massive or embedded grains	Blue	5½–6	In syenites with nepheline and feldspar	Sodalite 244
Botroyoidal or stalatitic masses	White, irid.	5–6	Light; G = 2.0–2.3; cavities in volcanic ash	Opal 235
Brilliant xls	Clear, green	5–5½	With zeolites in cracks and cavities of volcanic rocks	Datolite 279
Massive or disseminated grains	Green-white	5½	Fluoresces, in marbles; alteration of zinc ores	Willemite 275
Prismatic xls, granular massive	Pink, gray	5–6	In skarns with epidote, garnet, and diopside	Scapolite 242
Usually massive	White, gray	5½–6	Greasy; with feldspar in syenites and pegmatites	Nepheline 243
Reniform and stalactitic masses	Blue to green	6	In veins in altered volcanic rocks	Turquois 218
Reniform and stalactitic, with crystalline surface	Green, gray	6–6½	With zeolites in cracks and cavities of volcanic rocks	Prehnite 258
Prismatic xls, vertical striations	Red-brn, black	6–6½	In metamorphosed aluminous rocks, pegmatites	Rutile 181
Isolated grains in dolomitic marble	Yellow, brown	6–6½	In skarn with epidote, diopside, and garnet	Chondrodite 274
Square prismatic xls, columnar	Green-yellow	6½	In skarn with diopside, epidote, and garnet	Vesuvianite 271
Reniform surfaces, prismatic xls	Brown-black	6–7	G = 7.0; with tourmaline, wolframite in quartz veins and pegmatites	Cassiterite 182
Disseminated grains, massive granular	Green	6½–7	With feldspar, pyroxene in peridotite, gabbro, and basalt; in skarns	Olivine 273
Tough compact masses	Green, white	6½–7	In blue schists with glaucophane or pumpellyite	Jade 260, 264
Hexagonal xls, massive	Clear, white	7	Conchoidal fracture; in most rocks except those with olivine or nepheline	Quartz 229

Nonmetallic Luster: Hardness Greater than 5½ Less Than 7 (cont.)

Habit	Color	H	Remarks and Occurrence	Name, Page
			b. Shows no prominent cleavage	
Massive fine-grained	White to brown	7	Waxy; concretions or nodules in sediments or volcanics	Chalcedony 233
Elongate triangular xls	Black, brown	7–7½	All colors; with quartz in veins and pegmatites	Tourmaline 267
Prismatic xls, cruciform twins	Brown-black	7–7½	With garnet in mica schists	Staurolite 283
Square prismatic xls	Brown-green	7½	Meta-aluminous schists; chiastolite has black cross	Andalusite 280
Embedded grains resembling quartz	Blue, gray	7–7½	Pleochroic; in schists and gneisses with biotite and andalusite	Cordierite 270

Nonmetallic Luster Hardness Greater Than 7

Habit	Color	H	Remarks and Occurrence	Name, Page
			a. Shows good cleavage	
Octahedrons, rounded dodecahedrons	Clear	10	Octahedral cleavage, adamantine luster; with pyrope and ilmenite in breccias	Diamond 146
In long, slender xls	Grn-gry, brown	6–7	Prismatic cleavage; fibrous masses in schists, gneisses	Sillimanite 281
Flattened striated xls	White, lgt grn	6½–7	Square prismatic cleavage; red flame; in pegmatites with quartz and feldspar	Spodumene 264
Prismatic xls	Clear, yellow	8	Platy cleavage; in pegmatites with muscovite and tourmaline	Topaz 279
			b. Shows no prominent cleavage	
Prismatic xls, reniform surfaces	Red-brn, black	6–7	G = 7.0; with tourmaline, wolframite in quartz veins and pegmatites	Cassiterite 182
Rounded xls, granular massive	Green, brown	6½–7	In peridotite, gabbros, basalt, as masses with chromite	Olivine 273
Tough compact masses	Green, white	6½–7	In blue schists with glaucophane or pumpellyite	Jade 260, 264
Dodecahedrons or trapezohedrons	Brown, green	6½–7½	With biotite and feldspar in schist, gneiss, and skarn	Garnet 277
Striated hexagonal prisms, massive	Clear, white	7	Conchoidal fracture; in most rocks, as veins and as quartzite	Quartz 229
Slendar triangular prismatic xls	Black, brown	7–7½	All colors; with quartz in veins and pegmatites	Tourmaline 267
Square prismatic xls, interior black cross	Brown, green	7½	With chlorite or biotite in meta-aluminous schists	Andalusite 280
Prismatic xls and cruciform twin	Brown, black	7–7½	With biotite and garnet in schists	Staurolite 283
Usually in small prismatic xls	Brown	7½	G = 4.7; with quartz in granites, pegmatites, and skarns	Zircon 275
Hexagonal prisms	Green, yellow	7½–8	In pegmatites with quartz and feldspar	Beryl 266
Commonly in octahedral xls	Red, blu, grn	8	In marbles and magnesia skarns	Spinel 173
Frequent as tabular striated xls	Yellow, green	8½	With quartz and feldspar in pegmatites	Chrysoberyl 177
Barrel-shaped hexagonal xls	Gray to clear	9	Adamantine; G = 4.0; with mica in aluminous schists	Corundum 180

APPENDIX I

Common Minerals Arranged According to Prominent Elements

Aluminum

Corundum, aluminum oxide
Spinel, magnesium, aluminum oxide
Cryolite, aluminum, sodium fluoride
Turquois and wavellite, aluminum phosphates
Amblygonite, aluminum, lithium phosphate

Antimony

Native antimony
Stibnite, antimony sulfide

Arsenic

Native arsenic
Realgar and orpiment, arsenic sulfides

Barium

Barite, barium sulfate
Witherite, barium carbonate

Bismuth

Native bismuth

Calcium

Fluorite, calcium fluoride
Calcite and aragonite, calcium carbonates
Dolomite, calcium magnesium carbonate
Apatite, calcium phosphate
Anhydrite, calcium sulfate
Gypsum, hydrated calcium sulfate
Scheelite, calcium tungstate

Carbon

Diamond
Graphite

Cobalt

Skutterudite and cobaltite, cobalt arsenides

Copper

Native copper
Chalcocite, copper sulfide
Bornite and chalcopyrite, copper and iron sulfide

	Tetrahedrite, antimony and copper sulfide
	Cuprite, cuprous oxide
	Malachite and azurite, copper hydroxycarbonates
	Chrysocolla, copper silicate
Gold	Native gold
	Calaverite, gold telluride
	Sylvanite, gold–silver telluride
Hydrogen	Ice (and water)
Iron	Native iron
	Pyrrhotite, iron sulfide
	Pyrite and marcasite, iron disulfides
	Arsenophyrite, iron sulfarsenide
	Hematite, iron oxide
	Magnetite, magnetic iron oxide
	Franklinite, iron, zinc, manganese oxide
	Chromite, iron, chromium oxide
	Goethite and limonite, hydrated iron oxides
	Siderite, iron carbonate
Lead	Galena, lead sulfide
	Pyromorphite, lead phosphate
	Mimetite, lead arsenate
	Vanadinite, lead vanadate
	Cerussite, lead carbonate
	Anglesite, lead sulfate
	Wulfenite, lead molybdate
Magnesium	Brucite, magnesium hydroxide
	Magnesite, magnesium carbonate
	Dolomite, calcium magnesium carbonate
Manganese	Pyrolusite, manganite, and romanechite, managanese oxides
	Rhodonite, manganese silicate
	Rhodochrosite, manganese carbonate
Mercury	Native mercury
	Cinnabar, mercury sulfide
Molybdenum	Molybdenite, molybdenum disulfide
	Wulfenite, lead molybdate
Nickel	Millerite, nickel sulfide
	Nickeline, nickel arsenide
	Garnierite, nickel silicate
Phosphorus	Apatite, calcium phosphate
	Pyromorphite, lead phosphate
	Amblygonite, lithium phosphate
	Wavellite and turquois, aluminum phosphate
Platinum	Native platinum
Potassium	Sylvite, potassium chloride

Silver	Native silver
	Acanthite, silver sulfide
	Pyrargyrite, silver and antimony sulfide
	Proustite, silver and arsenic sulfide
	Chlorargyrite, silver chloride
Strontium	Celestite, strontium sulfate
	Strontianite, strontium carbonate
Sodium	Halite or rock salt, sodium chloride
	Borax, sodium borate
Silicon	Quartz, silicon dioxide
	Opal, hydrated silicon dioxide
	The silicate minerals, too numerous to list here, are grouped together in this book in the section "Silicates" in Chapter 7
Tin	Cassiterite, tin dioxide
Titanium	Rutile, anatase, and brookite, all titanium dioxide
	Ilmenite, iron titanium oxide
Tungsten	Wolframite, iron, manganese tungstate
	Scheelite, calcium tungstate
Uranium	Uraninite, uranium oxide
Zinc	Sphalerite, zinc sulfide
	Zincite, zinc oxide
	Franklinite, iron, zinc, manganese oxide
	Willemite and hemimorphite, zinc silicates
	Smithsonite, zinc carbonate

Most Important Minerals for a Small Collection

In the following list, the names of the minerals that it is most important for the young mineralogist to have in a collection are printed in SMALL CAPITAL letters. To these are added, in ordinary type, a number of others which are also important, but less so; they may well be present in the mineral collection of the school or university.

GOLD in quartz	Molybdenite
SILVER	Skutterudite
COPPER	TETRAHEDRITE
SULFUR	CUPRITE
GRAPHITE	Zincite
An ore of silver	CORUNDUM
Chalcocite	HEMATITE
Bornite	Spinel
GALENA	MAGNETITE
SPHALERITE	Franklinite
CHALCOPYRITE	CHROMITE
PYRRHOTITE	CASSITERITE
Nickeline	RUTILE
MILLERITE	Goethite
CINNABAR	Manganite (or pyrolusite)
Orpiment	Brucite
STIBNITE	HALITE
PYRITE	Cryolite
MARCASITE	FLUORITE
ARSENOPYRITE	CALCITE (several varieties)

DOLOMITE	CHABAZITE
Magnesite	NATROLITE
SIDERITE	Analcime
Rhodochrosite	TALC
SMITHSONITE	SERPENTINE
ARAGONITE	Garnierite
Strontianite	APOPHYLLITE
Witherite	Chlorite
CERUSSITE	PREHNITE
MALACHITE	MUSCOVITE
Azurite	BIOTITE
APATITE	Lepidolite
PYROMORPHITE	AMPHIBOLE (several varieties)
Amblygonite	PYROXENE (several varieties)
Wavellite	SPODUMENE
BARITE	RHODONITE
CELESTITE	TOURMALINE
Anglesite	BERYL
Anhydrite	Hemimorphite
GYPSUM	Olivine
Wulfenite	Willemite
QUARTZ (several varieties)	GARNET
OPAL	Vesuvianite
ORTHOCLASE	EPIDOTE
ALBITE	Zircon
Oligoclase	Datolite
Labradorite	Topaz
Nepheline	ANDALUSITE
Scapolite	Kyanite
Heulandite	STAUROLITE
STILBITE	Titanite

If the student limits himself to *small* specimens, a collection including the species mentioned will not occupy a great deal of space and, if desired, can be purchased at no great cost. Additional specimens can be obtained from time to time by exchange or purchase.

Of the minerals in the list above, the following are most desirable for the blowpipe and other chemical tests described in Chapter 5:

stibnite, molybdenite, an ore of silver, cinnabar, chalcopyrite, tetrahedrite, cuprite or malachite, galena, pyromorphite, cassiterite, rutile, pyrite, arsenopyrite, hematite or siderite, millerite, rhodonite, sphalerite, corundum, cryolite, fluorite, calcite, apatite, brucite, barite, celestite, orthoclase, amphibole (actinolite), garnet (almandite), tourmaline, and natrolite. Suitable fragments of the needed purity of these minerals can be purchased for a very small expenditure of money.

Recommended in addition to these are a mineral containing lithium (spodumene or amblygonite), one containing cobalt (skutterudite), one containing chromium (chromite), one containing vanadium (vanadinite), and one containing uranium (uraninite, or autunite).

APPENDIX III

Minerals Listed in Order of Increasing Specific Gravity

Mineral	G	Page	Mineral	G	Page
Ice	0.92	171	Colemanite	2.42	213
Borax	1.7	212	Leucite	2.45–2.50	243
Opal	1.9–2.2	236	Serpentine	2.5–2.65	251
Kernite	1.95	212	Microline	2.54–2.57	239
Ulexite	1.96	213	Nepheline	2.55–2.65	243
Sylvite	1.99	192	Orthoclase	2.57	239
Chrysocolla	2.0–2.4	258	Sanidine	2.57	239
Bauxite	2.0–2.5	189	Kaolinite	2.6	249
Sulfur	2.05	145	Chalcedony	2.6	233
Chabazite	2.05–2.15	246	Cordierite	2.6–2.66	270
Stilbite	2.1–2.2	246	Turquois	2.6–2.8	218
Sodalite	2.15–2.30	244	Chlorite	2.6–2.9	257
Halite	2.16	191	Albite	2.62	240
Heulandite	2.18–2.2	246	Plagioclase	2.62–2.76	241
Chrysotile	2.2–2.4	251	Oligoclase	2.65	240
Graphite	2.23	149	Scapolite	2.65–2.74	242
Natrolite	2.25	248	Quartz	2.66	231
Analcime	2.27	248	Andesine	2.68	240
Apophyllite	2.3–2.4	257	Talc	2.7–2.8	252
Gypsum	2.32	226	Laboradorite	2.71	240
Wavellite	2.33	218	Calcite	2.72	197
Brucite	2.4	186	Bytownite	2.73	240
Gibbsite	2.4	189	Beryl	2.75–2.80	267
Lazurite	2.4–2.45	244	Anorthite	2.76	240

Mineral	G	Page
Muscovite	2.76–2.88	254
Pyrophyllite	2.8	250
Prehnite	2.8–2.9	258
Wollastonite	2.8–2.9	266
Lepidolite	2.8–3.0	256
Datolite	2.8–3.0	279
Biotite	2.8–3.2	255
Dolomite	2.85	204
Phlogopite	2.86	255
Anhydrite	2.89–2.98	223
Aragonite	2.95	207
Cryolite	3.0–3.29	194
Magnesite	3.0–3.2	202
Boehmite	3.0	189
Nephrite	3.0	260
Tourmaline	3.0–3.25	269
Amblygonite	3.0–3.1	227
Tremolite	3.0–3.3	260
Amphiboles	3.0–3.3	260
Chondrodite	3.1–3.2	274
Zoisite	3.1–3.3	272
Apatite	3.15–3.20	214
Spodumene	3.15–3.20	264
Andalusite	3.16–3.20	281
Fluorite	3.2	195
Hornblende	3.2	261
Diopside	3.2–3.3	262
Augite	3.2–3.4	263
Jadeite	3.2–3.4	264
Pyroxenes	3.2–3.5	262
Enstatite	3.2–3.5	262
Sillimanite	3.23	281
Olivine	3.27–3.37	274
Vesuvianite	3.35–3.45	272
Epidote	3.35–3.45	272
Diaspore	3.4	189
Hemimorphite	3.4–3.5	271
Titanite	3.4–3.55	283
Topaz	3.4–3.6	280
Rhodonite	3.4–3.7	265
Rhodochrosite	3.45–3.6	203
Diamond	3.5	147
Realgar	3.5	162

Mineral	G	Page
Orpiment	3.5	162
Spinel	3.5–4.1	173
Kyanite	3.56–3.66	282
Garnet	3.58–4.37	277
Pyrope	3.58	278
Grossular	3.59	278
Limonite	3.6–4.0	187
Staurolite	3.65–3.75	284
Chrysoberyl	3.65–3.8	177
Strontianite	3.7	208
Romanechite	3.7–4.7	190
Azurite	3.77	211
Siderite	3.85	202
Andradite	3.86	278
Willemite	3.9–4.2	275
Uvarovite	3.90	278
Celestite	3.95–3.97	221
Sphalerite	3.9–4.1	155
Chalcopyrite	4.1–4.3	156
Malachite	3.9–4.0	210
Corundum	4.0	180
Spessartine	4.19	278
Rutile	4.2	182
Manganite	4.3	186
Goethite	4.37	187
Almandine	4.37	278
Smithsonite	4.4	203
Witherite	4.3	208
Enargite	4.43–4.45	157
Barite	4.5	220
Pyrrhotite	4.6	159
Stibnite	4.6	162
Chromite	4.6	176
Tetrahedrite	4.6–5.1	158
Molybdenite	4.7	169
Ilmenite	4.7	179
Zircon	4.7	275
Pyrolusite	4.75	184
Marcasite	4.9	166
Pyrite	5.0	165
Bornite	5.07	152
Magnetite	5.2	174
Franklinite	5.2	175

Mineral	G	Page
Hematite	5.26	178
Millerite	5.5	160
Zincite	5.5–5.7	173
Chlorargyrite	5.5	193
Proustite	5.55	170
Chalcocite	5.6	151
Arsenic	5.7	144
Pyrargyrite	5.85	170
Scheelite	5.9–6.1	223
Cuprite	6.0	172
Arsenopyrite	6.1	168
Anglesite	6.2–6.4	222
Cobaltite	6.33	166
Skutterudite	6.5	166
Pyromorphite	6.5–7.1	216
Cerussite	6.55	209
Wulfenite	6.7–6.9	224

Mineral	G	Page
Vanadinite	6.7–7.1	217
Cassiterite	6.8–7.1	183
Wolframite	7.0–7.5	225
Acanthite	7.3	150
Kamacite	7.3–8.2	143
Galena	7.5	152
Nickeline	7.8	159
Sylvanite	8.0–8.2	169
Cinnabar	8.1	161
Copper	8.9	141
Uraninite	9–9.7	185
Calavarite	9.35	169
Bismuth	9.8	144
Silver	10.1–11.1	139
Gold	15.6–19.3	137
Platinum	14–21.3	142

Index